Optimization Techniques

Neural Network Systems Techniques and Applications

Edited by **Cornelius T. Leondes**

VOLUME 1. Algorithms and Architectures

VOLUME 2. Optimization Techniques

VOLUME 3. Implementation Techniques

VOLUME 4. Industrial and Manufacturing Systems

VOLUME 5. Image Processing and Pattern Recognition

VOLUME 6. Fuzzy Logic and Expert Systems Applications

VOLUME 7. Control and Dynamic Systems

Optimization Techniques

Edited by

Cornelius T. Leondes
Professor Emeritus
University of California
Los Angeles, California

VOLUME 2 OF

Neural Network Systems Techniques and Applications

ACADEMIC PRESS
San Diego London Boston New York Sydney Tokyo Toronto

This book is printed on acid-free paper. ∞

Academic Press
a division of Harcourt Brace & Company
525 B Street, Suite 1900, San Diego, California 92101-4495, USA
http://www.apnet.com

Academic Press Limited
24-28 Oval Road, London NW1 7DX, UK
http://www.hbuk.co.uk/ap/

Library of Congress Card Catalog Number: 97-80441

International Standard Book Number: 0-12-443862-8

PRINTED IN THE UNITED STATES OF AMERICA
97 98 99 00 01 02 ML 9 8 7 6 5 4 3 2 1

Contents

Contributors xv
Preface xvii

Optimal Learning in Artificial Neural Networks: A Theoretical View

Monica Bianchini, Paolo Frasconi, Marco Gori, and Marco Maggini

I. Introduction 1
II. Formulation of Learning as an Optimization Problem 4
A. Static Networks 7
B. Recurrent Neural Networks 8
III. Learning with No Local Minima 10
A. Static Networks for Pattern Classification 11
B. Neural Networks with "Many Hidden Units" 22
C. Optimal Learning with Autoassociators 23
D. Recurrent Neural Networks 25
E. On the Effect of the Learning Mode 32
IV. Learning with Suboptimal Solutions 33
A. Local Minima in Neural Networks 34
B. Symmetrical Configurations 40
C. Network Saturation 41
D. Bifurcation of Learning Trajectories in Recurrent Neural Networks 42
V. Advanced Techniques for Optimal Learning 44
A. Growing Networks and Pruning 44
B. Divide and Conquer: Modular Architectures 45
C. Learning from Prior Knowledge 45

VI. Conclusions 45
References 47

Orthogonal Transformation Techniques in the Optimization of Feedforward Neural Network Systems

Partha Pratim Kanjilal

I. Introduction 53
II. Mathematical Background for the Transformations Used 55
A. Singular Value Decomposition 55
B. QR Factorization 56
C. QR with Column Pivoting Factorization and Subset Selection 56
D. Modified QR with Column Pivoting Factorization and Subset Selection 57
E. Remarks 58
III. Network-Size Optimization through Subset Selection 58
A. Basic Principle 58
B. Selection of Optimum Set of Input Nodes 59
C. Selection of Optimum Number of Hidden Nodes and Links 60
IV. Introduction to Illustrative Examples 61
V. Example 1: Modeling of the Mackey–Glass Series 62
VI. Example 2: Modeling of the Sunspot Series 65
A. Principle of Modeling a Quasiperiodic Series 65
B. Sunspot Series Model 66
VII. Example 3: Modeling of the Rocket Engine Testing Problem 71
VIII. Assessment of Convergence in Training Using Singular Value Decomposition 74
IX. Conclusions 76
Appendix A: Configuration of a Series with Nearly Repeating Periodicity for Singular Value Decomposition-Based Analysis 76
Appendix B: Singular Value Ratio Spectrum 77
References 77

Sequential Constructive Techniques

Marco Muselli

I. Introduction 81

II. Problems in Training with Back Propagation 82
A. Network Architecture Must Be Fixed *a Priori* 83
B. Optimal Solutions Cannot Be Obtained in Polynomial Time 85

III. Constructive Training Methods 85
A. Dynamic Adaptation to the Problem 87
B. High Training Speed 87

IV. Sequential Constructive Methods: General Structure 88
A. Sequential Decision Lists for Two-Class Problems 89
B. Sequential Decision Lists for Multiclass Problems 96
C. General Procedure for Two-Class Problems 98
D. General Procedure for Multiclass Problems 100

V. Sequential Constructive Methods: Specific Approaches 105
A. Halfspace Choice Set 106
B. Hyperplane Choice Set 117

VI. Hamming Clustering Procedure 123

VII. Experimental Results 125
A. Exhaustive Learning 128
B. Generalization Tests 132

VIII. Conclusions 139
References 140

Fast Backpropagation Training Using Optimal Learning Rate and Momentum

Xiao-Hu Yu, Li-Qun Xu, and Yong Wang

I. Introduction 145

II. Computation of Derivatives of Learning Parameters 148
A. Derivativcs of the Learning Rate 149
B. Derivatives of the Learning Rate and Momentum 151

III. Optimization of Dynamic Learning Rate 154
A. Method 1: Learning Rate Search with an Acceptable δE 154
B. Methods 2 and 3: Using a Newton-like Method to Compute μ 156
C. Method 4: Using the Higher-Order Derivatives of μ 156
IV. Simultaneous Optimization of μ and α 158
A. Method 5: Using the First Two Partial Derivatives 159
V. Selection of the Descent Direction 160
VI. Simulation Results 161
VII. Conclusion 168
References 172

Learning of Nonstationary Processes

V. Ruiz de Angulo and Carme Torras

I. Introduction 175
II. *A Priori* Limitations 177
III. Formalization of the Problem 178
IV. Transformation into an Unconstrained Minimization Problem 179
V. One-to-One Mapping *D* 182
VI. Learning with Minimal Degradation Algorithm 183
VII. Adaptation of Learning with Minimal Degradation for Radial Basis Function Units 186
VIII. Choosing the Coefficients of the Cost Function 188
IX. Implementation Details 190
A. Advance Rate 190
B. Stopping Criterion 191
C. Initial Hidden-Unit Configuration 191
X. Performance Measures 191
XI. Experimental Results 194
A. Scaling Properties 194
B. Solution Quality for Different Coefficient Settings 194

C. Computational Savings Derived from the Application of Learning with Minimal Degradation 197
D. Learning with Minimal Degradation versus Back Propagation 198

XII. Discussion 200
A. Influence of the Back Propagation Advance Rate on Forgetting 200
B. How to Prepare a Network for Damage or the Relation of Learning with Minimal Degradation with Fault Tolerance 201
C. Relation of Learning with Minimal Degradation with Pruning 204

XIII. Conclusion 204
References 206

Constraint Satisfaction Problems

Hans Nikolaus Schaller

I. Constraint Satisfaction Problems 209

II. Assessment Criteria for Constraint Satisfaction Techniques 213
A. P and NP Problems, Complexity Theory 213
B. Scaleability and Large-Scale Problems, Empirical Complexity 214
C. Parallelization 216
D. Design Principles for Computer Architectures 217
E. Examples of Constraint Satisfaction Problems 218
F. Summary 220

III. Constraint Satisfaction Techniques 221
A. Global Search 222
B. Local Search 223
C. Neural Networks 226

IV. Neural Networks for Constraint Satisfaction 227
A. Hopfield Networks 228
B. Neural Algorithms and the Strictly Digital Neural Network 231
C. Neural Computing Networks 233

D. Guarded Discrete Stochastic Net 234
E. Boltzmann Machine 234
F. *K*-Winner-Take-All 235
G. Dynamic Barrier Neural Network and Rolling Stone Neural Network 236

V. Assessment 240
A. *N*-Queens Benchmark 240
B. Comparison of Neural Techniques 241
C. Comparison of All Techniques 241
D. Summary 243

References 244

Dominant Neuron Techniques

Jar-Ferr Yang and Chi-Ming Chen

I. Introduction 249

II. Continuous Winner-Take-All Neural Networks 252

III. Iterative Winner-Take-All Neural Networks 256
A. Pair-Compared Competition 256
B. Fixed Mutually Inhibited Competition 259
C. Dynamic Mutual-Inhibition Competition 262
D. Mean-Threshold Mutual-Inhibition Competition 263
E. Highest-Threshold Mutual-Inhibition Competition 264
F. Dynamic Thresholding Competition 265
G. Simulation Results 267

IV. *K*-Winners-Take-All Neural Networks 268
A. Continuous *K*-Winners-Take-All Competition 268
B. Interactive Activation *K*-Winners-Take-All Competition 269
C. Coarse-Fine Mutual-Inhibition *K*-Winners-Take-All Competition 270
D. Dynamic Threshold Search *K*-Winners-Take-All Competition 270
E. Simulation Results 272

V. Conclusions 273
References 274

CMAC-Based Techniques for Adaptive Learning Control

Chun-Shin Lin, Ching-Tsan Chiang, and Hyongsuk Kim

I. Introduction 277

II. Neural Networks for Learning Control 278

A. Nonlinear Controller: Identification of Inverse Plant and Its Usage 278

B. Model Reference Adaptive Controller 280

C. Learning a Sequence of Control Actions by Back Propagation through Time 280

D. Neural Networks for Adaptive Critic Learning 283

III. Conventional Cerebellar Model Articulation Controller 284

A. Scheme 284

B. Application Example of Cerebellar Model Articulation Controller 287

IV. Advanced Cerebellar Model Articulation Controller-Based Techniques 290

A. Cerebellar Model Articulation Controller with Weighted Regression 290

B. Cerebellar Model Articulation Controller with General Basis Functions 293

V. Structure Composed of Small Cerebellar Model Articulation Controllers 298

A. Neural Network Structure with Small Cerebellar Model Articulation Controllers 298

B. Learning Rules 299

C. Example: Function Approximation 301

VI. Conclusions 302

References 303

Information Dynamics and Neural Techniques for Data Analysis

Gustavo Deco

I. Introduction 305

II. Statistical Structure Extraction: Parametric Formulation by Unsupervised Neural Learning 307

A. Basic Concepts of Information Theory 309
B. Independent Component Analysis 311
C. Nonlinear Independent Component Analysis 318
D. Linear Independent Component Analysis 322
E. Dual Ensemble Theory for Unsupervised and Supervised Learning 323

III. Statistical Structure Extraction: Nonparametric Formulation 326
A. Statistical Independence Measure 328
B. Statistical Test 331

IV. Nonparametric Characterization of Dynamics: The Information Flow Concept 337
A. Information Flow for Finite Partitions 339
B. Intrinsic Information Flow (Influential Partition) 342

V. Conclusions 345
References 349

Radial Basis Function Network Approximation and Learning in Task-Dependent Feedforward Control of Nonlinear Dynamical Systems

Dimitry Gorinevsky

I. Introduction 353

II. Problem Statement 357
A. Control Formulation 357
B. Example: Control of Two-Link Flexible Arm 360
C. Discretized Problem 362
D. Problems of Task-Dependent Feedforward Control 365

III. Radial Basis Function Approximation 366
A. Exact Radial Basis Function Interpolation 367
B. Radial Basis Function Network Approximation 369
C. Recursive Identification of the Radial Basis Function Model 370
D. Radial Basis Function Approximation of Task-Dependent Feedforward 372

IV. Learning Feedforward for a Given Task 373
A. Learning Control as On-Line Optimization 374

B. Robust Convergence of the Learning Control Algorithm 375
C. Finite-Difference Update of the Gradient 376

V. On-Line Learning Update in Task-Dependent Feedforward 378
A. Approximating System Sensitivity 378
B. Local Levenberg–Marquardt Update 379
C. Update of Radial Basis Function Approximation in the Feedforward Controller 380

VI. Adaptive Learning of Task-Dependent Feedforward 382
A. Affine Radial Basis Function Network Model of the System Mapping 382
B. Adaptive Update Algorithm 384
C. Discussion 386
D. Application Example: Learning Control of Flexible Arm 387

VII. Conclusions 391
References 391

Index 395

Contributors

Numbers in parentheses indicate the pages on which the authors' contributions begin.

Monica Bianchini (1), Dipartimento di Sistemi e Informatica, Università degli Studi di Firenze, 3-50139 Florence, Italy

Chi-Ming Chen (249), Department of Electrical Engineering, Kao Yuan College of Technology and Commerce, Luchu, Kaohsiung, Republic of China

Ching-Tsan Chiang (277), Department of Electrical Engineering, University of Missouri—Columbia, Columbia, Missouri 65211

Gustavo Deco (305), Siemens AG, Corporate Research and Development, Munich 81739, Germany

Paolo Frasconi (1), Dipartimento di Sistemi e Informatica, Università degli Studi di Firenze, 3-50139 Florence, Italy

Marco Gori (1), Dipartimento di Ingegneria dell'Informazione, Università degli Studi di Siena, 56-Siena, Italy

Dimitry Gorinevsky (353), Measurex Devron, Inc., North Vancouver, British Columbia V7J 3S4, Canada

Partha P. Kanjilal (53), Department of Electronics and Electrical Communication Engineering, Indian Institute of Technology, Kharagpur 721-302, India

Hyongsuk Kim (277), Department of Control and Instrumentation Engineering, Chonbuk National University, Republic of Korea

Chun-Shin Lin (277), Department of Electrical Engineering, University of Missouri—Columbia, Columbia, Missouri 65211

Marco Maggini (1), Dipartimento di Ingegneria dell'Informazione, Università degli Studi di Siena, 56-Siena, Italy

Marco Muselli (81), Istituto per i Circuiti Elettronici, Consiglio Nazionale delle Ricerche, 161 49 Genoa, Italy

Vicente Ruiz de Angulo (175), Institut de Ròbotica i Informàtica Industrial (CSIC-UPC), Edifici NEXUS, 08034 Barcelona, Spain

H. Nikolaus Schaller (209), DSJ TRI, D-80798 Munich, Germany

Carme Torras (175), Institut de Ròbotica i Informàtica Industrial (CSIC-UPC), Edifici NEXUS, 08034 Barcelona, Spain

Yong Wang (145), Department of Radio Engineering, National Communications Research Laboratory, Southeast University, Nanjing 210018, China

Li-Qun Xu (145), Intelligent Systems Research, Advanced Applications and Technology, BT Laboratories, Ipswich IP5 7RE, England

Jar-Ferr Yang (249), Department of Electrical Engineering, National Cheng Kung University, Tainan 70101, Taiwan

Xiao-Hu Yu (145), Department of Radio Engineering, National Communications Research Laboratory, Southeast University, Nanjing 210018, China

Preface

Inspired by the structure of the human brain, artificial neural networks have been widely applied to fields such as pattern recognition, optimization, coding, control, etc., because of their ability to solve cumbersome or intractable problems by learning directly from data. An artificial neural network usually consists of a large number of simple processing units, i.e., neurons, via mutual interconnection. It learns to solve problems by adequately adjusting the strength of the interconnections according to input data. Moreover, the neural network adapts easily to new environments by learning, and can deal with information that is noisy, inconsistent, vague, or probabilistic. These features have motivated extensive research and developments in artificial neural networks. This volume is probably the first rather comprehensive treatment devoted to the broad area of optimization techniques, including systems structures and computational methods. Techniques and diverse methods in numerous areas of this broad subject are presented. In addition, various major neural network structures for achieving effective systems are presented and illustrated by examples in all cases. Numerous other techniques and subjects related to this broadly significant area are treated.

The remarkable breadth and depth of the advances in neural network systems with their many substantive applications, both realized and yet to be realized, make it quite evident that adequate treatment of this broad area requires a number of distinctly titled but well-integrated volumes. This is the second of seven volumes on the subject of neural network systems and it is entitled *Optimization Techniques*. The entire set of seven volumes contains

Volume 1: *Algorithms and Architectures*
Volume 2: *Optimization Techniques*
Volume 3: *Implementation Techniques*
Volume 4: *Industrial and Manufacturing Systems*
Volume 5: *Image Processing and Pattern Recognition*
Volume 6: *Fuzzy Logic and Expert Systems Applications*
Volume 7: *Control and Dynamic Systems*

The first contribution to Volume 2 is "Optimal Learning in Artificial Neural Networks: A Theoretical View," by Monica Bianchini, Paolo Frasconi, Marco Gori, and Marco Maggini. The effectiveness of neural network systems emulating intelligent behavior and in solving many significant applied problems is strictly related to the learning algorithms intended to determine the optimal or near optimal values of the neural network systems weight elements. This contribution is a rather comprehensive treatment of techniques and methods for optimal learning (weight determination), and it provides a unified view of these techniques as well as a presentation of the state of the art in this broad and fundamental area. This contribution treats the issues and techniques related to the problem of local minima of the cost function that might be utilized in the process of determining neural network systems weights. Some rather significant links with the computational complexity of learning are presented, and various techniques for determining optimum neural network systems weights are presented. A number of rather illuminating illustrative examples are included in this contribution.

The next contribution is "Orthogonal Transformation Techniques in the Optimization of Feedforward Neural Network Systems," by Partha Pratim Kanjilal. Orthogonal transformation techniques can be utilized to identify the dominant modes in any information set. As this implies the realization of a neural network system of reduced or minimum order (minimum complexity), it is the basic motivation behind the use of orthogonal transformation techniques in optimizing (achieving the minimal complexity of) neural network systems. This contribution is a rather comprehensive treatment of the techniques and methods that are utilized in this important area, with illustrative examples to show the substantive effectiveness of the techniques presented.

The next contribution is "Sequential Constructive Techniques," by Marco Muselli. The theoretical and practical problems associated with the backpropagation algorithm have led to continual advances in learning techniques for this significant problem. Among these techniques is a new class of learning algorithms called sequential restoration methods. This highly effective method allows the treatment of training sets that contain several thousand samples. This contribution is a rather comprehensive treatment of the techniques and methods involved, with numerous substantive examples.

The next contribution is "Fast Backpropagation Training Using Optimal Learning Rate and Momentum," by Xiao-Hu Yu, Li-Qun Xu, and Yong Wang. This contribution presents a family of fast backpropagation (BP) learning algorithms for supervised training of neural networks. The achievement of rapid convergence rate is the result of using systematically optimized dynamic learning rate (and momentum, if required). This is in contrast to both the standard BP algorithm, in which a constant learning rate and momentum term is adopted, as well as other *ad hoc* or heuristics based methods. The main features of these algorithms are the attempts to explore the derivative information of the error surface (cost function) with respect to the learning rate and momentum to a certain necessary order rather than to obtain the Hessian matrix of synaptic weights, which is normally very costly to compute. This contribution is a rather comprehensive treatment of the methods and techniques for fast backpropagation neural network system learning methods. This contribution includes illustrations of the application of the techniques presented to several benchmark problems as well as comparisons to other well-studied classic algorithms. The highly effective performance of the techniques are made quite clear by these examples in terms of both fast convergence rate and robustness to weights initialization.

The next contribution to this volume is "Learning of Nonstationary Processes," by V. Ruiz de Angulo and Carme Torras. The degradation in performance of an associative network over a training set when new patterns are trained in isolation is usually called forgetting or catastrophic interference. Applications entailing the learning of a time-varying function require the ability to quickly modify some input–output patterns while at the same time avoiding catastrophic forgetting. Learning algorithms based on the popularity of the repeated presentation of the learning set backpropagation are suited only to tasks admitting two separate phases: an off-line phase for learning and another phase for operation. This contribution is a rather comprehensive treatment of techniques for the use of neural network systems for learning nonstationary processes. Numerous illustrative examples are presented that clearly manifest the effectiveness of the techniques presented.

The next contribution is "Constraint Satisfaction Problems," by Hans Nikolaus Schaller. System optimization problems in which the variables involved are either continuous or discrete are rather straightforward, comparatively speaking, when compared with similar problems wherein the continuous or discrete system variables are required to satisfy constraints. This contribution is a rather comprehensive treatment of the utilization of neural network systems for the treatment of this class of problems which has many diverse and broad applications of substantial applied significance. Numerous illustrative examples of the techniques and methods presented are included as an important element of this contribution.

The next contribution is "Dominant Neuron Techniques," by Jar-Ferr Yang and Chi-Ming Chen. This chapter provides an integrated and intensive investigation of the fundamental issues in the design and analysis of unsupervised learning neural networks for resolving which neuron (or neurons) has the maximum preference. The exploration of dominant neuron and K neurons (noted subsequently) can be related to the techniques for winner-take-all (WTA) and K winners-take-all (KWTA) problems, respectively. Generally, the KWTA neural network performs a selection of the K competitors whose activations are larger than the remaining $(M - K)$ ones (as noted in this contribution). When $K = 1$, the KWTA network devolves to the WTA process, in which the neuron with the maximum activation is determined. Hence, the KWTA network can be treated as a generalization of the WTA network. Well-known neural networks such as Grossberg's competitive learning, adaptive resonance theory, fuzzy associative memory, learning vector quantizers, and their various versions all require a WTA neural network. WTA methods have applications to such other diverse areas as classification applications, error correction systems, fuzzy associative memory systems, Gaussian classifiers, nearest match content addressable memory, signal processing, and the building of many complex systems. This contribution is a rather comprehensive treatment of the dominant neuron techniques of WTA and KWTA methods, with illustrative examples.

The next contribution is "CMAC-Based Techniques for Adaptive Learning Control," by Chun-Shin Lin, Ching-Tsan Chiang, and Hyongsuk Kim. This chapter treats the cerebellar model of articulation controller (CMAC) and CMAC-based techniques, which are often used in learning control applications. The CMAC was first developed by Albus in the mid-1970s for robot manipulator control and functional approximation. The CMAC is an efficient table lookup technique. Its most attractive characteristic is that learning always converges to the result with a least square error and the convergence is fast. The CMAC technique did not receive much attention until the mid-1980s when researchers started developing strong interests in neural networks. CMAC is now considered one type of neural network with major applications in learning control. Several illustrative examples are included which clearly manifest the significance and substantive effectiveness of CMAC systems.

The next contribution to this volume is "Information Dynamics and Neural Techniques for Data Analysis," by Gustavo Deco. One of the most essential problems in the fields of neural networks and nonlinear dynamics is the extraction and characterization of the statistical structure underlying an observed set of data. In the context of neural networks, the problem is posed as the data-based learning of a parametric form of the statistical dependences behind the data. In this parametric formulation, the goal is to model the observed process. On the other hand, an *a priori* requirement for the extraction of statistical structures is the detection of their existence and their characterization. For time series, for example, it is useful to know if the dynamics that originates the observed values is stationary or nonstationary, and if the time series is deterministic or stochastic, and to be able to distinguish between white noise, colored noise, Markov processes, and chaotic and nonchaotic determinism. The detection and characterization of such dependences should therefore be previously performed in a nonparametric fashion in order to be able *a posteriori* to model the process in a parametric form. The basic problem is of a statistical nature and therefore information theory offers the ideal theoretical framework for a mathematical formulation. This contribution is a rather substantive treatment of a detailed and unifying formulation of the theory of parametric and nonparametric structure extraction with a view toward establishing a consistent theoretical framework for the extremely important problem of discovering the knowledge implicit in empirical data. The significant implications are manifest by considering only a few of the many significant applications, including, biological data such as EEGs, financial data such as the stock market, etc. Illustrative examples are included.

The final contribution to this volume is "Radial Basis Function Network Approximation and Learning in Task-Dependent Feedforward Control of Nonlinear Dynamical Systems," by Dimitry Gorinevsky. This contribution considers intelligent control system architectures for task-level control. The problem is to compute feedforward control for a sequence of control tasks. Each task can be compactly described by a task parameter vector. The control update is performed in a discrete time: from task to task. This contribution considers an innovative controller architecture based on radial basis function (RBF) approximation of nonlinear mappings. The more advanced of these architectures enable on-line learning update for optimization of the system performance from task to task. This learning update can be considered as a generalization of the well-known learning (repetitive) control approach. Unlike repetitive control, which is only applicable to a single task, the proposed algorithms work for a parametric family of such tasks. As an example, a task-level feedforward control of a flexible articulated arm is considered. A vibration-free terminal control of such an arm is achieved using a task-level algorithm that learns optimal task-dependent feedforward as the arm goes through a random sequence of point-to-point motions.

This volume on neural network system optimization techniques clearly reveals the effectiveness and significance of the techniques available, and with further development, the essential role they will play in the future.

The authors are all to be highly commended for their splendid contributions to this volume which will provide a significant and unique reference source for students, research workers, practitioners, computer scientists, and others on the international scene for years to come.

Cornelius T. Leondes

Optimal Learning in Artificial Neural Networks: A Theoretical View*

Monica Bianchini
Dipartimento di Sistemi e Informatica
Università degli Studi di Firenze
Florence, Italy

Paolo Frasconi
Dipartimento di Sistemi e Informatica
Università degli Studi di Firenze
Florence, Italy

Marco Gori
Dipartimento di Ingegneria dell'Informazione
Università degli Studi di Siena
Siena, Italy

Marco Maggini
Dipartimento di Ingegneria dell'Informazione
Università degli Studi di Siena
Siena, Italy

I. INTRODUCTION

In the last few years impressive efforts have been made in using connectionist models either for modeling human behavior or for solving practical problems. In the field of cognitive science and psychology, we have been witnessing a debate on the actual role of connectionism in modeling human behavior. It has been claimed [1] that, like traditional associationism, connectionism treats learning as basically a sort of statistical modeling and that it is not adequate for capturing

*This chapter is partially reprinted from M. Bianchini and M. Gori, *Neurocomputing* 13:313–346, 1996, courtesy of Elsevier Science–NL, Sara Burgerhartstraat 25, 1055 KV Amsterdam, the Netherlands; and partially from M. Bianchini, M. Gori, and M. Maggini, *IEEE Trans. Neural Networks* 5:167–177 (© 1994 IEEE), M. Bianchini, P. Frasconi, and M. Gori, *IEEE Trans. Neural Networks* 6:512–515 (© 1995 IEEE), M. Bianchini, P. Frasconi, and M. Gori, *IEEE Trans. Neural Networks* 6:749–756 (© 1995 IEEE), M. Maggini and M. Gori, *IEEE Trans. Neural Networks* 7:251–254 (© 1996 IEEE).

Optimization Techniques

the rich structure of most significant cognitive processes. As for the actual novelty of the recent renew of connectionist models, Fodor and Pylyshyn [1] look quite skeptical and state "We seem to remember having been through this argument before. We find ourselves with a gnawing sense of deja vu." A parallel debate has been taking place concerning the application of connectionist models to engineering (pattern recognition, artificial intelligence, motor control, etc.). The arguments addressed in these debates seem strictly related to each other and refer mainly to the peculiar kind of learning that is typically carried out in connectionist models, which seem not to take enough of the structure into account. Unlike other symbolic approaches to machine learning, which are based on "intelligent search" (see, e.g., [2]), in connectionist models the learning is typically framed as an optimization problem. After the seminal books by the PDP group, Minsky published an extended edition of *Perceptrons* [3] that contains an intriguing epilogue on PDP's novel issues. He pointed out that what the PDP group calls a "powerful new learning result is nothing more than a straightforward hill-climbing algorithm" and commented on the novelty of backpropagation by saying: "We have the impression that many people in the connectionist community do not understand that this is merely a particular way to compute a gradient and have assumed instead that Backpropagation is a new learning scheme that somehow gets around the basic limitation of hill-climbing" (see [3, p. 286]).[1]

Minsky's issues call for the need to give optimal learning a theoretical foundation. Because simple gradient descent algorithms get stuck in local minima, in principle, one has no guarantee of learning the assigned task. It may be argued that more sophisticated optimization techniques (see, e.g., [5, 6]) guarantee reaching the global minimum, but the computational burden can become exaggerated early for most practical problems. The computational burden is obviously related to the shape of the error surface and particularly to the presence of local minima. Hence, it turns out to be very interesting to investigate the presence of local minima and particularly to look for conditions that guarantee their absence.

Obviously, we do not claim that the absence of local minima identifies the limit of practically solvable problems, because the use of sophisticated optimization techniques can actually be valuable also in the presence of error surfaces with local minima. However, beyond that bound, troubles are likely to begin for any learning algorithm, whose effectiveness seems very difficult to assess in advance.

One primary goal of this chapter is that of reviewing the basic results known in the literature concerning the optimal convergence of supervised learning algorithms in a unified framework. In the case of batch mode, the optimal convergence

[1]The criticism raised by Minsky to the backpropagation (BP) learning scheme also involves the mapping capabilities of feedforward nets. For example, Minsky [3, p. 265] points out that the net proposed by Rumelhart *et al.* [4, pp. 340–341] for learning to recognize the symmetry has very serious problems of scaling up. It may happen that the bits needed for representing the weights exceed those needed for recording the patterns themselves!

is strictly related to the shape of the error surface and particularly to the presence of local minima. When using pattern mode (or other schemes), the optimal convergence cannot be framed as an optimization problem, unless we use a conveniently small learning rate that leads us to an approximation of the "true gradient descent."

We focus mainly on batch mode by investigating conditions that guarantee local minima free error surfaces for both static and dynamic networks. In the case of feedforward networks, local minima free error surfaces are guaranteed when the patterns are linearly separable [7] or when using networks with as many hidden units as patterns to learn [8, 9]. Analogous results hold for radial basis function networks for which the absence of local minima is gained under the condition of patterns separable by hyperspheres [10]. Roughly speaking, these results suggest to us that optimal learning is certainly achieved in the limit cases of "many input" and "many hidden unit" networks. In the first case, the assumption of using networks with many inputs makes the probability of linearly separable patterns very high.[2] In the second case, the result holds independently of the problem at hand, but the main drawback turns out to be the excessive number of hidden units that are necessary for dealing with most practical problems. In the case of dynamic networks, local minima free error surfaces are guaranteed when matching the *decoupling network assumptions* (DNAs). They are essentially related to the decoupling of sequences of different classes on at least one gradient coordinate. Unlike other sufficient conditions, DNAs seem more valuable in network design. Basically, for a given classification task, one can look for architectures that are well suited for learning. In the best case, such a search leads to the discovery of networks for which learning takes place with no local minima. When no optimal network is found that guarantees local minima free error surface, one can, in any case, exploit DNAs for discovering architectures that are well suited for the task at hand.

The theoretical results described for batch mode can partially be extended, at least for feedforward networks, to the case of pattern mode learning [13]. This duality, which also holds for "nonsmall learning rates," is quite interesting, because it suggests conceiving new learning algorithms that are not necessarily based on function optimization, but on smart weight updating rules acting similarly to pattern mode. In practice, particularly for large experiments, the learning process takes place on subsets of the learning environment selected heuristically. For example, "difficult patterns" are commonly presented more often than others.

We also discuss some examples of suboptimal learning in the framework of the theory developed for understanding local minima free error surfaces. In so doing, two different kinds of local minima are identified that depend on joint *spurious*

[2]This follows directly from the results found by Cover [11] and Brown [12] concerning the average number of random patterns with random binary desired responses that can be absorbed by an ADALINE.

choices of the neuron nonlinear function and the cost (*spurious local minima*), and on the relationship between network and data (*structural local minima*), respectively. We also discuss *premature saturation* and appropriate choices of the cost for avoiding getting stuck in configurations in which the neurons are saturated.

This chapter is organized as follows. In the next section, we give the formulation of learning as an optimization problem and define the notation used throughout the chapter. In Section III, we review the basic results on local minima free error surfaces, while in Section IV, we discuss problems of suboptimal learning. In Section V we give a brief sketch of approaches that have been currently pursued to overcome the local minima problem. Finally, some conclusions are drawn in Section VI.

II. FORMULATION OF LEARNING AS AN OPTIMIZATION PROBLEM

In this section, we define the formalism adopted throughout the chapter. Basically, we are interested in experiments involving neural networks that can be represented concisely by $\mathcal{E} \doteq \{\mathcal{N}, \mathcal{L}_e, E_T\}$, $\mathcal{N}$ being the network, $\mathcal{L}_e$ the learning environment, and E_T the cost index.

1. Network $\mathcal{N}$

We consider neural networks whose N neurons are grouped into sets called layers. With reference to the index l, we distinguish between the input layer ($l = 0$), the output layer ($l = L$), and the hidden layers ($0 < l < L$). The number of neurons per layer is denoted by $n(l)$, whereas each neuron of layer l is referred to by its index $i(l)$, $i(l) = 1, \ldots, n(l)$.

We assume that the network is fed at discrete time $0, \ldots, t-1, t, t+1, \ldots, T$ by a sequence of vectors. For each t and l, we consider

$$A_l(t) \doteq \left[a_{1(l)}(t), \ldots, a_{n(l)}(t)\right]',$$
$$X_l(t) \doteq \left[x_{1(l)}(t), \ldots, x_{n(l)}(t)\right]',$$

where $A_l(t) \in \mathbb{R}^{n(l)}$ and $X_l(t) \in \mathbb{R}^{n(l)}$ are the *activation* and the *output* vector, respectively. The following model is assumed for the activation:

$$a_{i(l)}(t) = \mathcal{F}\big(W_{i(l)}, X_0(t), \ldots, X_{l-1}(t); X_l(t-1), \ldots, X_L(t-1)\big), \quad (1)$$

where $W_{i(l)}$ is the weight vector associated with the neuron $i(l)$. The function $\mathcal{F}(\cdot)$ depends on the particular model of each neuron and defines the way of combining the inputs received from all other neurons or external inputs. The initial state of the network is referred to as $X_0(0), \ldots, X_L(0)$. The output of neuron $i(l)$ is related to its activation as follows:

$$x_{i(l)} = f\big(a_{i(l)}\big),$$

where $f(\cdot)\colon \mathbb{R} \to [\,\underline{d},\ \overline{d}\,]$ is a $\mathcal{C}^2$ function and $f'(a_{i(l)}) \neq 0$ in $\mathbb{R}$. For example, a "squashlike" function [4] satisfies these hypotheses.

2. Learning Environment $\mathcal{L}_e$

In this chapter, we deal with supervised learning and, therefore, we need to refer to the following collection of T input–output pairs:

$$\mathcal{L}_e \doteq \big\{(I(t), D(t)),\ \ I(t) \in \mathcal{I},\ \ D(t) \in \{\underline{d},\ \overline{d}\}^n,\ t = 1, \ldots, T\big\},$$

where $I(t)$ is the input, $D(t)$ the corresponding target, and $\mathcal{I}$ is the *input space*. Each component of $D(t)$ belongs to $\{\underline{d},\ \overline{d}\}$. All the targets are collected in the matrix $\mathcal{D} \doteq [D(1), \ldots, D(T)]' \in \{\underline{d}, \overline{d}\}^{T,n}$.

3. Cost Function E_T

For a given experiment $\mathcal{E}_e$, the output–target data fitting is estimated by means of the cost function

$$E_T \doteq \sum_{t=1}^{T} E_t = \sum_{t=1}^{T} d\big(X_L(t), D(t)\big),$$

where $d(\cdot)$ is a distance in $\mathbb{R}^n$. The choice of this function plays a very crucial role in practice and depends significantly on the problem at hand. A common choice, which simplifies the mathematical analysis, is that of considering the distance induced by an L_p norm ($1 \leq p \leq \infty$). In the case of $p = 2$, which is most frequently considered, the cost is given by

$$E_T^Q \doteq \tfrac{1}{2} \sum_{t=1}^{T} \sum_{j=1}^{n} \big[x_j(t) - d_j(t)\big]^2.$$

The use of different values for p has been evaluated by Hanson and Burr [14] and by Burrascano [15] in a number of different domain problems. It turns out that the noise in the target domain can be reduced by rising power values less than 2, whereas the sensitivity of partition planes to the geometry of the problem may be increased with increasing power values.

The choice of the cost follows several different criteria, which may lead to opposite requirements. We focus on the requirements deriving from the need to limit problems of suboptimal solutions. One important requirement is that the particular function choice should not give rise to spurious local minima that, as will be shown in Section IV, depend on the relationship between the cost and the neuron functions. As pointed out in the following, this can be achieved by using an error criterion that does not penalize the outputs "beyond" the target values. Suppose that the outputs are exclusively coded, that is, if t belongs to class j, then $d_{i(L)}(t) = \overline{d}$ for $i(L) = j$, and $d_{i(L)}(t) = \underline{d}$ otherwise. In order to deal with spurious local minima, the introduction of the following LMS (least mean square)-*threshold* error function[3] turns out to be useful:

$$E_T^{\mathrm{LMS}} \doteq \sum_{j=1}^{n(L)} \sum_{t \in j} \left[l_2\big(x_j(t) - \overline{d}\big) + \sum_{i(L) \neq j} l_1\big(x_{i(L)}(t) - \underline{d}\big) \right],$$

where $l_k(\cdot)\colon \mathbb{R} \to \mathbb{R},\ k = 1, 2$, are C^2 functions (except for $\alpha = 0$), such that

$$\begin{cases} l_1(\alpha) = 0, & \text{if } \alpha \le 0, \\ l_1(\alpha) > 0,\ l_1'(\alpha) > 0, & \text{if } \alpha > 0, \end{cases}$$

$$\begin{cases} l_2(\alpha) = 0, & \text{if } \alpha \ge 0, \\ l_2(\alpha) > 0,\ l_2'(\alpha) < 0, & \text{if } \alpha < 0, \end{cases}$$

and $'$ stands for differentiation with respect to α.

Another important requirement for the error function is that of limiting the "premature saturation" problems due to erroneous choice of the initial weights. As will be shown in Section IV.C, the following *relative cross-entropy metric* [17–20]:

$$E_T^{\mathrm{RE}} \doteq \sum_{t=1}^{T} \sum_{j=1}^{n} d_j(t) \ln \frac{d_j(t)}{x_j(t)} + \big(1 - d_j(t)\big) \ln \frac{1 - d_j(t)}{1 - x_j(t)} \tag{2}$$

has the property of significantly reducing premature saturation problems. Solla *et al.* [19] have also pointed out that using the logarithmic error function (2) yields significant reductions in learning times. It is worth mentioning that in the case of an exact solution $E_T^{Q} = E_T^{\mathrm{LMS}} = E_T^{\mathrm{RE}} = 0$, because $d_j(t) = x_j(t)$ for all t, j.

The learning experiments analyzed in this chapter involve networks $\mathcal{N}$ with the architecture fixed in advance. The studies on variable networks are beyond the scope of this work.

[3]This definition is just the extension of the function proposed in [16] for the case of single-output networks.

A. STATIC NETWORKS

In the previous general framework, static networks [4],[4] whose neuron activation does not involve past samples, commonly adopt the following choice of function $\mathcal{F}(\cdot)$:

$$a_{i(l)}(t) = \mathcal{F}\big(W_{i(l)}, X_{l-1}(t)\big) = w_{i(l)} + \sum_{j(l-1)=1}^{n(l-1)} w_{i(l),j(l-1)} x_{j(l-1)}(t)$$
$$= W'_{i(l)} X^e_{l-1}(t),$$

where $W_{i(l)} \in \mathbb{R}^{n(l-1)+1}$, $w_{i(l),j(l-1)}$ is the weight of the link between the neurons $j(l-1), i(l)$ and $w_{i(l)}$ is the neuron threshold. In this case, the function $\mathcal{F}$ is simply the dot product of $X^e_{l-1} \doteq [X'_{l-1}, 1]'$ and $W_{i(l)}$, with "1" accounting for bias. The output of neuron $i(l)$ is related to the activation by means of the squashing function[5]:

$$x_{i(l)} = f\big(a_{i(l)}\big) \doteq \frac{1}{1 + \exp(-a_{i(l)})}. \tag{3}$$

Another common choice of the function $\mathcal{F}(\cdot)$ leads to radial basis function networks (RBFs). The activation of the locally tuned units of these networks follows the equation:

$$a_{i(l)}(t) = \mathcal{F}\big(W_{i(l)}, X_{l-1}(t)\big) = w_{i(l)} + \frac{1}{\sigma^2_{i(l)}} \sum_{j(l-1)=1}^{n(l-1)} \big(x_{j(l-1)}(t) - w_{i(l),j(l-1)}\big)^2$$
$$= w_{i(l)} + \frac{\|X_{l-1}(t) - \overline{W}_{i(l)}\|^2}{\sigma^2_{i(l)}}, \tag{4}$$

where $W_{i(l)} = [\overline{W}'_{i(l)}, w_{i(l)}, \sigma_{i(l)}]' \in \mathbb{R}^{n(l-1)+2}$, $w_{i(l),j(l-1)}$ is the weight of the link between the neurons $j(l-1), i(l)$, $w_{i(l)}$ is the unit threshold, and $\sigma_{i(l)}$ is the "width" of the Gaussian. In this case, the output neuron function is

$$x_{i(l)} = \tilde{f}(a_{i(l)}) \doteq \exp(-a_{i(l)}).$$

The RBF networks also contain an output layer of linear or sigmoidal neurons.[6]

[4]In the literature and throughout the chapter, feedforward networks are also referred to as multilayered networks (MLNs).

[5]The symmetric squashing function $\tanh(u)$ can be used with analogous results.

[6]Unlike the RBFs proposed in [21], Eq. (4) includes a bias term. Let $L = 2$ be. Because $x_{i(1)} = \exp(-a_{i(1)}) = \exp(-w_{i(1)} - \|X_0(t) - \overline{W}_{i(1)}\|^2/\sigma^2_{i(1)}) = \exp(-w_{i(1)})\exp(-\|X_0(t) - \overline{W}_{i(1)}\|^2/\sigma^2_{i(1)})$, the contribution of hidden neuron $i(1)$ to output neuron $j(2)$ turns out to be $[w_{j(2),i(1)}\exp(-w_{i(1)})]\exp(-\|X_0(t) - \overline{W}_{i(1)}\|^2/\sigma^2_{i(1)})$. When denoting $\widehat{w}_{j(2),i(1)} \doteq [w_{j(2),i(1)}\exp(-w_{i(1)})]$, the RBF we consider becomes exactly the one assumed in [21].

For both MLNs and RBFs, the learning environment involves pairs of input–output vectors as follows:

$$\mathcal{L}_e \doteq \{(X_0(t), D(t)),\ X_0(t) \in \mathbb{R}^{n(0)},\ D(t) \in \mathbb{R}^{n(L)},\ t = 1, \ldots, T\},$$

where $X_0(t)$ is the input pattern and $D(t)$ is its corresponding target.

When using a feedforward network as an autoassociator [22], the target is set to the input, thus reducing the learning environment to

$$\mathcal{L}_e \doteq \{X_0(t) \in \mathbb{R}^{n(0)},\ t = 1, \ldots, T\}.$$

In order to understand the basic results proposed in this paper, additional notation is needed that allows us to deal with a compact vectorial formulation of the problem.

1. The vector $X_{i(l)} \doteq [x_{i(l)}(1), \ldots, x_{i(l)}(T)]' \in \mathbb{R}^T$, called the *output trace* of neuron $i(l)$, stores the output of neuron $i(l)$ for all the T patterns of the learning environment. The output trace for all the neurons of a given layer l is called the *layer output trace*. It is kept in the matrix $\mathcal{X}_l \doteq [X_{1(l)} \cdots X_{n(l)}] \in \mathbb{R}^{T,n(l)},\ 0 \le l \le L$.
2. $w_{i(l),j(l-1)}$ is the weight connecting neuron $j(l-1)$ to neuron $i(l)$. The associated matrix $\mathcal{W}_{l-1} \in \mathbb{R}^{n(l),n(l-1)}$ is referred to as the *layer weight matrix*. The symbol Ω denotes the weight space.
3. Let $y_{i(l)}(t) \doteq \partial E_t / \partial a_{i(l)}(t)$. We call the *delta trace* the vector $Y_{i(l)} \doteq [y_{i(l)}(1), \ldots, y_{i(l)}(T)]'$ and the *layer delta trace* the matrix $\mathcal{Y}_l \doteq [Y_{1(l)} \cdots Y_{n(l)}] \in \mathbb{R}^{T,n(l)}$. We denote by $\mathcal{S}_l^{\mathcal{Y}} \subset \mathbb{R}^{T,n(l)}$ the set of all $\mathcal{Y}_l$ generated when varying the weights in Ω. Moreover, let us define $\tilde{\mathcal{Y}}_l$ such that $\tilde{y}_{i(l)}(t) \doteq y_{i(l)}(t)/f'(a_{i(l)}(t))$.
4. We assume that there is no connection that jumps a layer. Therefore, for MLNs and autoassociators, for weights connecting layers $l-1$ to l the gradient can be represented by a matrix $\mathcal{G}_{l-1} \in \mathbb{R}^{n(l-1)+1,n(l)}$, whose generic element $g(j(l-1), i(l))$ is given by $\partial E_T / \partial w_{i(l),j(l-1)}$ if $j(l-1) \le n(l-1)$ and $\partial E_T / \partial w_{i(l)}$ if $j(l-1) = n(l-1)+1$. For the hidden layer of an RBF network, $\mathcal{G}_{l-1} \in \mathbb{R}^{n(l-1)+2,n(l)}$, where the $(n(l)+1)$th row accounts for the derivative w.r.t. $\sigma_{i(l)}$.

B. Recurrent Neural Networks

We are mainly interested in recurrent networks used for processing sequences. The computational style we consider is that of feeding the network by sequences of frames (*tokens*) and computing the activation on-line without waiting for state relaxation as in Hopfield networks [23] and Boltzmann machines [24].

Formally, a token $S_{F(t)}(t)$, with $t = 1, \ldots, T$, is a sequence of $F(t)$ frames (input vectors): $S_{F(t)}(t) \doteq \{X_0(f,t) \in \mathbb{R}^{n(0)},\ f = 1, \ldots, F(t)\}$. The number of frames composing a given token t is referred to as the *token length* $F(t) \leq F_{\max}$, where $F_{\max} = \max_{1 \leq t \leq T}\{F(t)\}$. The activation of each neuron is reset to its initial value after feeding the network with a new token.

Let us consider single-layer networks,[7] fed by the input vectors $X_0(f,t) \doteq [x_{0,1}(f,t), \ldots, x_{0,n(0)}(f,t)]'$. We are particularly interested in networks whose neuron activation evolves according to

$$\begin{aligned} a_{i(1)}(f,t) &= \mathcal{G}\big(W_{i(1)}, X_0(f,t), X_1(f-1,t)\big) \\ &= \big(W^1_{i(1)}\big)' X_1(f-1,t) + \big(W^0_{i(1)}\big)' X_0(f,t), \end{aligned}$$

where $W^1_{i(1)} \in \mathbb{R}^{n(1)}$ denotes the *feedback connections* and $W^0_{i(1)} \in \mathbb{R}^{n(0)}$ the *connections from external inputs*. The output of the neuron $i(1)$ is related to its activation by a squashing function [see, Eq. (3)]. Finally, among the processing units of the network, the *output neuron* plays the important role of coding the class of the sequences that feed the network, as we are interested in dealing with positive and negative tokens only.

The learning process is based on a set of supervised tokens, collected in the T input–target pairs:

$$\mathcal{L}_e \doteq \big\{(S_{F(t)}(t), D(t)),\ S_{F(t)}(t) \in \mathcal{S}_T,\ D(t) \in \mathbb{R},\ t = 1, \ldots, T\big\},$$

where $S_{F(t)}(t)$ is the input sequence, $D(t)$ is its corresponding target value for the output at $F(t)$, and $\mathcal{S}_T$ is the token space.

For recurrent neural networks the following notation needs to be defined:

1. $\mathcal{X}_{0,s}(t) \doteq [X_0(1,t) \cdots X_0(F(t),t)] \in \mathbb{R}^{n(0),F(t)}$ is called the *token trace*. Let $F^* \doteq \sum_{t=1}^{T} F(t)$. $\mathcal{X}_0 \doteq [\mathcal{X}_{0,s}(1) \cdots \mathcal{X}_{0,s}(T)] \in \mathbb{R}^{n(0),F^*}$ is called the *input trace*. It collects all the tokens of the learning environment.
2. Let us define $\mathcal{X}_{0,f}(f) \doteq [X_0(f,1) \cdots X_0(f,T)] \in \mathbb{R}^{n(0),T}$. If $f > F(t)$, then we assume $X_0(f,t) \doteq 0$. $\mathcal{X}_{0,f}(f)$ is referred to as the *frame trace*.
3. The matrix $\mathcal{X}_s(t) \doteq [X(0,t) \cdots X(F(t)-1,t)] \in \mathbb{R}^{n(1),F(t)},\ 1 \leq t \leq T$, is referred to as the *output token trace*. $\mathcal{X} \doteq [\mathcal{X}_s(1) \cdots \mathcal{X}_s(T)] \in \mathbb{R}^{n(1),F^*}$ is called the *neuron trace*. It collects the outputs of all the neurons of the learning environment.
4. The matrix $\mathcal{X}_f(f) \doteq [X(f,1) \cdots X(f,T)] \in \mathbb{R}^{n(1),T},\ 0 \leq f \leq F_{\max} - 1$, is called the *output frame trace*. For $f \geqslant F(t)$ we define $X(f,t) \doteq 0$.
5. Let us define $y_i(f,t) \doteq \partial E / \partial a_i(f,t)$. The $y_i(f,t)$ *delta error* can be collected in vectorial structures similar to those used for inputs and neuron outputs. Hence, $\mathcal{Y}_s(t) \in \mathbb{R}^{n(1),F(t)}$ is called the the *delta token trace*,

[7]Thus, for the sake of simplicity, we will discard the layer index.

$\mathcal{Y}_f(f) \in \mathbb{R}^{n(1),T}$ is referred to as the *delta frame trace*, and $\mathcal{Y} \in \mathbb{R}^{n(1),F^*}$ is the *delta trace*.

6. The gradient of the cost function $E_T(w^1_{i,j}, w^0_{k,l}; \mathcal{N}, \mathcal{L}_e)$ w.r.t. the weights $\mathcal{W}^1 \in \mathbb{R}^{n(1),n(1)}$ and $\mathcal{W}^0 \in \mathbb{R}^{n(1),n(0)}$ may be kept in the matrices $\mathcal{G}_{\mathcal{W}^1} \in \mathbb{R}^{n(1),n(1)}$ and $\mathcal{G}_{\mathcal{W}^0} \in \mathbb{R}^{n(0),n(1)}$, respectively. Notice that the transpose of these matrices must be used for weight updating.

III. LEARNING WITH NO LOCAL MINIMA

This section contains some theoretical results aimed at guaranteeing local minima free error surfaces under some hypotheses on networks and data. The identification of similar conditions ensures global optimization just by using simple gradient descent learning algorithms (*batch mode*). The interest in similar conditions is motivated by the comparison with the perceptron learning (PL) algorithm [3, 25, 26] and with ADALINE [27, 28] for which optimal learning is guaranteed under the assumption of linearly separable patterns.

Baldi and Hornik [29] proposed a first interesting analysis on local minima under the assumption of linear neurons. They proved that the attached cost function has only saddle points and a unique global minimum. As the authors pointed out, however, it does not seem easy to extend such an analysis to the case of nonlinear neurons. Sontag and Sussman [30] provided other conditions guaranteeing local minima free error surfaces in the case of single-layered networks of sigmoidal neurons. When adopting LMS-threshold cost functions, they proved the absence of local minima for linearly separable patterns. This is of remarkable interest, in that it allows us to get rid of spurious local minima arising with an improper joint selection of cost and squashing functions [31]. Shynk [32] showed that the perceptron learning algorithm may be viewed as a steepest-descent method by defining an appropriate performance function. In so doing, the problem of optimal convergence in perceptrons turns out to be closely related to that of the shape of such performance function.

However, although interesting, these analyses make no prediction in the case of networks with nonlinear hidden neurons.

Beginning from an investigation of small examples, Hush and Salas [33] gave some interesting qualitative indications on the shape of the cost surface. They pointed out that the cost surface is mainly composed of plateaus, which extend to infinity in all directions, and very steep regions. When the number of patterns is "small," they observed "stair-steps" in the cost surface, one for each pattern. When increasing the cardinality of the training set, however, the surface become smoother. Careful analyses on the shape of the cost surface, also supported by a detailed investigation of an example, were proposed by Gouhara *et al.* [34, 35]. They introduced the concepts of *memory and learning surface*. The learning surface is the surface attached to the cost function, whereas the memory surface is the region in the weight space that represents the solution to the problem of mapping the patterns onto the target values. One of their main conclusions is that the learning process "...has the tendency to descend along the memory surfaces because of the valley–hill shape of the learning surface." They also suggest what the effect of the P and S symmetries[8] [37] is on the shape of the learning surface.

In the next sections, we give a detailed review of studies that addresses the problem of local minima for networks with nonlinear hidden layers from a theoretical point of view.

A. STATIC NETWORKS FOR PATTERN CLASSIFICATION

In this section, all our analyses and conclusions rely upon the following assumption:

ASSUMPTION 1. *The entire training set can be learned exactly.*

This hypothesis can be met when using a network with just one hidden layer, provided that it is composed of a sufficient number of neurons [38–41]. According to more recent research, when using the hard-limiting neurons ($\text{sgn}(\cdot)$ function), the perfect mapping of all the training patterns can also be attained by using at least $T - 1$ hidden neurons [42]. It may be argued that this architectural requirement is unreasonable in most interesting problems dealing with redundant information. On the other hand, for many problems of this kind (e.g., pattern recognition), the architectures that are commonly selected simply by *trial and error* give errors that are very close to 0, and it is likely even to find examples showing perfect mapping (see, e.g., [43, 44]).

1. Feedforward Networks

We begin by imposing the condition for finding stationary points in the cost function. On the basis of the definitions given in the previous section and on the

[8]P and S symmetries are weight transformations that do not affect the network output. The P symmetry can act on any vector W_i of input weights of a given hidden neuron i. The vectors of the hidden neurons of an assigned layer can be permuted in any order, because their global contribution to the upper layer is not affected at all. The S symmetry acts for symmetric squashing functions such that $f(a) = -f(-a)$. In this case, a transformation of the weights can be created which inverses the sign of all the input and output connections of a neuron. More recently, Chen *et al.* [36], have proven that when using P and S symmetries, there are $n!2^n$ different assignments with the output.

backpropagation rule,[9] the gradient of the cost can be written as

$$\mathcal{G}_{l-1} = \left(\mathcal{X}^e_{l-1}\right)'\mathcal{Y}_l, \qquad l = 1, \ldots, L, \tag{5}$$

where $\mathcal{X}^e_{l-1} \doteq [\mathcal{X}_{l-1}\ \Pi] \in \mathbb{R}^{T,n(l-1)+1}$, and $\Pi \doteq [1, \ldots, 1]' \in \mathbb{R}^T$.

The following theorem introduces some hypotheses primarily concerning the network architecture, but also the relationship between the network and the learning environment. Basically, the theorem gives a sufficient condition for local minima free error surfaces in the case of pyramidal networks, commonly used in pattern recognition.

THEOREM 1. *The cost function $E_T^{\text{LMS}}(w_{i,j}; \mathcal{N}, \mathcal{L}_e)$ is local minima free if the network $\mathcal{N}$ and the associated learning environment $\mathcal{L}_e$ meet the following PR1 (pattern recognition) hypotheses:*

1. $n(l+1) \le n(l)$, $l = 1, \ldots, L-1$ *(pyramidal hypothesis).*
2. *The weight layer matrices $\mathcal{W}_l$, $l = 1, \ldots, L-1$, are full-rank matrices.*
3. $\text{Ker}\left[\mathcal{X}_0'\right] \cap \mathcal{S}_1^{\mathcal{Y}} = \{0\}$.

Proof Sketch (see [50] for more details). Because of PR1.3, $\mathcal{G}_1 = (\mathcal{X}^e_0)'\mathcal{Y}_1 = 0$ implies $\mathcal{Y}_1 = 0$. According to the backpropagation step, $\tilde{\mathcal{Y}}_l = \mathcal{Y}_{l+1}\mathcal{W}_l$. From $\mathcal{Y}_l = 0$, $\tilde{\mathcal{Y}}_l = 0$ follows and, consequently, $0 = \tilde{\mathcal{Y}}_l = \mathcal{Y}_{l+1}\mathcal{W}_l$, from which $\mathcal{Y}_{l+1} = 0$ because of PR1.2. Because $\mathcal{Y}_1 = 0$, $\mathcal{Y}_L = 0$ follows by induction on l. Finally, $\mathcal{Y}_L = 0$ implies $E_T^{\text{LMS}} = 0$. ■

A few remarks concerning the hypotheses of this theorem help to clarify its meaning. First, we notice that the pyramidal assumption does not involve the input layer ($l = 0$). This hypothesis appears as a natural consequence of the task accomplished by the neurons in networks devoted to classification, because the more a hidden layer gets close to the output, the more the information is compressed. This structure is often adopted in many practical experiments, and particularly for classification (see, e.g., [43, 44, 51–54]). Second, as also pointed out in [29], the hypothesis concerning the rank of the weight layer matrices $\mathcal{W}_l$ is quite reasonable.

Finally, the PR1.3 hypothesis involves both the network and the learning environment. Unfortunately, it is quite hard to understand its practical meaning as it requires knowledge of $\mathcal{S}_1^{\mathcal{Y}}$, that is, the set of all the $\mathcal{Y}_1$ generated when varying the weights in Ω. The computation of $\mathcal{S}_1^{\mathcal{Y}}$ seems to be very hard with no assump-

[9] As pointed out by le Cun [45], to some extent, the basic elements of backpropagation can be traced back to the famous book of Bryson and Ho [46]. A more explicit statement of the algorithm has been proposed by Werbos [47], Parker [48], le Cun [49], and the members of the PDP group [4]. Although many researchers have contributed in different ways to the development and proposition of different aspects of BP, there is no question that Rumelhart and the PDP group are given credit for the current high diffusion of the algorithm.

tion on the problem at hand. Basically, this condition involves both the network and the learning environment very closely, thus stating formally the intuitive feeling that the presence of local minima depends heavily on the mutual relationship between the given problem and the architecture chosen for its solution.

We can think of Theorem 1 as a first general attempt to investigate the presence of stationary points in the error surface in the case of pyramidal networks. From this general point of view, the problem is very complex and the role of this theorem is essentially that of moving all the difficulties to condition PR1.3. A case in which PR1.3 holds is when all the patterns are linearly independent, because, in that case, $\mathrm{Ker}[\mathcal{X}_0'] = 0$. It is worth mentioning that if $\mathrm{Ker}[\mathcal{X}_0'] = 0$ holds, then the PR1.3 hypothesis only involves the learning environment. This is a very desirable property but, on the other hand, when the patterns are linearly independent the number of patterns T cannot be greater than $n(0)$. This is a very serious restriction, because the number of inputs dramatically limits the cardinality of the learning environment. However, as it will be shown later, this condition can be extended to more significant practical cases.

Theorem 1 can be easily restated to provide a necessary and sufficient condition guaranteeing local minima free error surfaces.

COROLLARY 1. *Let us consider experiments based on pyramidal networks matching* PR1.2 *and learning environments satisfying Assumption* 1. *The associated cost function $E_T^{\mathrm{LMS}}(w_{i,j}; \mathcal{N}, \mathcal{L}_e)$ is local minima free if and only if, for all the stationary points $\mathcal{W}^o$, $\mathcal{Y}_1(\mathcal{W}^o) = 0$ holds.*

Proof. If $\mathcal{Y}_1(\mathcal{W}^o) = 0$ holds for all the stationary points $\mathcal{W}^o$, then $E_T^{\mathrm{LMS}}(w_{i,j}; \mathcal{N}, \mathcal{L}_e)$ is local minima free using the same arguments of the proof of Theorem 1. On the other hand, if $E_T^{\mathrm{LMS}}(w_{i,j}; \mathcal{N}, \mathcal{L}_e)$ has only one global minimum, $E_T^{\mathrm{LMS}} = 0$ implies $\mathcal{Y}_L(\mathcal{W}^o) = 0$ for this point, from which $\mathcal{Y}_1(\mathcal{W}^o) = 0$ follows, because of the recursive application of the backpropagation relationship $\tilde{\mathcal{Y}}_{l-1} = \mathcal{Y}_l \mathcal{W}_{l-1}$ and because $\tilde{\mathcal{Y}}_{l-1} = 0$ presupposes $\mathcal{Y}_l = 0$. ■

This corollary deserves attention primarily for the intriguing relationship with Rosenblatt's PL algorithm [26] and ADALINE [27] which it suggests. A close relationship between learning in multilayered and single-layered networks comes out because, under the PR1.1 and PR1.2 hypotheses, the search for global minima of the cost in the case of multilayered networks is restricted to inspecting $\mathcal{Y}_1(\mathcal{W}^o) = 0$ only, exactly as in single-layered networks. This also makes clear that the additional problem coming out in the analysis of multilayered networks is that of providing a description of $\mathcal{Y}_1(\mathcal{W}^o)$, that is, of space $\mathcal{S}_1^{\mathcal{Y}}$.

In order to discover meaningful conditions with a straightforward and practical interpretation, we propose investigating the case of patterns that are separable by

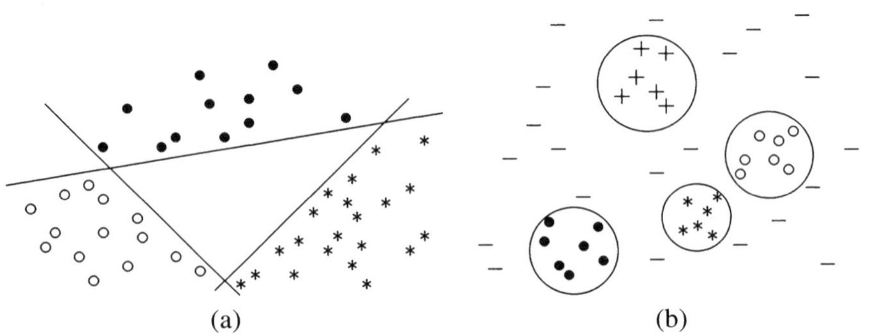

Figure 1 Separable patterns: (a) linearly separable patterns; (b) patterns separated by hyperspheres. Reprinted with permission from *Neurocomputing* 13, 1996; courtesy of Elsevier Science–NL.

a family of surfaces $\Phi_c(\cdot),\ c = 1, \ldots, C$; that is,

$$\begin{aligned} \Phi_c(X_0(t)) &< 0, \qquad \forall t \text{ in class } c,\\ \Phi_c(X_0(t)) &> 0, \qquad \text{otherwise.} \end{aligned}$$

For example, if $\Phi_c(X_0(t)) \doteq A_c' X_0^e(t),\ A_c \in \mathbb{R}^{n(0)+1}$, the patterns are *linearly separable* (see Fig. 1a), whereas if $\Phi_c(X_0(t)) = \|X_0(t) - \mathcal{C}_c\| - r_c$ the patterns are *separable by hyperspheres*, where $\mathcal{C}_c$ and r_c are the center and the radius of the hypersphere, respectively. In the last case, all "positive examples" of different classes in the learning environment belong to regions bounded by hyperspheres, whereas all eventual negative examples, which do not belong to the assumed classes, are in the complementary domain (see Fig. 1b).

The following theorem deals with the simplest case of linearly separable patterns and specializes the results given in Theorem 1 under this new assumption.

THEOREM 2. *The cost function $E_T^{\mathrm{LMS}}(w_{i,j}; \mathcal{N}, \mathcal{L}_e)$ is local minima free if the network and the learning environment satisfy the following PR2 hypotheses:*

- *Network*

  1. *The network has only one hidden layer ($L = 2$).*
  2. *The network has C outputs where C is the number of classes.*
  3. *Full connections are assumed from the input to the hidden layer. The hidden layer is divided into C sublayers, $H_1, \ldots, H_c, \ldots, H_C$, and connections are only permitted from any sublayer to the associated output unit (see Fig. 2). The sublayer H_c contains $n_c(1)$ neurons.*

- *Output coding*
  Exclusive coding is used for the output.
- *Learning environment*
  All the patterns of $\mathcal{L}_e$ are linearly separable.

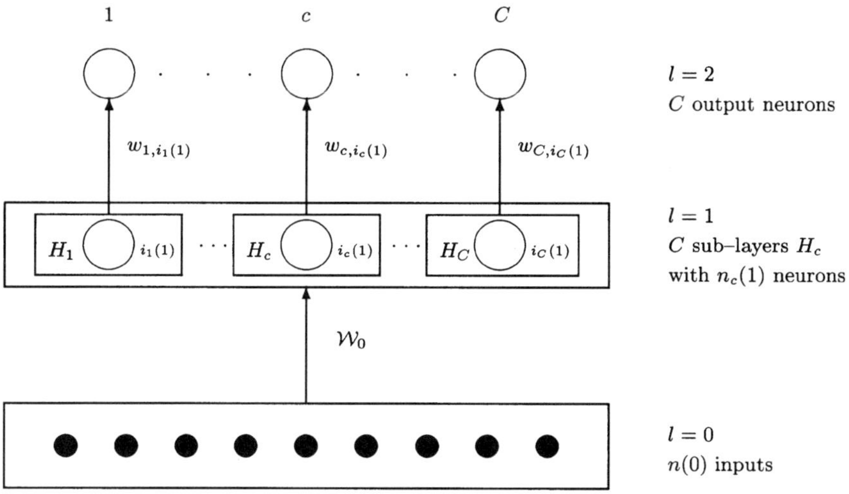

Figure 2 Network architecture with multiple outputs.

Proof Sketch (see [50] for more details). Because of PR2.3, $\mathcal{Y}_1(\omega)$ is composed of elements of the same sign for patterns of the same class. Let us consider $\mathcal{G}_1 = (\mathcal{X}_0^e)'\mathcal{Y}_1 = 0$ and $A \in \mathbb{R}^{n(0)+1}$. The equation $A'(\mathcal{X}_0^e)'\mathcal{Y}_1 = 0$ has, at least, the same solutions as $\mathcal{G}_1 = 0$. When considering $\mathcal{Y}_1$'s sign and the hypothesis of linearly separable patterns, it follows that $(A'(\mathcal{X}_0^e)')\mathcal{Y}_1$ is composed of terms with the same sign. As a result, $\mathcal{Y}_1 = 0$, which, in turn, implies that E_T^{LMS} is local minima free, because of Corollary 1. ■

The hypothesis on the architecture is not very restrictive. No output interaction is assumed; that is, the outputs are computed independently of each other. This hypothesis has also been adopted for proving the interpolation capabilities of MLNs in [38, 40, 41]. Plaut and Hinton [55] have shown that these architectures learn faster than those with fully connected layers. Jacobs *et al.* [56] have considered network assumption PR2.2 as a first step toward the conception of modular architectures that are usually well suited for high generalization. When keeping the pyramidal assumption, this architectural hypothesis can be removed at the price of introducing the assumption that $\mathcal{W}_1$ is a full-rank matrix [57].

The hypothesis of linearly separable patterns suggests a comparison with Rosenblatt's perceptron. It is well known that this hypothesis is also sufficient for guaranteeing, in the case of the simple perceptron, the convergence of the δ rule [3, 25, 26] to configurations where all the patterns are correctly classified. Nevertheless, this must not lead us to conclude that when dealing with linearly separable patterns, perceptrons and MLNs are equivalent. As pointed out in [7, 44], the generalization to new examples is significantly better for networks with a hidden layer. In the case of MLNs, the assumption of linearly separable patterns is only sufficient to guarantee the convergence of a gradient descent learning algorithm. Moreover, also in the presence of local minima, one still has a chance to perform global optimization. Finally, there are cases in which backpropagation gets stuck in local minima, but the resulting suboptimal solutions are still useful in practice, whereas the PL algorithm for the perceptron oscillates.[10] As a result, we can state that the superiority of MLNs with respect to single-layer perceptrons is not only a matter of experiments, but that it can be established on the basis of theoretical results. It is still an open research problem to identify sharper sufficient conditions guaranteeing local minima free cost surfaces.

2. Radial Basis Functions

The results stated in Theorems 1 and 2 involve classic sigmoidal neurons with activation computed by the dot product of weights and inputs. When looking into the details of the proofs of these theorems, it is quite easy to realize that they are essentially based on the network architecture that is responsible for the factorization of the gradient stated by Eq. (5). This factorization is gained by using backpropagation and has nothing to do with the special neuron with which we are dealing.

These remarks suggest extending the previous results to other multilayered networks based on different neurons. From the different choices following Eq. (1), the radial basis function networks [21, 59] seem the most interesting. We consider multilayered architectures with a hidden layer of *locally tuned* units [21] and an output layer of *ordinary* processing units [4].[11] The multilayered architecture of radial basis function networks makes it possible to give the gradient a factorization that looks analogous to Eq. (5).

- For the output layer, the use of the BP computing scheme makes it possible to determine the stationary points, as for MLN networks, by means of

$$\mathcal{G}_2 = \left(\mathcal{X}_1^e\right)'\mathcal{Y}_2 = 0.$$

- For locally tuned hidden neurons, the use of backpropagation leads to

$$\mathcal{G}_{i(1)} = 0 \Rightarrow \tilde{\mathcal{X}}'_{0,i(1)} Y_{i(1)} = 0, \qquad i(1) = 1, \ldots, n(1),$$

[10] This undesirable behavior, however, is not found for single-layered networks trained by LMS [28]. For Rosenblatt's perceptron, there is a generalization of the PL algorithm, called the pocket algorithm [58], that avoids cycling in the case of nonlinearly separable patterns.

[11] The RBFs proposed in [21] have linear outputs. The assumption made in this chapter, however, does not change the essence of the analysis, which can also be carried out under the hypothesis of linear outputs.

where

$$\tilde{\mathcal{X}}'_{0,i(1)} \doteq \begin{bmatrix} -2\dfrac{\mathcal{X}'_0 - \mathcal{M}'_T(\overline{W}'_{i(1)})}{\sigma^2_{i(1)}} \\ -2\dfrac{\mathcal{A}'_{i(1)}}{\sigma_{i(1)}} \\ \Pi \end{bmatrix}$$

and $\mathcal{M}_T(\cdot)$ is the *T-matrix-replica operator*, which creates a matrix with T rows all equal to the argument $\overline{W}'_{i(1)}$. $\mathcal{A}'_{i(1)} \doteq [A_{i(1)}(1), \ldots, A_{i(1)}(T)]' \in \mathbb{R}^T$, where $A_{i(1)}(t) \doteq \|X_0(t) - \overline{W}_{i(1)}\|^2/\sigma^2_{i(1)}$, $t = 1, \ldots, T$, derives from differentiation with respect to Gaussian widths, and $\Pi \doteq [1, \ldots, 1]' \in \mathbb{R}^T$ derives from biases.

We can provide a natural extension of Theorems 1 and 2, stated for sigmoidal networks, to this case.

THEOREM 3. *The cost function $E_T^{\mathrm{LMS}}(w_{i,j}; \mathcal{N}, \mathcal{L}_e)$ is local minima free if the network $\mathcal{N}$ and the associated learning environment $\mathcal{L}_e$ satisfy the following PR1-bis hypotheses:*

1. $n(2) \le n(1)$ *(pyramidal hypothesis).*
2. *The weight layer matrix $\mathcal{W}_1$ is full row rank.*
3. *The following relationships between the kernel of matrices $\tilde{\mathcal{X}}'_{0,i(1)}$ and $\mathcal{S}^{\mathcal{Y}}_{i(1)}$ hold:*

$$\mathrm{Ker}\big[\tilde{\mathcal{X}}'_{0,i(1)}\big] \cap \mathcal{S}^{\mathcal{Y}}_{i(1)} = \{0\}, \qquad \forall i(1) = 1, \ldots, n(1).$$

Proof (see [10]). ■

Nevertheless, as in the case of MLNs, in order to discover more significant conditions, we propose a geometric hypothesis on the data. The presence of locally tuned processing units leads us to consider inputs separated by hyperspheres. This assumption turns out to be dual with respect to that of linearly separable patterns for inner product–based neurons.

THEOREM 4. *The cost function $E_T^{\mathrm{LMS}}(w_{i,j}; \mathcal{N}, \mathcal{L}_e)$ is local minima free if the hypotheses of Theorem 2 are satisfied for the network $\mathcal{N}$ and the output coding, whereas the patterns of the learning environment $\mathcal{L}_e$ are separated by hyperspheres (PR2-bis hypotheses).*

Proof.

• If the *PR*2-bis hypotheses hold, then

$$\tilde{\mathcal{X}}'_{0,i(1)} Y_{i(1)} = 0, \qquad i(1) = 1, \ldots, n(1), \tag{6}$$

only admits the solution $Y_{i(1)} = 0$. In fact, if the patterns are separable by hyperspheres, for each $W_{i(1)} \in \mathbb{R}^{n(0)}$, there exist C vectors $\Phi'_c(W_{i(1)}) = (\tilde{\Phi}'(W_{i(1)}), 1, \beta(W_{i(1)}))' \in \mathbb{R}^{n(0)+2}$, where

$$\tilde{\Phi}(W_{i(1)}) = \frac{2(W_{i(1)} - \mathcal{C}_c)}{\sigma_{i(1)}}, \qquad c = 1, \ldots, C,$$

and

$$\beta(W_{i(1)}) = \frac{2}{\sigma^3_{i(1)}} [r_c^2 - \|\mathcal{C}_c - W_{i(1)}\|^2], \qquad c = 1, \ldots, C,$$

such that

$$\begin{aligned} \operatorname{sgn}[\Phi'_c \tilde{\mathcal{X}}'_{0,i(1)}] &= \operatorname{sgn}\Big[-\frac{1}{\sigma^2_{i(1)}} \big\{ \tilde{\Phi}'\big(\mathcal{X}_0(t) - \mathcal{M}_T(W_{i(1)})\big) \\ &\qquad + \|\mathcal{X}_0(t) - W_{i(1)}\|^2 \big\} + \beta \Big] \\ &= \big[-e_1^1 | -e_1^2 | \cdots | e_1^c | \cdots | -e_1^C \big]. \end{aligned} \tag{7}$$

The solution $Y_{i(1)}$ of Eq. (6) must necessarily satisfy the following equations:

$$\Phi'_c \tilde{\mathcal{X}}'_{0,i(1)} Y_{i(1)} = 0, \qquad \forall c = 1, \ldots, C.$$

Hence, for each class c and for each neuron $i_c(1)$ of the hidden sublayer H_c, the following equality must hold:

$$\Phi'_c \tilde{\mathcal{X}}'_0 Y_{i_c(1)} = 0.$$

Because of (7) and the $Y_{i_c(1)}$ sign structure, it easily follows that $Y_{i_c(1)} = 0$, $\forall i_{c(1)} = 1, \ldots, n_{c(1)}$ and $c = 1, \ldots, C$. Consequently $Y_{i(1)} = 0$ is the unique solution of (6).

• Let us assume the existence of a stationary point whose weights connecting the neurons of a hidden sublayer H_c to a corresponding output c are all null, that is,

$$w_{c,i_c(1)} = 0, \qquad \forall i_c(1) \in H_c. \tag{8}$$

Then this stationary point is not a local minimum.

Condition (8) implies that for $i(2) = c$ we get

$$y_{i(2)}(t) = f'(w_c)\big(f(w_c) - d_c(t)\big) \neq 0, \tag{9}$$

where w_c is the bias for output c, $d_c(t)$ is the target for pattern t, and f is the squash function of output neurons.

From the network hypotheses, it follows that the Hessian matrix $\mathcal{H}$ has the following block-diagonal structure

$$\mathcal{H} = \text{diag}[\mathcal{H}_1, \ldots, \mathcal{H}_c, \ldots, \mathcal{H}_C], \tag{10}$$

where $\mathcal{H}_c \in \mathbb{R}^{(n(0)+2)n_c(1)+n_c(1)+1,(n(0)+2)n_c(1)+n_c(1)+1}$ is the submatrix associated with the subnetwork defined by the input layer, hidden sublayer H_c, and output c. This submatrix is partitioned as follows:

$$\mathcal{H}_c = \begin{bmatrix} \mathcal{H}_c(1,1) & \mathcal{H}_c(1,0) \\ \mathcal{H}_c(0,1) & \mathcal{H}_c(0,0) \end{bmatrix},$$

where $\mathcal{H}_c(1,1) \in \mathbb{R}^{n_c(1)+1,n_c(1)+1}$ and $\mathcal{H}_c(0,0) \in \mathbb{R}^{(n(0)+2)n_c(1),(n(0)+2)n_c(1)}$ are generated by the weights connecting the hiddens to the outputs and the inputs to the hiddens, respectively, whereas $\mathcal{H}_c(0,1) = \mathcal{H}_c(1,0)' \in \mathbb{R}^{(n(0)+2)n_c(1),n_c(1)+1}$ represents the cross-contribution of these weights.

We observe that condition (8) implies that the delta trace $Y_{i_c(1)} \in \mathbb{R}^T$, $\forall i_c(1) \in H_c$, is identically null. Hence, from the BP equations, we obtain that $\mathcal{H}_c(0,0)$ is the null matrix. Now, the generic element of $\mathcal{H}_c(0,1)$ has the following expression:

$$\frac{\partial^2 E_T^{\text{LMS}}}{\partial w_{i_c(1),j(0)} \partial w_{i(2),i_c(1)}} = 2 \sum_{t=1}^{T} \frac{(x_{j(0)}(t) - w_{i_c(1),j(0)})}{\sigma_{i_c(1)}^2} \tilde{f}'(a_{i_c(1)}(t)) y_{i(2)}(t),$$

where $i(2) = c$, $j(0)$ denotes the generic input, and $\tilde{f}$ is the locally tuned output function. Hence, any subcolumn $\mathcal{H}_c^v(0,1) \in \mathbb{R}^{n(0)+2}$ of $\mathcal{H}_c(0,1)$ can be written as

$$\mathcal{H}_c^v(0,1) = \tilde{\mathcal{X}}'_{0,i(1)} \begin{bmatrix} \tilde{f}'(a_{i_c(1)}(1)) y_c(1) \\ \cdots \\ \tilde{f}'(a_{i_c(1)}(T)) y_c(T) \end{bmatrix} \doteq \tilde{\mathcal{X}}'_{0,i(1)} \widehat{Y}_{c(2)},$$

where $\widehat{Y}_{c(2)} \in \mathbb{R}^T$ has the sign structure explained by the right-hand side of Eq. (7). From the previous item and condition (9), we get that $\mathcal{H}_c^v(0,1)$ is not identically null. Therefore, the matrix $\mathcal{H}_c$ has the following structure:

$$\mathcal{H}_c = \begin{bmatrix} P & D' \\ D & 0 \end{bmatrix},$$

where the matrix P is assumed positive definite and D is not the null matrix. By applying Sylvester's theorem (see, e.g., [60, p. 104]), it can be obtained that $\mathcal{H}_c$ has both positive and negative eigenvalues, and, hence, the proof directly follows from (10).

In conclusion, if the PR2-bis hypotheses hold, the stationary points defined by the condition of zero weight for the connections between any hidden sublayer to the corresponding output are not local minima. Therefore, only the case in which the weights connecting each hidden sublayer with the corresponding output are not all null remains to be considered. In this case, the proof easily follows from a direct application of Theorem 3 because the PR2-bis network conditions satisfy the PR1.1-bis and PR1.2-bis hypotheses, and the PR2-bis learning environment and output coding conditions satisfy the PR1.3-bis hypothesis. ■

Some remarks are worth mentioning concerning this result.

Remark 1 (Output coding). It is quite easy to see that Theorem 4 also holds for networks with only one output and *positive examples* belonging to a region delimited by a given hypersphere. In that case, the class c is simply coded by 1, whereas $\bar{c}$ is coded by 0. This extension also holds in the case of C exclusively coded classes and a set of *negative examples* that do not belong to the assumed classes.

Remark 2 (Beyond hyperspheres?). When choosing more general processing units for the hidden layer, one may wonder if the results put forward in Theorem 4 also hold for different separation surfaces. It can be shown that this is indeed the case for more general radial basis functions in which the locally tuned processing units follow the equation:

$$a_{i(1)}(t) = \big(X_0(t) - W_{i(1)}\big)' Q_{i(1)} \big(X_0(t) - W_{i(1)}\big),$$

$Q_{i(1)}$ being a symmetric matrix associated with neuron $i(1)$. This extension is related to *input preprocessing*, which allows us to obtain complex separation surfaces with just single-layer networks (linear machines; see, e.g., [25, pp. 29–30]). Notice that the quadratic preprocessing for linear machines *is not equivalent* to radial basis functions for at least a couple of reasons. First, with linear machines and quadratic preprocessing, the quadratic separation between patterns of different classes is a necessary condition for learning their parameters. That is, if the separation condition does not hold, no solution can be found. However, with radial basis functions, the condition established by Theorem 4 *is only sufficient* and, therefore, solutions can still be discovered also in the case in which the separation condition is not met. Second, the generalization to new examples is generally different. With RBFs, the prior knowledge on a given problem can be exploited much better than for linear machines and this may lead us to significant improvements in generalization.

Remark 3 (Forcing limited hyperspheres). In many cases, for discriminating patterns of different classes, the optimization algorithms may discover hyperspheres with "large" σ. That means the same pattern discrimination could have been accomplished with significantly smaller hyperspheres. This is due to the fact that, in these cases, we do not restrict the space of the possible solutions with "negative examples" that do not belong to the assigned classes. Unfortunately, in most relevant applications, the definition of the set of negative examples is almost impossible. We can, however, constrain the radii of the hyperspheres and learn under such hypotheses. Theorem 4 also holds under these constraints. It suffices to assume $\sigma = B/(1 + \exp(-\sigma_p))$ and to consider the cost as a function of σ_p. In so doing, σ is constrained into the interval $[-B, +B]$, which can be properly selected with some prior knowledge of the problem at hand.

Remark 4 (Relationships with hybrid learning). It is well known that pattern mode weight updating departs, to some extent, from exact gradient descent of the cost function E_T. The adoption of "small" learning rates,[12] however, reduces arbitrarily the difference with respect to batch mode [4]. Following the gradient descent scheme in pattern mode, the locally tuned processing unit weights are updated according to

$$W_{i(1)}(r+1) = W_{i(1)}(r) + \frac{2\varepsilon y_{i(1)}(t)}{\sigma_{i(1)}^2}(X_0(t) - W_{i(1)}(r)), \quad \forall i(1) = 1, \ldots, n(1),$$

where $W_{i(1)} \doteq [w_{i(1),1(0)}, \ldots, w_{i(1),n(0)}]'$. This equation is essentially the same as that used for performing the updating of codebook vectors in learning vector quantization (LVQ)[61]. In particular, $2\varepsilon y_{i(1)}(t)/\sigma_{i(1)}^2$ plays the role of the coefficient $\alpha(t)$ and is consistent with the requirement of being asymptotically null. The basic difference, however, is that with BP optimization schemes any pattern of the learning environment affects, in principle, any vector $W_{i(1)}$ (codebook vector), whereas with LVQ just the closest patterns to $X_0(t)$ in the Euclidean metric are taken into account (one vector in LVQ1, two vectors in LVQ2 and LVQ3). However, with BP optimization schemes, the updating of the codebook vectors depends strongly on the distance $\|X_0(t) - W_{i(1)}\|$. For a given pattern $X_0(t)$, no codebook vectors $W_{i(1)}$ such that $a_{i(1)}(t) \approx 0$, in practice, are updated because, in that case, $y_{i(1)} \approx 0$. One more remarkable difference is that with BP schemes one may have two or more codebook vectors reacting to a given pattern almost in the same way. These remarks also suggest that hybrid learning, as proposed in [61], may turn out to be less affected by suboptimal solutions. The self-organization step provides a first tuning of the codebook vectors such that they go away from each other. In so doing, in practice, the number of codebook vectors reacting to a given pattern decreases. The resulting effect is that of linking the reaction of patterns to the processing units. As a result, for a given processing unit the optimization takes place in a subset of the learning environment. Hence, hybrid learning

[12]The term "small" is strictly related to the dimension of the learning environment.

performs a sort of *divide and conquer*, and is likely to exceed ordinary BP optimization starting with random weights. The conclusion is that, in many practical cases, hybrid learning can be very successful, particularly if BP optimization is used as second step instead of simple LMS.

B. Neural Networks with "Many Hidden Units"

In the previous section, we investigated the presence of local minima without considering the influence of the number of hidden units.

Let us consider a two-layered static network with linear output units[13] and let E_T^Q be the cost function. The following theorem puts forward the relationship between local minima and the number of hidden neurons.

THEOREM 5. *The cost function $E_T^Q(w_{i,j}; \mathcal{N}, \mathcal{L}_e)$ is local minima free[14] provided that the number of hidden units $n(1) = T - 1$.*

Proof Sketch (see [62] for more details). Let $\mathcal{W}^o$ be a stationary point for the cost function. The proof of this result can be obtained by considering the following two cases:

- $\mathcal{X}_1^e$ is nonsingular when computed in $\mathcal{W}^o$.
  Notice that, under this assumption, $\mathcal{X}_1^e = [\mathcal{X}_1, \Pi] \in \mathbb{R}^{T,T}$ is a square nonsingular matrix, and the condition $\mathcal{G}_2 = (\mathcal{X}_1^e)'\mathcal{Y}_2 = 0$ implies directly $\mathcal{Y}_2 = 0$ which, in turn, yields $E_T^Q(w_{i,j}; \mathcal{N}, \mathcal{L}_e) = 0$.
- $\mathcal{X}_1^e$ is singular in $\mathcal{W}^o$.
  In this case, the proof needs some more technical arguments, based on the fact that, for any small neighborhood of $\mathcal{W}_1^o$, there exists $\mathcal{W}_1^n$ lying in this neighborhood such that, the resulting $\mathcal{X}_1^{e,n}$ is nonsingular. ■

This result gives PDP's claims on the role of the hidden units a clear theoretical foundation. On the other hand, the assumption on the number of hidden units is completely unrealistic in most practical problems dealing with redundant information. We are confident, however, that Poston and Yu's result [8, 9] can be extended to data distribution "clustered" on some centers. In those cases, one could look for conditions involving networks with as many hidden units as centers and relax to "quasi-optimal" configurations.

PR and Poston and Yu's conditions refer to two limit cases involving/not involving data structure assumption, respectively. PR conditions represent an attempt to break the joint relationship between Ker$[\mathcal{X}_0']$ and $\mathcal{S}_1^{\mathcal{Y}}$ (see Theorem 1). The assumption on pattern separation allows us to perform such a break and state

[13]The same conclusions can also be drawn in the case of sigmoidal output neurons.

[14]The theorem holds also when using LMS-threshold cost functions.

the absence of local minima without being involved in careful analyses on $\mathcal{S}_1^{\mathcal{Y}}$. The knowledge on $\mathcal{S}_1^{\mathcal{Y}}$ that is exploited in the PR conditions is very limited and only concerns the sign of $\mathcal{Y}_1$ [7, 10].

Poston and Yu's condition avoids the problem of describing $\mathcal{Y}_1$ by giving a very general result that is independent of the problem with which we are dealing. This is, at the same time, the strength and the weakness of this result, which is only of theoretical interest in most interesting practical problems.

Open research problems are the extension of both PR and Poston and Yu's conditions and the exploration of possible integration of the ideas on which they are based.

C. Optimal Learning with Autoassociators

The results given in the previous sections concern static networks. It has been pointed out (see, e.g., [61]) that multilayered networks can also be used in a self-organizing style and that they are very useful for compression of information [22] and for speech verification [63]. The case of linear neurons has been analyzed carefully from a theoretical point of view and it is well known that linear autoassociators are very successful as autoassociative memories [64]. One significant limitation of linear autoassociators is that they do not perform input clustering. On the contrary, *nonlinear autoassociators* deal with clustering very well [65].

Concerning the learning process, however, linear autoassociators behave significantly better. It has recently been shown [29] that linear autoassociators produce local minima free error surfaces, whereas, as pointed out in [66, 67], there is no such guarantee in the nonlinear case.

Let us choose the quadratic cost function:

$$E_T^Q \doteq \frac{1}{2}\sum_{t=1}^{T} E_t = \frac{1}{2}\sum_{t=1}^{T}\sum_{j=1}^{n(L)} \left[x_{j(0)}(t) - x_{j(L)}(t)\right]^2,$$

and let $\mathcal{N}$ be an autoassociator with linear outputs. A general result can be derived by inspecting solely the null gradient condition w.r.t. the weights connecting the last layer. This result can be established for a network of any number of layers, but is reported here only for the case of one hidden layer, typically used in practice.

THEOREM 6. *Let $\mathcal{N}$ be a network used as an autoassociator for the learning environment $\mathcal{L}_e$. Under this assumption, the following condition:*

$$\mathcal{X}_2'\mathcal{X}_2 = \mathcal{X}_2'\mathcal{X}_0 \tag{11}$$

holds at the end of the learning.

Proof Sketch (see [65] for more details). The equation $\mathcal{W}_1\mathcal{X}_1'\mathcal{Y}_2 = 0$ has, at least, the same solutions as $\mathcal{G}_2 = \mathcal{X}_1'\mathcal{Y}_2 = 0$. Because the network has linear outputs, $\mathcal{Y}_2 = \mathcal{X}_2 - \mathcal{D} = \mathcal{X}_2 - \mathcal{X}_0$. As a result, $\mathcal{W}_1\mathcal{X}_1'\mathcal{Y}_2 = 0$ implies $\mathcal{X}_2'(\mathcal{X}_2 - \mathcal{X}_0) = 0$. ■

Therefore, at the end of the learning process, this condition imposes the equality of the *output coordinate correlation* $\mathcal{X}_2'\mathcal{X}_2$ and the *input–output coordinate correlation* $\mathcal{X}_2'\mathcal{X}_0$. In this chapter, Eq. (11) is referred to as the *end-of-learning* condition. Notice that this condition is also met for pyramidal networks, as no hypothesis has been done on $\mathcal{W}_1$.

Remark 5 (The end-of-learning condition). In order to understand better the meaning of condition (11), let us consider the generic input $i(0)$ and the associated output $i(2)$. With reference to these units, the *end-of-learning* condition becomes $\|X_{i(2)}\|^2 = \langle X_{i(2)}, X_{i(0)}\rangle$ and, if we sum up w.r.t. all input (output) units, $\sum_{i=1}^{n} \|X_{i(2)}\|^2 = \sum_{i=1}^{n}\langle X_{i(2)}, X_{i(0)}\rangle$ holds. Because the scalar product and the induced norm are defined in $\mathbb{R}^T$, the sums w.r.t. the pattern coordinate i and the pattern itself t may be exchanged, thus obtaining $\sum_{t=1}^{T} \|X_2(t)\|^2 = \sum_{t=1}^{T} \langle X_2(t), X_0(t)\rangle$. This relationship has a very intriguing geometric meaning, which comes out while considering the network transformation $\mathcal{N}_a$: $\mathbb{R}^n \to \mathbb{R}^n$: $\mathcal{N}_a(X_0(t)) \to X_2(t)$, from which $\sum_{t=1}^{T} \|\mathcal{N}_a(X_0(t))\|^2 = \sum_{t=1}^{T}\langle\mathcal{N}_a(X_0(t)), X_0(t)\rangle$ follows. In the case of linear hidden units, the operator $\mathcal{N}_a$ is linear and the associated matrix N_a satisfies the relationship $N_a^2 = N_a$. N_a is, therefore, a *projection matrix* and represents the optimal solution. By analogy with linear algebra, the operator $\mathcal{N}_a$ is referred to as a *projection operator*. Moreover, if we exploit the identity $\|X_2(t) - X_0(t)\|^2 = \|X_2(t)\|^2 - 2X_0(t)X_2(t) + \|X_0(t)\|^2$, it becomes $\sum_{t=1}^{T} \|X_0(t)\|^2 = \sum_{t=1}^{T} \|\delta(t)\|^2 + \sum_{t=1}^{T} \|X_2(t)\|^2$, where $\delta(t) \doteq X_0(t) - X_2(t)$. A concise way of expressing the preceding equation is that of introducing the norm $\langle X_l\rangle \doteq (1/T)\sum_{t=1}^{T} \|X_l(t)\|^2$. Using this symbolism, the projection condition sounds like $\langle X_2, X_2\rangle < \langle X_0, X_0\rangle$. It becomes clear that the "efficiency" of the learning depends on the "energy" lost in $\sum_{t=1}^{T} \|\delta(t)\|^2$. Because the matrix $\mathcal{X}_2 \in \mathbb{R}^{T,n(2)}$ has rank $n(1)$, at most, for an efficient learning a dimensionality reduction of the input pattern must take place.

Remark 6 (The end-of-learning condition for classification tasks). Notice that Eq. (11) is properly derived for networks acting as autoassociators though it can similarly be restated when considering classification tasks. We still need to consider a linear output, but any target $\mathcal{D}$ can be assumed. In that case, Eq. (11) becomes $\mathcal{X}_2'\mathcal{X}_2 = \mathcal{X}_2'\mathcal{D}$.

D. Recurrent Neural Networks

In this section, we discuss conditions guaranteeing local minima free error surfaces for recurrent neural networks. The analysis that we put forward has many relationships with the previous results given for static networks because it relies on the structure of the gradient equations which, because of backpropagation through time (BPTT) [4, 68]), resemble Eq. (5).

The analysis that we propose leads us to identify some optimal conditions guaranteeing a local minima free cost function. These assumptions, referred to as the decoupling network assumptions (DNAs), involve both the network and the learning environment and are based on a new idea that, so far, has not been explored in the case of static networks.

Let us begin by considering the similarities with static networks that emerge very clearly when dealing with the gradient equations. In order to give the gradient a formal expression, let us consider the learning environment

$$\mathcal{L}_e \doteq \big\{\big(S_{F(t)}(t), d(t)\big),\ S_{F(t)}(t) \in \mathcal{S}_T,\ d(t) \in \{d^-, d^+\},\ t = 1, \dots, T\big\},$$

where $[d^-, d^+] \subset [\underline{d}, \overline{d}\,]$. It can be partitioned into the following sets:

$$\mathcal{C}^+ \doteq \{t \in \mathcal{L}_e\colon\ d(t) = d^+\},$$
$$\mathcal{C}^- \doteq \{t \in \mathcal{L}_e\colon\ d(t) = d^-\},$$

collecting the positive and the negative tokens, respectively.

With the adopted formalism, we can compute the gradient of the error function by the following vectorial equations:

$$\begin{aligned}
\mathcal{G}_{\mathcal{W}^0} &= \sum_{f=1}^{F_{\max}} \mathcal{X}_{0,f}(f)\mathcal{Y}_f(f)' = \sum_{t=1}^{T} \mathcal{X}_{0,s}(t)\mathcal{Y}_s(t)' = \mathcal{X}_0\mathcal{Y}',\\
\mathcal{G}_{\mathcal{W}^1} &= \sum_{f=1}^{F_{\max}} \mathcal{X}_f(f-1)\mathcal{Y}_f(f)' = \sum_{t=1}^{T} \mathcal{X}_s(t)\mathcal{Y}_s(t)' = \mathcal{X}\mathcal{Y}'.
\end{aligned} \tag{12}$$

1. Role of the Input Structure

Let us focus on the role of the input structure. This analysis is motivated by the results obtained in the case of feedforward networks, where the presence of many inputs is likely to limit the convergence problems due to local minima. We focus attention on the frame structure regardless of the dynamic relationships among frames within the sequences. The following theorem gives a first insight into the role of the input structure.

THEOREM 7. *If* rank $\mathcal{X}_0 = F^*$, *then the cost function* $E_T^{\mathrm{LMS}}(w^1_{i,j}, w^0_{k,l}; \mathcal{N}, \mathcal{L}_e)$ *has no local minima.*

Proof (see [69]).[15] ■

This condition is hardly met in practice because it requires the adoption of networks with an exaggerated number of inputs. In particular, the condition is likely to hold provided that $n(0) > F^* - 1$. On the other hand, this theorem does not fully exploit the network structure and the stated result looks interesting only from a theoretical point of view. The theorem formalizes the "intuitive" feeling that the absence of local minima is strongly related to the number of inputs of each frame. Apart from the rank deficiency of the matrix $\mathcal{X}_0$, which is not likely to hold if $n(0) > F^* - 1$, the choice of enough inputs guarantees the absence of local minima, no matter what problem we consider.[16]

One may wonder if more interesting results can be obtained when assuming some structural properties on the learning environment. Because of the network dynamics, such a structure can involve both the single frames and their sequential relationships. The next theorem gives a result that only involves the frame structure.

THEOREM 8. *The cost function $E_T^{\mathrm{LMS}}(w_{i,j}^1, w_{k,l}^0; \mathcal{N}, \mathcal{L}_e)$ has no local minima if the network $\mathcal{N}$ and the learning environment $\mathcal{L}_e$ satisfy the following hypotheses:*

- *Network*
  The matrix $\mathcal{W}^1$ is composed of nonnegative weights.
- *Learning environment*
  All the frames of $\mathcal{L}_e$ are linearly separable into two classes, depending on the token to which they belong.

Proof (see [69]). ■

Remark 7 (Network architecture). In practice, the assumption on the $w_{i,j}^1$'s sign is not restrictive. In fact, for the case of symmetric squashing functions, a mapping from a general network with no constraints on the weight sign to a network with all nonnegative weights is always possible, which makes their input–output response equivalent [69]. Notice that no constraint is placed on the $w_{i,j}^0$'s sign. Architectures for which the theorem's constraints hold have already been shown to be very useful for applications to automatic speech recognition [70].

Remark 8 (Learning environment). The assumption of linearly separable frames is certainly more reasonable than that proposed in Theorem 7. Notice, however, that only one hyperplane must separate all the frames of different

[15]The proof of this theorem, as well as that of Theorem 8, follows arguments closely related to those given for proving the theoretical results for static networks.

[16]Notice that we implicitly assume that there exists at least one solution with null cost for the given problem. As a consequence, this also places some indirect assumptions on the network architecture.

classes. Therefore, a comparison with the case of feedforward networks shows that this condition is still quite restrictive, as it would suggest dealing with linearly separable tokens.

The given results do not provide a very useful bound because they are likely to hold for networks requiring too many inputs for most interesting practical problems. From a theoretical point of view, however, they are remarkable for their clear explanation of the role of the input dimension and geometrical structure on the shape of the cost. In order to get some more general and useful results, one must be able to exploit the network architecture with its nonlinear dynamics. Basically, neither Theorem 7 nor Theorem 8 exploits the dynamic relationship between different frames, thus neglecting the most significant structural assumptions of a given problem.

2. Decoupling Network Assumptions

In order to deal with the sequential relationships of different frames, we need the introduction of the following *network unfolding matrix*. Let $\mathcal{N}_Y \in \mathbb{R}^{n(1), F_{\max}}$ be defined as follows:

1. $\mathcal{N}_Y(i, F_{\max}) \doteq \delta_{i,n(1)}$,[17] $\forall i\colon\ 1 \le i \le n(1)$.
2. $\forall j\colon\ 1 \le j < F_{\max}, \forall i\colon\ 1 \le i \le n(1)$,

$$\mathcal{N}_Y(i, j) \doteq \begin{cases} 1, & \text{if } \exists \hat{i}\colon\ 1 \le \hat{i} \le n(1),\ w^1_{\hat{i},i} \ne 0, \text{ and } \mathcal{N}_Y(\hat{i}, j+1) = 1, \\ 0, & \text{otherwise.} \end{cases}$$

$\mathcal{N}_Y$ is only associated with the network $\mathcal{N}$ and, particularly, only with the neuron connection matrix $\mathcal{W}^1 \in \mathbb{R}^{n(1),n(1)}$. The number of columns of $\mathcal{N}_Y$ is the maximum token length $F_{\max}$. Each token $t\colon\ 1 \le t \le T$ of the learning environment $\mathcal{L}_e$ can be associated with a network unfolding matrix. In particular, for each token t, the following unfolding matrix can be defined as

$$\widehat{\mathcal{Y}}_s(t) \doteq \Phi\big(\mathcal{N}_Y, F(t)\big),$$

where $\Phi(\cdot)$ is the operator that extracts the last $F(t)$ columns of $\mathcal{N}_Y$. From the gradient equation (12), we thus define the *gradient contribution matrix* for token t as

$$\widehat{\mathcal{G}}_t(i, j) \doteq \begin{cases} 1, & \text{if } [\mathcal{X}_{0,s}(t)\widehat{\mathcal{Y}}_s(t)'](i, j) \ne 0, \\ 0, & \text{otherwise,} \end{cases}$$

where $\widehat{\mathcal{G}}_t(i, j)$ is nonzero if token t contributes to the gradient element $\mathcal{G}_{\mathcal{W}^0}(i, j)$. Finally, for each element of $\widehat{\mathcal{G}}_t$, we define the *token contribution set* w.r.t. the

[17] $\delta_{i,j}$ is the Kronecker symbol, $\delta_{i,j} = 1$ for $i = j$ and $\delta_{i,j} = 0$ otherwise.

learning environment $\mathcal{L}_e$, $\lambda(i, j|\mathcal{L}_e) \doteq \{t \in \mathcal{L}_e\colon \widehat{\mathcal{G}}_t(i, j) = 1\}$, which collects all the tokens that contribute to the corresponding element of $\mathcal{G}_{\mathcal{W}^0}$.

In the following, we assume that each sequence contributes, at least, to one element of the matrix $\mathcal{G}_{\mathcal{W}^0}$. Formally, this is stated by

$$\bigcup_{i,j} \lambda(i, j|\mathcal{L}_e) = \mathcal{L}_e. \tag{13}$$

Using the preceding definitions, we can now introduce the concept of *decoupling* for the classes $\mathcal{C}^+$ and $\mathcal{C}^-$ w.r.t. a gradient component $\mathcal{G}_{\mathcal{W}^0}(\hat{i}, \hat{j})$. We say that the gradient component $\mathcal{G}_{\mathcal{W}^0}(\hat{i}, \hat{j})$ is decoupled w.r.t. the classes $\mathcal{C}^+$ and $\mathcal{C}^-$ provided that $\lambda(\hat{i}, \hat{j}|\mathcal{L}_e) = \mathcal{C}^+$ or $\lambda(\hat{i}, \hat{j}|\mathcal{L}_e) = \mathcal{C}^-$ holds.

Let us consider the case in which $\mathcal{G}_{\mathcal{W}^0}(\hat{i}, \hat{j})$ is decoupled w.r.t. $\mathcal{C}^+$ ($\mathcal{C}^-$) and $\lambda(\hat{i}, \hat{j}|\mathcal{L}_e) \subset \mathcal{C}^-$ strictly ($\lambda(\hat{i}, \hat{j}|\mathcal{L}_e) \subset \mathcal{C}^+$). In order to extend the preceding definition of decoupling, a simple algorithm can be conceived that recursively checks if some sequences in $\mathcal{L}_e$ can be decoupled by, at least, one of the gradient elements.

ALGORITHM 1 (Gradient decoupling test).

1. *Initialize*:
   $k \leftarrow 0;\ \Delta_0^+ \leftarrow \mathcal{C}^+,\ \Delta_0^- \leftarrow \mathcal{C}^-$.
2. **If** $\Delta_k^+ = \emptyset$ *and* $\Delta_k^- = \emptyset$ **then** *stop*.
3. **If** $\exists (i_k, j_k)$: $\mathcal{G}_{\mathcal{W}^0}(i_k, j_k)$ *is decoupled w.r.t.* Δ_k^+

   then $\Delta_{k+1}^- \leftarrow \Delta_k^- \setminus \lambda(i_k, j_k|\mathcal{L}_e)$ *and* $\Delta_{k+1}^+ \leftarrow \Delta_k^+$;

   else

   if $\exists (i_k, j_k)$: $\mathcal{G}_{\mathcal{W}^0}(i_k, j_k)$ *is decoupled w.r.t.* Δ_k^-

   then $\Delta_{k+1}^+ \leftarrow \Delta_k^+ \setminus \lambda(i_k, j_k|\mathcal{L}_e)$ *and* $\Delta_{k+1}^- \leftarrow \Delta_k^-$;

   else *stop*.
4. $k \leftarrow k + 1$ *and goto step* 3.

DEFINITION 1. The matrix $\mathcal{G}_{\mathcal{W}^0}$ is decoupled w.r.t. the classes $\mathcal{C}^+$ and $\mathcal{C}^-$ if Algorithm 1 terminates for $k = \hat{k}$ and $\Delta_{\hat{k}}^+ = \Delta_{\hat{k}}^- = \emptyset$.

Remark 9. It is quite easy to prove that if there exists $(\hat{i}, \hat{j})$ such that the gradient component $\mathcal{G}_{\mathcal{W}^0}(\hat{i}, \hat{j})$ is decoupled w.r.t. $\mathcal{C}^+$ and $\mathcal{C}^-$, then Algorithm 1 terminates with $\Delta_{\hat{k}}^+ = \Delta_{\hat{k}}^- = \emptyset$ and, therefore, $\mathcal{G}_{\mathcal{W}^0}$ is decoupled w.r.t. $\mathcal{C}^+$ and $\mathcal{C}^-$, too. It suffices to choose $(i_1, j_1) = (\hat{i}, \hat{j})$ at step 3 of the algorithm. The consequence is $\Delta_1^+ = \emptyset$ ($\Delta_1^- = \emptyset$); that is, the next steps will involve only elements in Δ_1^- (Δ_1^+). Because of (13), after $\hat{k}$ steps the algorithm necessarily ends with $\Delta_{\hat{k}}^+ = \Delta_{\hat{k}}^- = \emptyset$.

THEOREM 9. *The cost function* $E_T^{\mathrm{LMS}}(w_{i,j}^1, w_{k,l}^0; \mathcal{N}, \mathcal{L}_e)$ *has no local minima if the network* $\mathcal{N}$ *and the learning environment* $\mathcal{L}_e$ *satisfy the following DNA hypotheses:*

- *Network*
  The matrix $\mathcal{W}^1$ *is composed of nonnegative weights.*
- *Output coding*
  The supervision is only placed on the neuron $n(1)$ *at the end of each token.*
- *Learning environment*

  1. *The network is fed by nonnegative inputs;*
  2. *The gradient component* $\mathcal{G}_{\mathcal{W}^0}(\hat{i}, \hat{j})$ *is decoupled w.r.t. the classes* $\mathcal{C}^+$ *and* $\mathcal{C}^-$.

Proof. The proof of the theorem is based on the implications of the condition $\mathcal{G}_{\mathcal{W}^0} = 0$. Because weight constraints[18] are only assumed on the neuron weight matrix $\mathcal{W}^1$, this condition must certainly hold for any optimal solution.

- *For all* t, $E_t = 0$ *if and only if* $y_n(F(t), t) = 0$.

It follows directly from the BPTT relationships

$$y_n(F(t), t) = \begin{cases} f'(a_n(F(t), t)) l_2'(x_n(F(t), t) - d^+) & \text{if } t \in \mathcal{C}^+, \\ f'(a_n(F(t), t)) l_1'(x_n(F(t), t) - d^-) & \text{if } t \in \mathcal{C}^-, \end{cases}$$

$$y_i(F(t), t) = 0, \qquad i = 1, \ldots, n-1, \tag{14}$$

$$y_i(f, t) = f'\big(a_i(f, t)\big) \sum_k w_{k,i}^1 y_k(f+1, t), \qquad f < F(t),$$

and the definitions of E_t and $y_n(F(t), t)$.

- *If the matrix* $\mathcal{W}^1$ *is composed of nonnegative elements and there exists a neuron* i: $1 \le i < n$, *such that* $y_i(f, t) = 0$ *and* $\mathcal{N}_Y(i, f) = 1$, *then* $y_n(F(t), t) = 0$.

According to the BPTT backward step (14),

$$y_i(f, t) = f'\big(a_i(f, t)\big) \sum_k w_{k,i}^1 y_k(f+1, t)$$

holds. Because of the assumption $\mathcal{N}_Y(i, f) = 1$, a path connecting neuron i to the output n exists in the unfolded network associated with $\mathcal{N}$. Because $\mathcal{W}^1$ has nonnegative weights, along that path the weights of $\mathcal{W}^1$ are certainly positive. As a result, the proof follows from Eq. (14) by induction on f.

[18] A simple implementation of the nonnegativity constraints on $\mathcal{W}^1$ can be achieved by the introduction of *hyperparameters* $\phi_{i,j}$ such that $w_{i,j}^1 = \phi_{i,j}^2$.

• *Let the matrix $\mathcal{W}^1$ be composed of nonnegative weights. For a token t, if $y_n(F(t), t) > 0$ [$y_n(F(t), t) < 0$], then $\mathcal{Y}_s(t)$ elements are positive (negative) for all coordinates (i, f), where the correspondent element of $\widehat{\mathcal{Y}}_s(t)$ is* 1.
The proof can easily be obtained by induction on f by using the backward BPTT relationship (14) and considering the hypothesis on the sign of $\mathcal{W}^1$. The hypothesis on $\widehat{\mathcal{Y}}_s(t)$ just allows us to identify neurons and frames where the backpropagation takes place. If this assumption does not hold for indices i, f, then $y_i(f, t) = 0$.

Assume that the DNA hypotheses hold. We prove that $\forall t \in \mathcal{L}_e \Rightarrow E_t = 0$. Let us execute Algorithm 1 step by step. At the beginning, $\Delta_0^+ = \mathcal{C}^+$ and $\Delta_0^- = \mathcal{C}^-$ hold. Because of the hypothesis $\Delta_{\hat{k}}^- = \Delta_{\hat{k}}^+ = \emptyset$, there exists (i_0, j_0) such that $\mathcal{G}_{\mathcal{W}^0}(i_0, j_0)$ is decoupled w.r.t. $\mathcal{C}^+$ or $\mathcal{C}^-$. As a consequence, all the tokens $\hat{t} \in \lambda(i_0, j_0|\mathcal{L}_e)$ belong to the same class, causing the corresponding final delta errors $y_n(F(\hat{t}), \hat{t})$ to have all the same sign. Then all the delta errors $y_{i_0}(f, \hat{t})$, $\forall i, f$ have the same sign. Hence, the null gradient condition

$$\mathcal{G}_{\mathcal{W}^0}(i_0, j_0) = \sum_{\hat{t} \in \lambda(i_0, j_0|\mathcal{L}_e)} \sum_{f=1}^{F(\hat{t})} x_{j,0}(f, \hat{t}) y_i(f, \hat{t}) = 0,$$

implies that

$$\forall f(\hat{t}): x_{j_0,0}\big(\hat{t}, f(\hat{t})\big) > 0, \qquad \mathcal{N}_Y\big(i_0, f(\hat{t})\big) = 1 \Rightarrow y_{i_0}\big(f(\hat{t}), \hat{t}\big) = 0.$$

Therefore, $y_n(F(\hat{t}), \hat{t}) = 0$ follows, which, in turn, implies $E_{\hat{t}} = 0$.

Because $E_{\hat{t}} = 0$, we can consider all the tokens collected in $\lambda(i_0, j_0|\mathcal{L}_e)$ as correctly classified, thus reducing the learning environment.

Let us assume by induction on k that the application of Algorithm 1 implies $E_t = 0$ for all the tokens considered up to step $k - 1$ and choose (i_k, j_k) such that $\mathcal{G}_{\mathcal{W}^0}(i_k, j_k)$ is decoupled w.r.t. Δ_k^+ or Δ_k^-. If all the tokens in $\lambda(i_k, j_k|\mathcal{L}_e)$ are of the same class, we can proceed as before; otherwise consider the case for which $\mathcal{G}_{\mathcal{W}^0}(i_k, j_k)$ is decoupled w.r.t. Δ_k^+.[19] From this assumption, it follows that each token $t \in \mathcal{C}^+$, $t \in \lambda(i_k, j_k|\mathcal{L}_e)$ was eliminated in a previous execution of step 3 of Algorithm 1 and, therefore, $E_t = 0$. The tokens that actually contribute to element $\mathcal{G}_{\mathcal{W}^0}(i_k, j_k)$ are only from class $\mathcal{C}^-$. If we impose the condition $\mathcal{G}_{\mathcal{W}^0}(i_k, j_k) = 0$, then we deduce that $E_t = 0$ also for these tokens. Because $\Delta_{\hat{k}}^- = \Delta_{\hat{k}}^+ = \emptyset$, $\forall t \in \mathcal{L}_e \Rightarrow E_t = 0$ and, finally, $E = 0$. ■

Remark 10 (DNA and Architectural Design). The hypotheses concerning the network architecture and the output coding are the same as those of Theorem 8 and have already been discussed, whereas the conditions on the learning environment are different. The first assumption involving the input sign is not restrictive

[19] In the case in which $\mathcal{G}_{\mathcal{W}^0}(i_k, j_k)$ is decoupled w.r.t. Δ_k^-, we can proceed in the same way.

at all and can always be met in practice under simple linear translation of all the data. On the other hand, the practical implications of the last condition are more difficult to evaluate directly. However, the analysis of the network unfolding matrix suggests that the decoupling test (condition 2 on the learning environment) is likely to succeed for networks having few connections. Obviously, the choice of similar networks requires a sort of prior knowledge of the task at hand. More interestingly, the role of the DNAs can go beyond the simple test. The DNA conditions can be used to design the network architecture in order to avoid the presence of local minima [69].

EXAMPLE 1. In this example, we show how to choose the network architecture to meet the DNA conditions for the following task:

- Consider the set of the binary tokens for which $F(t) = 3p,\ p = 1, 2, \ldots, p_{\max}$. Classify these strings so that the positive strings are those for which $X_0(f, t) = 0,\ f \neq 3k,\ k = 1, 2, \ldots, p_{\max}$, whereas all the others are negative.

Because the positive strings do not generate the whole Euclidean space for each sequence length, it is possible to choose a vector that is orthogonal to these sequences. If we construct an unfolding matrix $\mathcal{N}_Y$ having this vector as a row, then the corresponding gradient component will be decoupled w.r.t. classes $\mathcal{C}^+$ and $\mathcal{C}^-$. In particular, the vector $[\ldots, 1, 1, 0, 1, 1, 0]$ meets this requirement. If we choose such a vector as the first row of the unfolding matrix (see Fig. 3b), then $\lambda(1, 1|\mathcal{L}_e) = \mathcal{C}^-$.

Because of the structure of the problem, an unfolding matrix is required in which the columns are repeated with period 3. Notice that the previous row suggested for the unfolding matrix meets this requirement. As a design choice, let us assume that the network has a *ring structure* as in Fig. 3a. In particular, the network is composed of a *ring* of three neurons containing the output neuron and

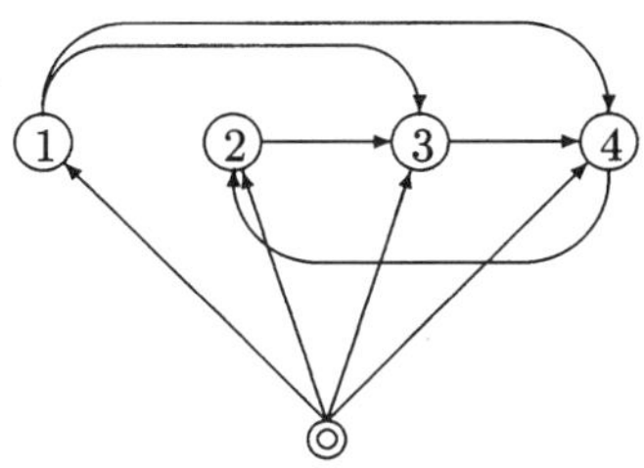

(a)

$$\mathcal{N}_Y = \begin{bmatrix} \cdots & 0 & 1 & 1 & 0 & 1 & 1 & 0 \\ \cdots & 0 & 1 & 0 & 0 & 1 & 0 & 0 \\ \cdots & 0 & 0 & 1 & 0 & 0 & 1 & 0 \\ \cdots & 1 & 0 & 0 & 1 & 0 & 0 & 1 \end{bmatrix}$$

(b)

Figure 3 Example of DNA application: (a) network architecture; (b) network unfolding matrix. Reprinted with permission from *IEEE Trans. Neural Networks* 5(2), 1994; courtesy of IEEE.

a single *control neuron* properly connected to the ring. These design choices lead us to define the network unfolding matrix depicted in Fig. 3b. Using the $\mathcal{N}_Y$ definition, the connections from the control neuron to each neuron of the ring turn out to be automatically specified (see Fig. 3).

Several experiments were carried out in order to get some comparisons between the network created using the DNA design criteria (DNA network) and some fully connected networks having one input, n fully connected hidden neurons, and one output connected to all the hidden units (1–n–1 networks; see [69]). In all cases, the DNA network exhibited a perfect generalization even when trained with few examples and the convergence behavior was significantly better. These experimental results confirm the importance of choosing the "right architecture" for solving a given problem. The design criteria we have proposed are very successful for both speeding up the convergence and improving the generalization, because they lead us to choose an architecture tuned to the task at hand and such that the associated cost function is local minima free.

The same design criteria can be extended to a class of analogous problems, whereas the basic idea is likely to be useful in general. Of course, there are problems for which this simple design scheme does not allow us to reach a complete decoupling. In those cases, one may introduce a sort of decoupling index and conceive searching algorithms, in the space of the network architectures, aimed to optimize such an index. In so doing, the ordinary learning step based on optimization in the weight space would be preceded by a searching step producing a network architecture tuned to the problem at hand.

E. On the Effect of the Learning Mode

All the analyses that have been carried out so far concern the shape of the error surface and are independent of the learning algorithm used. This makes the previous results very attractive from a theoretical point of view, because there is no doubt that any learning algorithm has to deal with the shape of the error surface which, to some extent, gives us a sort of index of complexity. Moreover, if one uses *batch mode* with gradient descent optimization techniques, then the previous results on the absence of local minima sound like results on optimal learning.

We should not neglect, however, that in many experiments, particularly those based on redundant data, the *pattern mode* weight updating turns out to be more efficient than batch mode. If we place the learning in artificial neural networks in the framework of function optimization, then the use of learning modes different from batch mode looks quite heuristic and appears not to have any theoretic foundations. All we can do is to realize that the smaller the learning rate is, the slighter pattern mode departs from correct batch mode gradient descent. On the other hand, if pattern mode is just an approximation of batch mode, there is neither theoretical nor practical interest in its application. Pattern mode and other weight updating schemes are themselves interesting and worthy of exploration.

The extension of Rosenblatt's analyses on the optimal convergence of the perceptron does not appear a simple task, but we feel that both the practical and the theoretical exploration of learning modes different from batch are very important. The results given in the previous sections suggest that progress in the field is based on the ability of optimization algorithms to go beyond the "border of local minima" efficiently. The conditions that we have reported are a first attempt to draw such a border. Beyond that border, however, the results rely heavily on the capability of our learning algorithm to perform global optimization. It is the difficulty of this general problem that suggests alternative learning modes.

Recently, Gori and Maggini [13] have proven that a feedforward network with one hidden layer and one output unit, learning with pattern mode, converges to an optimal solution if the patterns are linearly separable. Notice that this result holds independently of the learning rate, which is also the case in which pattern mode is not just an approximation of the correct function optimization performed by batch mode.

IV. LEARNING SUBOPTIMAL SOLUTIONS

In this section, we explore cases in which the learning process may not produce the optimal solution. There are several reasons for which a learning algorithm can fail to discover the optimal solution. When using batch mode, the presence of local minima in the cost function is the direct flag of potential failures. With other modes, the algorithm's behavior becomes difficult to understand only on the basis of the shape of the cost function, although it can still be useful. For example, if we use pattern mode on a large database, a potential problem is that the use of too large learning rates may lead to updating the weights only on the basis of the "recently seen" patterns. This forgetting behavior is not the only problem one has to face. Numerous different problems may emerge from special updating techniques, depending on the choice of the learning parameters. As pointed out by Lee *et al.* [71], a very remarkable problem is *premature saturation*, that is, saturation of neuron outputs in configurations that are far away from optimal solutions. Premature saturation causes the learning to cross very flat regions (plateaus) from which it may escape only if there is enough patience and computational power available. In the case of recurrent networks, this problem may become very serious when dealing with "long" sequences, because of backpropagation through time of the errors. Moreover, another source of troubles for "long sequences" is bifurcation of the learning trajectories [72], commonly found by researchers in experiments on inductive inference of regular grammars.

A. Local Minima in Neural Networks

In this section, we propose some artificial examples in which the associated error surface is populated by local minima or other stationary points. These simple examples have been conceived to clarify the mechanisms behind the creation of local minima.

As for their relevance in most interesting practical applications, one should not forget that these problems have a significantly different structure typically due to the data redundancy. As pointed out by Sontag [16], "It is entirely possible that 'real' problems—as opposed to mathematically constructed ones—will not share these pathologies." Thus, it becomes even more urgent to extensively characterize all the features that would cause "real" problems to be incorrectly faced by neural networks.

In the following, we will try to give a detailed exhibition of all these features. Then several examples are proposed and referred to having local minima in the error surface. Their analysis makes clear the mutual role of networks and data. Most serious local minima are essentially due to dealing with "difficult" problems: these minima depend on the structure of the problem (*structural local minima*) and on the fitness of the network to the assigned data. Moreover, *spurious local minima* may arise from an inappropriate joint choice of $\mathcal{N}$, $\mathcal{L}_e$, and E_T (e.g., squashing and cost functions, target values).

1. Spurious and Structural Local Minima

a. Spurious Local Minima

There have been some efforts to understand the BP behavior in feedforward networks with no hidden layer. Even if this case may also be approached with the perceptron learning algorithm [26] or in the framework of ADALINE [28][20] it provides a testing ground for hypotheses on the local minima structure of the cost function in more general cases. Brady *et al.* [31] give examples illustrating that with a linearly separable training set, a network performing gradient descent may get stuck in a solution that fails to separate the data, thus leading to the pessimistic conclusion that BP fails where perceptron succeeds. Nevertheless, the analysis of these cases reveals that those *spurious* local minima are due to an *improper* joined choice of the cost function, the nonlinear neuron functions, and the target values. A quick glance makes it clear that these examples only hold when choosing targets different from the asymptotic squashing function values. As pointed out in [30], using instead an LMS-threshold cost function (see Section II), where values "beyond" the targets are not penalized, these counterexamples cease to exist, whereas a convergence theorem that closely parallels that of perceptrons holds [7, 57] also

[20]The choice of the algorithm depends on the use of hard-limiting or linear neurons.

for networks with a hidden layer. The spurious local minima suggested in [31] are only present in the case in which the target values differ from the asymptotic limits $\underline{d}$, $\overline{d}$ of the squashing function $f(\cdot)$. They are essentially due to the fact that guaranteeing $\mathcal{Y}_1$'s sign is no longer possible (see Theorem 2 and [7] for further details).

b. Structural Local Minima

If we look at the problem of supervised learning in general, the shape of the cost function depends on several elements. Keeping fixed the pattern of connectivity, we have seen that squashing and cost functions still play quite an important role. As a result, different choices of the cost may lead to optimization problems with different minima. Most importantly, the optimization problem at hand is closely related to the mapping performed by the network. Consequently, the network architecture and the learning environment play a very fundamental role. The data are fixed, because they are an input from the problem at hand, whereas the network architecture is ordinarily tailored to the problem itself. Sontag and Sussman [16] have proposed an example of a local minimum in a single-layered network which is remarkably different from those cited previously, as it involves the problem structure. They observed that, intuitively, the existence of local minima is due to the fact that the error function is the superposition of functions whose minima are at different points. In the case of linear response units, all these functions are convex, so no difficulties arise because a sum of convex functions is still convex. In contrast, sigmoidal units give rise to nonconvex functions, which give no guarantee for the summed function to have a unique minimum. In order to exhibit such a behavior, it is necessary to obtain terms whose minima are far apart and to control the second derivative so that the effect of such minima is not cancelled by the other terms. In the example given in [16], this is done via a network with one output neuron without threshold, which is incapable of correctly learning the proposed set of patterns. Other "difficult" problems where the networks were, in fact, capable of mapping the given data are given in [7, 50, 73, 74].

2. Examples of Local Minima in Feedforward Networks

We now discuss some examples of MLNs where BP gets stuck in local minima. Basically, these examples belong to the two different classes described previously, depending on the fact that the local minima are associated with the cost and squashing function chosen, or are intrinsically related to the network and the learning environment.

EXAMPLE 2. The first example is taken from [31], where a single-layered sigmoidal network is considered. The following linearly separable learning environment is selected for minimizing the quadratic cost

$$\begin{aligned}\mathcal{L}_e &= \{([x_0, x_1]', d)\} \\ &= \{([-1, 0]', 0.1), ([1, 0]', 0.9), ([0, 1]', 0.9), ([0, -5]', 0.9)\}. \end{aligned} \tag{15}$$

For the sake of simplicity, the explicit dependence of x_0, x_1, and d on t has been omitted. It turns out that there exists a nonseparating local minimum. The presence of this minimum is due to the fact that the asymptotic values $(\underline{d}, \overline{d})$ are not used as targets as required for the quadratic cost function. If asymptotic values were used, then local minima no longer hold. In particular, as previously pointed out, this kind of local minimum is due to the fact that, for a given pattern, $Y_{i(1)}$'s may change their sign, depending on the weight configuration, whereas this sign cannot change when using asymptotic targets. This change of the sign is a source of spurious combinations of terms in the gradient equations that give rise to local minima. The $Y_{i(1)}$'s "sign structure" is associated with that of $\mathcal{Y}_2$, and depends strictly on the asymptotic target assumption. Brady *et al.* [31] have proposed other similar examples concerning linearly separable patterns. The common characteristic in these examples is that the patterns are not of the same "importance" [e.g., in Eq. (15) the last pattern has a module sensibly greater than the others]. It is quite intuitive that for such a set of patterns, a strong learning decision (e.g., asymptotic target values) must be used for BP to work properly.

EXAMPLE 3. Sontag and Sussman [16] have proposed an example of a local minimum in a single-layered network which has no bias and which has the symmetric squash function as activation function. The quadratic cost function is minimized with reference to the following linearly separable learning environment

$$\begin{aligned}\mathcal{L}_e = \{&([1, 1, 1, -1, -1]', 1),\ ([1, 1, -1, 1, -1]', 1),\\ &([1, -1, 1, -1, 1]', 1),\ ([-1, 1, 1, -1, 1]', 1),\\ &([-1, 1, 1, 1, -1]', 1),\ ([-1, -1, -1, 1, 1]', 1),\\ &([-1, -1, 1, -1, 1]', 1),\ ([-1, 1, -1, 1, -1]', 1),\\ &([1, -1, -1, 1, -1]', 1),\ ([1, -1, -1, -1, 1]', 1),\\ &([1, 1, 1, 1, 1]', 1)\}.\end{aligned}$$

This example is different with respect to the one discussed previously in that it assumes asymptotic target values. In this case, the presence of the local minimum is due to the fact that the chosen network (without bias) is not able to learn the patterns exactly. Hence, it turns out that Assumption 1 is no longer satisfied.

EXAMPLE 4. Let us consider the standard XOR Boolean function. Following [74], it can be shown that there is a manifold of local minima with cost

$E_T = 0.5$. Another particular local minimum weight configuration is that having null all the weights. This fact makes it clear how situations in which a sort of symmetry involving the combined effect of weights and data may cause local minima for nonlinearly separable patterns.

EXAMPLE 5. In this example, we consider the XOR5 Boolean function [XOR, plus the training pattern $([0.5, 0.5]', 0)$; see Fig. 4a]. Obviously, in this case, Theorem 2 cannot also be applied because the patterns are not linearly separable. Depending on the initial weights, the gradient can get stuck in points where the cost is far from being null. The presence of these local minima is intuitively re-

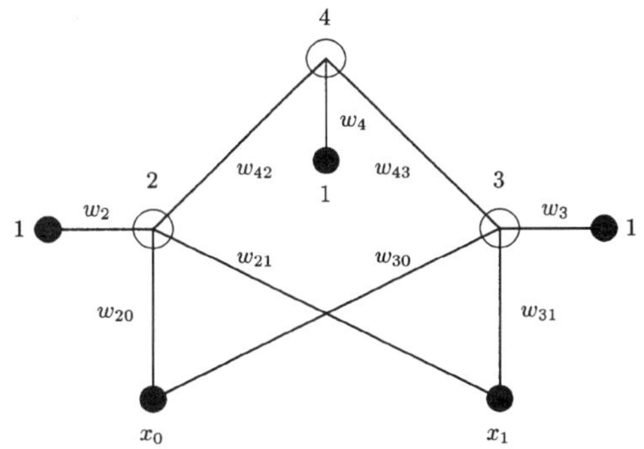

Pattern	x_0	x_1	Target
A	0	0	0
B	1	0	1
C	1	1	0
D	0	1	1
E	0.5	0.5	0

(a)

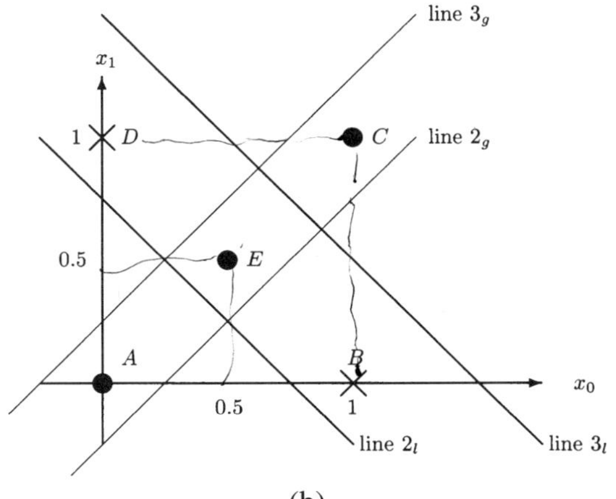

(b)

Figure 4 (a) Network and learning environment of Example 5; (b) separation lines of the local and global minima configurations. For simplicity, neurons are progressively numbered, with no regard to the layered structure.

lated to the symmetry of the learning environment. Experimental evidence of the presence of local minima is given in Fig. 5. We ran this experiment several times with different initial weights. We found that we can be trapped in these minima no matter what gradient descent algorithm is used. From a geometric point of view, these minima are related to the position of the separation lines identified by vec-

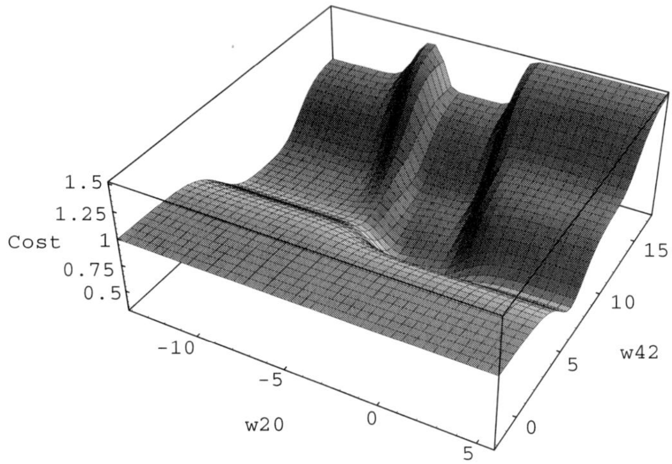

(a)

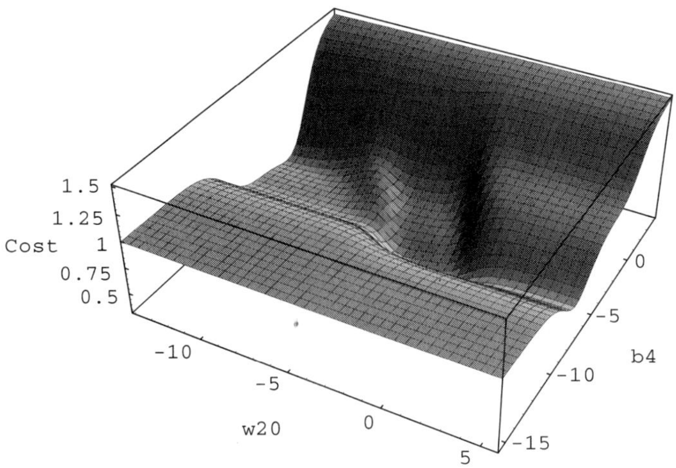

(b)

Figure 5 Cost surfaces as functions of (a) w_{20}, w_{42} and (b) w_{20}, w_4, respectively, for Example 5.

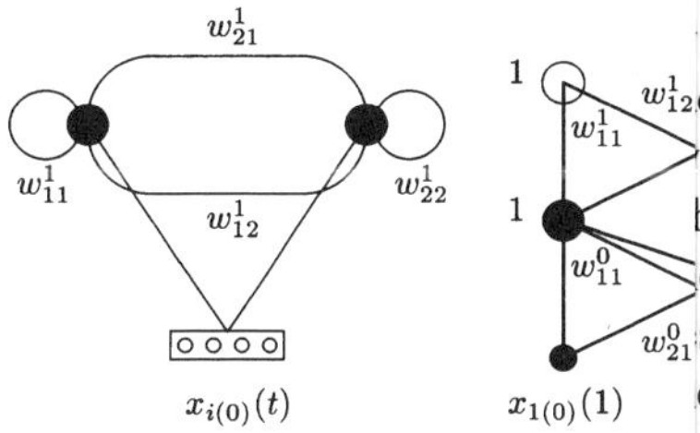

Figure 6 (a) The recurrent network; (b) th
Reprinted with permission from *IEEE Trans.* N

biases act at the right time. Finally, t
output neuron.

EXAMPLE 7. We can also consi
posed in [7] by simply adding the tok

$\frac{1}{2}\ \frac{1}{2}\ 1\ 0$

to (16). As for feedforward network
minima on actual learning of the wei
more token makes the problem of lea

The method used for mapping the
works can obviously be extended to
networks. As a result, any learning ta
to one for recurrent networks fed by
same associated cost function. It is q
with $F_{\max} = 2$, because the initial sta
of the associated unfolded network,
with three or more frames. In these
ferent layers of the unfolded feedfor
finding local minima increases, thus
networks even more difficult.

B. SYMMETRICAL CONFIG

The XOR5 example, given in Sec
from problems where a sort of sym
data. Similar considerations suggest

with equal or null weights. Of course,
ct on the learning depends on the data,
ays associated with stationary points,
h.

e case of feedforward networks with
sholds in the output layer. In that case,
e delta error $\mathcal{Y}_l$ is not null only at the
there are no thresholds in the output
weights are null, the backpropagation
$\forall l = 1, \ldots, L-1$. As a result, $\mathcal{G}_l =$

ggest neither to learn from symmetric
s.

also arise when learning with "high"
erred to as *premature saturation* (see,
h weights are essentially due to neuron
rror backpropagation very hard. Obvi-
on saturation is strictly related to neu-
the higher is the probability of neuron
e initial weights randomly distributed
in of the unit to which the connection
ments of the previous section suggest
a nor too small. Drago and Ridella [77]
t determining the relationship between
m magnitude of the weights—and the
has been shown to be very useful for
n their analysis, they also show how to
computer simulations.
re saturation on the output units is a
t function. Indeed, LMS is a wrong
statistical model for classification and
n likelihood principle. In this case, as-
s variable, the negative log-likelihood
oss-entropy metric. The main differ-
dinary quadratic cost is that the erro-
not lead to plateaus, but to very high
output is erroneously saturated (e.g.,
∞. It is worth mentioning that the

large plateaus associated with the quadratic cost do not represent local minima and, consequently, do not attract the learning trajectory toward suboptimal solutions. However, the computational burden for escaping from similar configurations may be really exaggerated and serious problems due to limited numerical precision may also arise. When using the relative cross-entropy metric, the repulsion from the previous erroneous configurations is much more effective, because there are no plateaus but surfaces with high gradient and underflow errors are likely to be avoided. Saturation problems emerge also when learning with radial basis functions in the case in which the Gaussian centers are randomly placed and the associated variance σ is "small." The hybrid learning scheme suggested in [21] handles this problem very effectively thanks to a proper weight initialization.

The neuron saturation is significantly more serious in recurrent networks, particularly when applied to capture long-term dependencies. This problem can be understood if we bear in mind the BPTT scheme for gradient computation. The network time unfolding for long sequences leads to vanishing the gradient, and, consequently, it seems very hard to keep track of long-term dependencies [76]. Recent attempts to deal with these very serious problems can be found in [78–80].

D. Bifurcation of Learning Trajectories in Recurrent Neural Networks

Let us consider the problem of learning long-term dependencies using recurrent networks. A very serious problem arises that depends on the different qualitative dynamic behavior taking place in recurrent networks acting on long sequences. It has been proven that, depending on the network weights, the network dynamics can change significantly (see, e.g., [72]). For example, depending on the weight of the self-loop connection, a recurrent network can exhibit a *forgetting behavior*, or *information latching* [81].

This different dynamic behavior can be understood by considering a single neuron having a self-loop connection as follows:

$$\begin{aligned}
a_{i(1)}(t) &= \sum_{j(1)\in\mathcal{N}} w^1_{i(1),j(1)} x_{j(1)}(t-1) + \sum_{k(0)\in\mathcal{I}} w^0_{i(1),k(0)} x_{k(0)}(t) \\
&\doteq w^1_{i(1),i(1)} x_{i(1)}(t-1) + \mathcal{I}_{i(1)}(t), \\
x_{i(1)}(t) &= f\big(a_{i(1)}(t)\big) = \tanh\big[a_{i(1)}(t)\big],
\end{aligned}$$

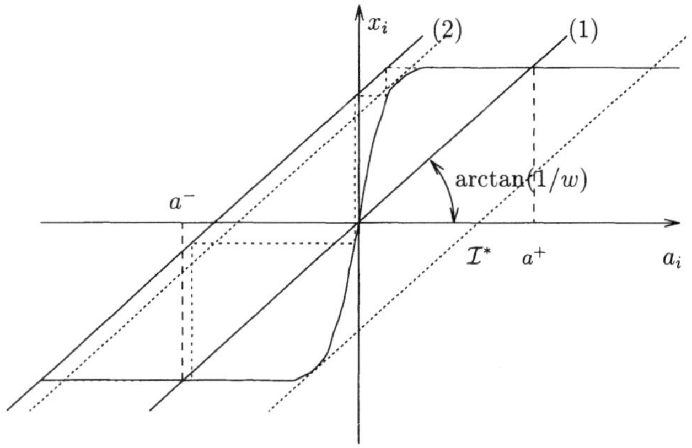

Figure 7 Equilibrium points for a neuron with a self-loop connection.

where

$$\mathcal{I}_{i(1)}(t) \doteq \sum_{j(1)\in\mathcal{N}} w^1_{i(1),j(1)}\big(1-\delta_{i(1),j(1)}\big)x_{j(1)}(t-1) + \sum_{k(0)\in\mathcal{I}} w^0_{i(1),k(0)}x_{k(0)}(t).$$

Let us investigate the possibility of latching the information of a given state.

DEFINITION 2. We say that a given dynamic hidden neuron latches the information at t_0, represented by its activation $a_i(t_0)$ provided that the following relationship holds:

$$x_i^b(t) \doteq \operatorname{sgn}\big(a_i(t)\big) = \operatorname{sgn}\big(a_i(t_0)\big), \qquad \forall i, t, t_0\colon t \geqslant t_0.$$

It is quite easy to realize that, depending on the value of $w^1_{i,i}$, the neuron output will reach one of the three equilibrium points depicted in Fig. 7.

As a result, if $w^1_{i,i} = 1$, a very small change will lead to different equilibrium points, which results in a very different dynamic behavior. In the case in which the neuron output reaches the "zero," there is a sort of *forgetting behavior*, whereas in the other two cases an event is latched independently of the length of the sequence.

Similar but more detailed analyses are drawn in [72] for the case of continuous recurrent networks and lead us to conclude that there are bifurcation points in the learning trajectories. Like the saturation problem, the bifurcation of the learning trajectories is due to dealing with sequences of unbounded length.[22] Facing saturation and bifurcation is an open research problem.

[22]In practice, "long" sequences.

V. ADVANCED TECHNIQUES FOR OPTIMAL LEARNING

The theoretical limitations that have been shown in the previous sections suggest looking for learning techniques capable of dealing more effectively with local minima and generalization to new examples. Of course, there are many different approaches for coping with this problem that are making neural network learning a very multifaceted discipline. The following brief review is not supposed to cover the many different techniques recently proposed in the literature, but simply to offer a sketch of some ideas that look promising.

A. Growing Networks and Pruning

As pointed out in the previous section, the requirement of reaching good convergence for any learning algorithm, no matter how it is conceived, and of performing high generalization to new examples leads to a sort of uncertainty principle. In order to face this problem, several researchers have proposed *pruning* algorithms, in which we begin training a network larger than "necessary," and then continue by pruning connections that, to some extent, do not affect to the learning process significantly. As a result, the networks turn out to be tuned to the task at hand with a consequent improvement of the generalization to new examples.[23]

Pruning algorithms may be grouped into two broad categories [82]: *sensitivity* and *penalty-term methods*. The *sensitivity methods* modify a trained network with a given structure by estimating the sensitivity of the error function with respect to the removal of an element (weight or unit), and then remove the element with the least effect (see, e.g., [83–86]). The *penalty-term methods* modify the cost function so that backpropagation based on that function drives unnecessary weights to zero. Even if the weights are not actually removed, the network acts like a smaller one (see, e.g., [87–92]).

Rather than beginning from large networks and pruning subsequently unnecessary connections, an alternative approach is that of using a small network that grows gradually for facing problems of optimal convergence. Successful techniques based on this idea have been proposed in [58, 93–98]. Unlike pruning methods, those based on growing networks can often be guaranteed to converge to an optimal solution, but, the reverse of the coin is that the resulting networks may be too large with a consequent poor generalization to new examples.

[23] Small networks also have the advantage of being cheaper to build whereas their operations are easier to understand.

In this chapter, we have addressed
oretical point of view. The focus is
the shape of the cost function. In pa
guarantee local minima free error surf
PR conditions are based on the hypo
tern space, whereas Poston and Yu's
functions, no matter what examples a
hidden units as patterns. These cond
problem, but, unfortunately, both of t
seem limited by their restrictive assun
conditions appear severely limited b
units. Bridging PR and Poston and Y
necessary and sufficient conditions fo
open research problem.

In the light of our theoretical fram
optimal learning due to the presence c
mature saturation, and also bifurcatio

The theoretical analyses on local n
interesting themselves, but also give
machine learning. Basically, the dec
for a given problem, some networks
ing. In particular, there are cases in
the associated cost function is local r
can introduce an index accounting fo
perform a search, in the space of the
such an index [107]. In so doing, the
function optimization in the weight sp
selecting an architecture that is likel
integration of search and optimizatio
neural network approaches to machin

Finally, the presence of local minin
algorithm will fail to discover an opti
ence as a boundary beyond which tro
to begin.

We are confident that the theoretic
the doors to more thorough analyses in
could shed light on the computational

cognitive architecture: a critical analysis. *Con-*
achine Learning, an Artificial Intelligence Ap-
o, CA, 1983.
led Edition. MIT Press, Cambridge, MA, 1988.
ams. Learning internal representations by error
(D. E. Rumelhart and J. L. McClelland, Eds.),
ridge, MA, 1986.
Lecture Notes in Computer Sciences, 1987.
lobal Random Search. Kluwer Academic, Dor-
ninima in backpropagation. *IEEE Trans. Pattern*
992.
l minima and backpropagation. In *International*
Vol. 2, pp. 173–176. IEEE, New York, 1991.
ave local minima? *IEEE Trans. Neural Networks*
ng without local minima in radial basis function
–756, 1995.
of linear threshold devices. Ph.D. Thesis, Stan-
964.
old systems and their storage capacities. Ph.D.
ford, CA, 1964.
of pattern mode backpropagation. *IEEE Trans.*
ck-propagation: learning in connectionist mod-
nces in Neural Information Processing Systems
the generalized delta rule. *IEEE Trans. Neural*
can give rise to spurious local minima even for
ems 3:91–106, 1989.
es. *Artificial Intelligence* 40:185–234, 1989.
of probability distributions by neural networks.
Systems (D. Anderson, (Ed.), pp. 52–61, 1988.
d learning in layered neural networks. *Complex*
ed on heuristic arguments. In *International Joint*
DC, Vol. 1, pp. 565–568. IEEE, New York, 1990.
works of locally-tuned processing units. *Neural*
internal representation of gray scale images: an
h Annual Cognitive Science Society Conference,
ems with emergent collective computational abil-
58, 1982.

[24] D. H. Ackley, G. E. Hinton, and T. J. Sejnowski. A learning algorithm for Boltzmann machines. *Cognitive Sci.* 9:147–169, 1985.

[25] N. J. Nilsson. *Learning Machines*. McGraw-Hill, New York, 1965. Reissued as *Mathematical Foundations of Learning Machines*. Morgan Kaufmann, San Mateo, CA, 1990.

[26] F. Rosenblatt. *Principles of Neurodynamics: Perceptrons and the Theory of Brain Mechanism*. Spartan Books, Washington, DC, 1962.

[27] B. Widrow and M. Hoff. Adaptive switching circuits. In *1960 IRE WESCON Convention Record*, Vol. 4, pp. 96–104. IRE, New York, 1960.

[28] B. Widrow. 30 years of adaptive neural networks: perceptron, Madaline, and backpropagation. *IEEE Trans. Neural Networks* 78:1415–1442, 1990.

[29] P. Baldi and K. Hornik. Neural networks and principal component analysis: learning from examples without local minima. *Neural Networks* 2:53–58, 1989.

[30] E. Sontag and H. Sussman. Backpropagation separates when perceptrons do. In *International Joint Conference on Neural Networks*, Washington, DC, Vol. 1, pp. 639–642. IEEE, New York, 1989.

[31] M. Brady, R. Raghavan, and J. Slawny. Back-propagation fails to separate where perceptrons succeed. *IEEE Trans. Circuits Systems* 36:665–674, 1989.

[32] J. Shynk. Performance surfaces of a single-layer perceptron. *IEEE Trans. Neural Networks* 1:268–274, 1990.

[33] D. Hush and J. Salas. Improving the learning rate of back-propagation with the gradient reuse algorithm. In *IEEE International Conference on Neural Networks*, San Diego, Vol. 1, pp. 441–447, IEEE, New York, 1988.

[34] K. Gouhara, N. Kanai, and Y. Uchikawa. Experimental learning surface and learning process in multilayer neural networks. Technical Report, Naguya University, Naguya, Japan, 1993.

[35] K. Gouhara and Y. Uchikawa. Memory surface and learning surfaces in multilayer neural networks. Technical Report, Naguya University, Naguya, Japan, 1993.

[36] A. M. Chen, H. Lu, and R. Hecht-Nielsen. On the geometry of feedforward neural network error surfaces. *Neural Comput.* 5:910–927, 1993.

[37] F. Jordan and G. Clement. Using the symmetries of multi-layered networks to reduce the weight space. In *International Joint Conference on Neural Networks*, Seattle, Vol. 2, pp. 391–396. IEEE, New York, 1991.

[38] G. Cybenko. Approximation by superpositions of a sigmoidal function. *Math. Control Signals Systems* 3:303–314, 1989.

[39] K. Funahashi. On the approximate realization of continuous mappings by neural networks. *Neural Networks* 2:183–192, 1989.

[40] R. Hecht-Nielsen. Theory of the backpropagation neural network. In *International Joint Conference on Neural Networks*, Washington, DC, Vol. 1, pp. 593–605, IEEE, New York, 1989.

[41] K. Hornik, M. Stinchcombe, and H. White. Multilayer feedforward networks are universal approximators. *Neural Networks* 2:359–366, 1989.

[42] S. C. Huang and Y. F. Huang. Bounds on the number of hidden neurons in multi-layer perceptrons. *IEEE Trans. Neural Networks* 2:47–55, 1991.

[43] Y. Bengio, P. Cosi, and R. De Mori. Phonetically-based multi-layered networks for vowel classification. *Speech Comm.* 9:15–29, 1990.

[44] Y. le Cun. Generalization and network design strategies. In *Connectionism in Perspective*, pp. 143–155. North-Holland, Amsterdam, 1989.

[45] Y. le Cun. A theoretical framework for backpropagation. In *The 1988 Connectionist Models Summer School* (D. Touretzky, G. E. Hinton, and T. Sejnowski, Eds.), pp. 21–28. Morgan Kaufman, San Mateo, CA, 1988.

[46] A. Bryson and Y. C. Ho. *Applied Optimal Control*. Blaisdell, New York, 1969.

[47] P. Werbos. Beyond regression: new tools for prediction and analysis in the behavioral sciences. Ph.D. Thesis, Harvard University, Cambridge, MA, 1974.

[48] D. Parker. Learning logic. Technical Report TR-47, Center for Computational Research in Economics and Management Science, MIT, Cambridge, MA, 1985.

[49] Y. le Cun. Learning processes in an asymmetric threshold network. In *Disordered Systems and Biological Organization* (F. F. Soulie, E. Bienenstock, and G. Weisbuch, Eds.), pp. 233–240. Springer-Verlag, Les Houches, France, 1986.

[50] M. Gori and A. Tesi. Some examples of local minima during learning with backpropagation. In *Parallel Architectures and Neural Networks*, Vietri sul Mare, Italy, 1990.

[51] H. Bourlard and C. Wellekens. Speech pattern discrimination and multi-layered perceptrons. *Comput. Speech Language* 3:1–19, 1989.

[52] J. Elman and D. Zipser. Learning the hidden structure of speech. *J. Acoust. Soc. Amer.* 83:1615–1626, 1988.

[53] Y. Bengio, R. De Mori, and M. Gori. Learning the dynamic nature of speech with backpropagation for sequences. *Pattern Recognition Lett.* 13:375–386, 1992.

[54] A. Waibel, T. Hanazawa, G. E. Hinton, K. Shikano, and K. Lang. Phoneme recognition using time-delay neural networks. *IEEE Trans. Acoust. Speech Signal Process.* 37:328–339, 1989.

[55] D. C. Plaut and G. E. Hinton. Learning set of filters using back-propagation. *Comput. Speech Language* 2:35–61, 1987.

[56] R. A. Jacobs, M. I. Jordan, and A. Barto. Task decomposition through competition in a modular connectionist architecture: the what and where vision tasks. Technical Report, COINS, 1990.

[57] P. Frasconi, M. Gori, and A. Tesi. Backpropagation for linearly separable patterns: a detailed analysis. In *IEEE International Conference on Neural Networks*, San Francisco, Vol. 3, pp. 1818–1822. IEEE, New York, 1993.

[58] S. I. Gallant. Perceptron-based learning algorithms. *IEEE Trans. Neural Networks* 1:179–192, 1990.

[59] T. Poggio and F. Girosi. Networks for approximation and learning. *Proc. IEEE* 78:1481–1497, 1990.

[60] R. Bellman. *Introduction to Matrix Analysis*, 2nd ed. McGraw-Hill, New York, 1974.

[61] T. Kohonen. The self-organizing map. *Proc. IEEE* 78:1464–1480, 1990.

[62] X. Yu and G. Chen. On the local minima free condition of backpropagation learning. *IEEE Trans. Neural Networks* 6:1300–1303, 1995.

[63] M. Gori, L. Lastrucci, and G. Soda. Neural autoassociators for phoneme-based speaker verification. In *International Workshop on Automatic Speaker Recognition, Identification, and Verification*, Martigny, Switzerland, pp. 189–192, 1994.

[64] T. Kohonen. *Self-Organization and Associative Memory*, 3rd ed. Springer-Verlag, Berlin, 1989.

[65] M. Bianchini, P. Frasconi, and M. Gori. Learning in multilayered networks used as autoassociators. *IEEE Trans. Neural Networks* 6:512–515, 1995.

[66] H. Bourlard and Y. Kamp. Auto-association by multilayer perceptrons and singular value decomposition. *Biol. Cybernet.* 59:291–294, 1988.

[67] J. L. McClelland and D. E. Rumelhart. *Explorations in Parallel Distributed Processing*, Vol. 3. MIT Press, Cambridge, MA, 1988.

[68] R. J. Williams and J. Peng. An efficient gradient-based algorithm for on-line training of recurrent network trajectories. *Neural Comput.* 2:490–501, 1990.

[69] M. Bianchini, M. Gori, and M. Maggini. On the problem of local minima in recurrent neural networks. *IEEE Trans. Neural Networks* 5:167–177, 1994.

[70] P. Frasconi, M. Gori, M. Maggini, and G. Soda. Unified integration of explicit rules and learning by example in recurrent networks. *IEEE Trans. Knowledge Data Engineering* 7:340–346, 1995.

[71] Y. Lee, S. Oh, and M. Kim. The effect of weights on premature saturation in back-propagation learning. In *International Joint Conference on Neural Networks*, Seattle, Vol. 1, pp. 765–770, 1991.
[72] K. Doya. Bifurcations of recurrent neural networks in gradient descent learning. *Connectionist News Neuroprose*, 1993.
[73] J. M. McInerny, K. G. Haines, S. Biafore, and R. Hecht-Nielsen. Back propagation error surfaces can have local minima. In *International Joint Conference on Neural Networks*, Washington, DC, Vol. 2, p. 627. IEEE, New York, 1989.
[74] E. Blum. Approximation of boolean functions by sigmoidal networks, I, XOR and other two-variable functions. *Neural Comput.* 1:532–540, 1989.
[75] M. Gori. Apprendimento con supervisione in reti neuronali. Ph.D. Thesis, Università degli Studi di Bologna (in Italian), 1990.
[76] Y. Bengio, P. Frasconi, and P. Simard. Learning long-term dependencies with gradient descent is difficult. *IEEE Trans. Neural Networks* 5:157–166, 1994.
[77] C. Drago and S. Ridella. Statistically controlled activation weight initialization (SCAWI). *IEEE Trans. Neural Networks* 3:627–631, 1992.
[78] J. Schmidthubler. Learning complex, extended sequences using the principle of history compression. *Neural Comput.* 4:234–242, 1992.
[79] M. Gori, M. Maggini, and G. Soda. Scheduling of modular architectures for inductive inference of regular grammars. In *Workshop on Combining Symbolic and Connectionist Processing, ECAI '94*, Amsterdam, pp. 78–87, 1994.
[80] T. a. B. G. H. Lin, P. Tino, and C. L. Giles. Learning long-term dependencies in narx recurrent neural networks. Technical Report UMIACS-TR-95-78, University of Maryland, 1995.
[81] P. Frasconi, M. Gori, and G. Soda. Local feedback multi-layered networks. *Neural Comput.* 4:120–130, 1992.
[82] R. Reed. Pruning algorithms — a survey. *IEEE Trans. Neural Networks* 4:740–747, 1993.
[83] Y. le Cun, J. Denker, and S. Solla. Optimal brain damage. In *Advances in Neural Information Processing Systems* (D. Touretzky, Ed.), Vol. 2, pp. 598–605. Morgan Kaufmann, San Mateo, CA, 1990.
[84] M. Mozer and P. Smolensky. Skeletonization: a technique for trimming the fat from a network via relevance assessment. In *Advances in Neural Information Processing Systems* (D. Touretzky, Ed.), Vol. 1, pp. 107–115. Morgan Kaufmann, San Mateo, CA, 1989.
[85] E. Karnin. A simple procedure for pruning back-propagation trained neural networks. *IEEE Trans. Neural Networks* 1:188–197, 1990.
[86] B. Hassibi, D. Stork, and G. Wolff. Second order derivatives for network pruning: optimal brain surgeon. In *Advances in Neural Information Processing Systems* (S. J. Hanson, J. D. Cowan, and C. L. Giles, Eds.), Vol. 5. Morgan Kaufmann, San Mateo, CA, 1992.
[87] Y. Chauvin. A back-propagation algorithm with optimal use of hidden units. In *Advances in Neural Information Processing Systems* (D. Touretzky, Ed.), Vol. 1, pp. 519–526. Morgan Kaufmann, San Mateo, CA, 1989.
[88] C. Ji, R. Snapp, and D. Psaltis. Generalizing smoothness constraints from discrete samples. *Neural Comput.* 2:188–197, 1990.
[89] S. J. Nowlan and G. E. Hinton. Simplifying neural networks by soft weight-sharing. *Neural Comput.* 4:473–493, 1992.
[90] A. S. Weigend, D. E. Rumelhart, and B. A. Hubermann. Back-propagation, weight-elimination, and time series prediction. In *Connectionist Model Summer School* (D. Touretzky, J. Elmann, T. Sejnowski, and G. E. Hinton, Eds.), pp. 105–116, 1990.
[91] A. S. Weigend, D. E. Rumelhart, and B. A. Hubermann. Generalization by weight-elimination applied to currency exchange rate prediction. In *International Joint Conference on Neural Networks*, Seattle, Vol. 1, pp. 837–841, 1991.

[92] A. S. Weigend, D. E. Rumelhart, and B. A. Hubermann. Generalization by weight-elimination with application to forecasting. In *Advances in Neural Information Processing Systems* (R. Lippmann, J. Moody, and D. Touretzky, Eds.), Vol. 3, pp. 875–882, 1991.
[93] S. I. Gallant. Three constructive algorithms for network learning. In *Eighth Annual Conference of the Cognitive Science Society*, Amherst, MA, pp. 652–660. IEEE, New York, 1986.
[94] S. Gallant. *Neural Network Learning and Expert Systems*. MIT Press, Cambridge, MA, 1993.
[95] S. E. Fahlman and C. Lebiere. The cascade–correlation learning architecture. In *Advances in Neural Information Processing Systems* (D. S. Touretzky, Ed.), Vol. 2, pp. 524–532. Morgan Kaufmann, San Mateo, CA, 1990.
[96] M. Mézard and J. Nadal. Learning in feedforward layered networks: the Tiling algorithm. *J. Phys. A* 22:2191–2204, 1989.
[97] M. Frean. The upstart algorithm: a method for constructing and training feedforward neural networks. *Neural Comput.* 2:198–209, 1990.
[98] T. S. Chang and K. A. S. Abdel-Ghaffar. A universal neural net with guaranteed convergence to zero system error. *IEEE Trans. Acoust. Speech Signal Process.* 40:3022–3030, 1992.
[99] A. Waibel. Modular construction of time-delay neural networks for speech recognition. *Neural Comput.* 1:39–46, 1989.
[100] R. A. Jacobs. Task decomposition through competition in a modular connectionist architecture. Ph.D. Thesis, Department of Computer and Information Science, University of Massachusetts, 1990.
[101] R. A. Jacobs, M. I. Jordan, S. J. Nowlan, and G. E. Hinton. Adaptive mixture of local experts. *Neural Comput.* 3:79–87, 1991.
[102] R. A. Jacobs and M. I. Jordan. Hierarchical mixtures of experts and the EM algorithm. *Neural Comput.* 6:181–214, 1994.
[103] P. Dempster, N. Laird, and D. Rubin. Maximum-likelihood from incomplete data via the EM algorithm. *J. Roy. Statist. Soc.* 39:1–38, 1977.
[104] *Machine Learning* 7, 1991.
[105] *Artificial Intelligence* 46, 1990.
[106] R. Sun. *Integration Rules and Connectionism for Robust Commonsense Reasoning*. Wiley, New York, 1990.
[107] N. J. Nilsson. *Principles of Artificial Intelligence*. Tioga, Palo Alto, CA, 1980.
[108] J. Anderson and E. Rosenfeld (Eds.). *Neurocomputing: Foundations of Research*. MIT Press, Cambridge, MA, 1988.

Orthogonal Transformation Techniques in the Optimization of Feedforward Neural Network Systems

Partha Pratim Kanjilal
Department of Electronics and Electrical Communication Engineering
Indian Institute of Technology
Kharagpur 721-302, India

I. INTRODUCTION

Orthogonal transformation can be used to identify the dominant modes in any information set, which is the basic idea behind the application of orthogonal transformation techniques for the optimization of neural networks. In this chapter, singular value decomposition (SVD) and different forms of QR with column pivoting factorization are used for the optimization of the size of a feedforward neural network in terms of the optimum number of links and nodes; the prime objective is to improve representativeness and generalization ability [1] of the model. As corollaries to the application of orthogonal transformation for optimization, studies on (a) the compaction of the process information through orthogonal transformation followed by the operation of the neural network with the reduced set of transformed data and (b) the assessment of the convergence of the training process for the network using SVD are also included in this chapter.

In any method of modeling, overparameterization or redundancy in the structure is undesirable [2]. In case of an oversized neural network, if the training set of data are not noise-free (which is usually the case), the network will tend to learn the information along with the noise associated with the data, leading to poor validation results. It is well known that during the flow of information in any stage, collinearity among the set of information links can lead to identification problems [3–5]; any model should be parsimonious to be representative [2], hence the primary need for optimizing the size of the neural network model. For an optimum representation of the underlying process, there should be parsimony in both the configuration of the problem and the design of the model; that is, the neural network should be fed with the optimum number of inputs, and there should be an optimum number of links and nodes within the network.

There have been some studies concerning the optimization of the size of neural networks [6–8]. The approaches [9–11], largely based on statistical information criteria [12], can be complex and quite computation intensive; further the imprecision associated with such information criteria (e.g., [13]) may lead to improper modeling. Pruning-based methods [6, 8] often use *ad hoc* conditions to assess the importance of links within the network. The approaches employing eigenvalue-based optimization [8] can suffer from numerical ill-conditioning problems [15, 16].

In this chapter direct and robust methods for the optimization of homogeneous as well as nonhomogeneous feedforward neural networks in terms of essential links and nodes within the network are discussed. The three different approaches considered for the optimization of neural networks are based on (i) the singular value decomposition (SVD) [15], (ii) SVD followed by QR with column pivoting (QRcp) factorization, and (iii) the modified QR with column pivoting (m-QRcp) factorization coupled with the C_p statistic for assessment of optimality [17]. SVD is used for homogeneous network optimization. QRcp factorization (with SVD) and m-QRcp factorization (with C_p) are used for nonhomogeneous network optimization. Both QRcp and m-QRcp factorizations can also be used for selection of the optimum set of inputs to the network. All the transformations used are numerically robust and can have robust implementations.

In all cases, the optimization is performed in the linear sections of the network. The problem is configured as the subset selection problem in each case. Three-layer neural networks with a single hidden layer are considered for the present study. Three illustrative examples are considered: (i) the Mackey–Glass series representing the chaotic dynamics in controlled physiological systems [18]; (ii) the nonlinear data series of yearly averaged sunspot numbers [2, 19]; and (iii) the rocket engine testing process [20], which is a multi-input, single-output problem.

The organization of this chapter is as follows. The mathematical background for the orthogonal transformations used is presented in Section II. Section III explains the principles for the optimization of neural networks. The illustrative examples are presented and the results are discussed in Sections IV to VII. The convergence analysis during the training using SVD features in Section VIII.

II. MATHEMATICAL BACKGROUND FOR THE TRANSFORMATIONS USED

Orthogonal transformation can be very effectively used for data analysis, modeling, prediction, and filtering [15, 21, 5]. It can be used to convert data sets into relatively decorrelated sets of transform coefficients (or spectral components). The energy in the data, which represents the information content, remains conserved throughout the transformation, but the distribution of energy becomes more compact following the transformation. The process of transformation is linear and reversible.

Two popular classes of orthogonal transformation are singular value decomposition and QR factorization; the two special forms of QR factorizations used here are the QR with column pivoting factorization and the modified-QR with column pivoting factorization, both of which can be used for subset selection. The characteristic features of all these transformations are discussed here.

A. SINGULAR VALUE DECOMPOSITION

Singular value decomposition [15] of an $m \times n$ matrix $\mathbf{A}$ is given by $\mathbf{A} = \mathbf{U\Sigma V}^T$, where $\mathbf{U} = [\mathbf{u}_1, \ldots, \mathbf{u}_m] \in \mathbb{R}^{m\times m}$ and $\mathbf{V} = [\mathbf{v}_1, \ldots, \mathbf{v}_n] \in \mathbb{R}^{n\times n}$ are orthogonal matrices (i.e., $\mathbf{U}^T\mathbf{U} = \mathbf{U}\mathbf{U}^T = \mathbf{I}$, etc.); $\mathbf{U}^T\mathbf{A}\mathbf{V} = \mathbf{\Sigma} = [\text{diag}\{\sigma_1, \ldots, \sigma_p\}\colon 0] =\in \mathbb{R}^{m\times n}$, where $p = \min(m, n)$ and $\sigma_1 \geq \cdots \geq \sigma_p \geq 0$. $\sigma_1, \ldots, \sigma_p$ are the singular values of $\mathbf{A}$ which are nonnegative. $\mathbf{U}$ and $\mathbf{V}$ are the left and the right singular vector matrices, respectively. The left and the right singular vectors form a basis for the row space and the column space of $\mathbf{A}$.

The number of nonzero singular values give the rank of the matrix $\mathbf{A}$. In fact, SVD is the most numerically robust and precise method for the determination of the null space of a matrix. The smallest nonzero singular value of $\mathbf{A}$ gives the precise 2-norm distance of $\mathbf{A}$ from the set of all rank-deficient matrices.

The energy contained in $\mathbf{A}$ $(= \{\{a_{ij}\}\})$ is given by

$$Q_A = \sum_{i=1}^{m}\sum_{j=1}^{n} a_{ij}^2 = \|\mathbf{A}\|_F^2 = \sigma_1^2 + \sigma_2^2 + \cdots + \sigma_p^2.$$

Because

$$\mathbf{A} = \sum_{i=1}^{p} \mathbf{u}_i \sigma_i \mathbf{v}_i^T, \tag{1}$$

the energy in the ith decomposed mode, $\mathbf{u}_i\sigma_i\mathbf{v}_i$, is given by σ_i^2. If q of the p singular values are dominant, the prime information of $\mathbf{A}$ will be contained in

$$\mathbf{A} = \sum_{i=1}^{q} \mathbf{u}_i\sigma_i\mathbf{v}_i^T.$$

A nearly periodic series $\{x(k)\}$ of periodicity n can be arranged in the $m \times n$ matrix $\mathbf{A}$ such that successive n-long segments occupy the successive rows of the matrix; if σ_1 of $\mathbf{A}$ is significantly dominant with $\sigma_1 \gg \sigma_2$,

$$\mathbf{A} \simeq \mathbf{u}_1\sigma_1\mathbf{v}_1. \tag{2}$$

In such a case, $\mathbf{v}_1$ will represent the periodic pattern or the normalized distribution of the series over one period, and the successive elements of $\mathbf{u}_1\sigma_1$ will represent the scaling factors for the successive periodic segments of $\{x(k)\}$. Thus, SVD can be very effectively used for the characterization of periodic or nearly periodic series [5].

B. QR Factorization

QR factorization [15] of a data matrix $\mathbf{A}$ is expressed as $\mathbf{A} = \mathbf{QR}$, where

$$\mathbf{A} = [\mathbf{a}_1, \ldots, \mathbf{a}_j, \ldots, \mathbf{a}_N] \in \mathbb{R}^{m\times N},$$
$$\mathbf{Q} = [\mathbf{q}_1, \ldots, \mathbf{q}_i, \ldots, \mathbf{q}_N] \in \mathbb{R}^{m\times N};$$

$\mathbf{a}_j$ are the m-column vectors; $\mathbf{q}_i$ are orthonormal columns: $\mathbf{Q}^T\mathbf{Q} = \mathbf{I}$, and $\mathbf{R}$ is upper triangular.

The columns of $\mathbf{Q}$ span the same subspace as the columns of $\mathbf{A}$. The number of nonzero diagonal elements $\mathbf{R}_{ii}$ [where $i < \min(m, N)$) of $\mathbf{R}$ indicates the rank of $\mathbf{A}$. $|\mathbf{R}_{jj}| = 0$ implies that the jth column vector of $\mathbf{A}$ is redundant, as it has no component in the $\mathbf{q}_j$ vector space that is orthogonal to the $\mathbf{q}_i$ $(i \neq j)$ vector space.

C. QR with Column Pivoting Factorization and Subset Selection

QR with column pivoting (QRcp) factorization of any $m \times n$ matrix $\mathbf{A}$ involves the pivoting of the columns of the matrix $\mathbf{A}$ in order of maximum Euclidean norm in successive orthogonal directions while QR factorization is performed on the matrix. Subset selection is inherent with the pivoting or the rotation of the columns within QRcp factorization. The mechanism of the rotation of the columns can be explained as follows [5].

Given any $m \times n$ matrix $\mathbf{A} = [\mathbf{a}_1, \ldots, \mathbf{a}_i, \ldots, \mathbf{a}_n]$, with n m-column vectors, the column vector of $\mathbf{A}$ with $\max(\mathbf{a}_i^T \mathbf{a}_i)$ is first selected, which is swapped with $\mathbf{a}_1$. Let $\mathbf{q}_1 (= \mathbf{a}_1 / \|\mathbf{a}_1\|)$ be the unit vector in the direction of $\mathbf{a}_1$. The selected (or rotated) second vector is the one maximizing the norm $(\mathbf{a}_j - \mathbf{q}_1^T \mathbf{a}_j \mathbf{q}_1)^T (\mathbf{a}_j - \mathbf{q}_1^T \mathbf{a}_j \mathbf{q}_1)$, which is swapped with $\mathbf{a}_2$, and $\mathbf{q}_2$, the corresponding unit vector, is computed.

At the ith stage of selection, the rotated vectors $(\mathbf{a}_j^*)$ are

$$\mathbf{a}_j^* = \mathbf{a}_j - (\mathbf{q}_1^T \mathbf{a}_j \mathbf{q}_1 + \cdots + \mathbf{q}_{i-1}^T \mathbf{a}_j \mathbf{q}_{i-1}),$$

$i = 2$ to n, $j = i$ to n, and ith selected vector is the one maximizing $\mathbf{a}_j^{*T} \mathbf{a}_j^*$. The subsequent rotation within QR decomposition will be with respect to this vector and so on. The selection is continued for up to r stages, where r $[\leq \min(m, n)]$ may be the rank of $\mathbf{A}$ or may be prespecified. The sequence of successive selections of the columns of $\mathbf{A}$ is registered in the permutation matrix $\mathbf{P}$; $\mathbf{AP}$ will have the first r columns of $\mathbf{A}$ appearing in order of selection.

If $\mathbf{A}$ has q $(< p)$ dominant singular values with $\sigma_q \gg \sigma_{q+1}$, for increased numerical stability QRcp factorization may be performed on a $q \times n$ matrix $\mathbf{W}^T$ instead of $\mathbf{A}$ [15], where $\mathbf{W}$ constitutes of the first q columns of $\mathbf{V}$:

$$\mathbf{W} = [\mathbf{v}_1 \quad \mathbf{v}_2 \quad \cdots \quad \mathbf{v}_q].$$

If $\mathbf{W}^T = [\mathbf{W}_1^T \quad \mathbf{W}_2^T]$, where $\mathbf{W}_1$ is a $q \times q$ matrix, QRcp factorization of $\mathbf{W}^T$ will produce the $n \times n$ permutation matrix $\mathbf{P}$:

$$\mathbf{Q}^T [\mathbf{W}_1 \quad \mathbf{W}_2]^T \mathbf{P} = [\mathbf{R}_{11} \quad \mathbf{R}_{12}],$$

such that $\mathbf{R}_{11}$ is upper triangular and $\mathbf{Q}^T \mathbf{Q} = \mathbf{I}$. The selected subset is given by the first q columns of $\mathbf{AP}$.

D. Modified QR with Column Pivoting Factorization and Subset Selection

Modified QR with column pivoting (m-QRcp) factorization [17] can lead to optimal successive selection of the n $(< N)$ regressors in $\mathbf{A} \in \mathbb{R}^{m \times N}$, with respect to the output vector $\mathbf{y}$ in a linear regression problem (3) discussed in Section III.A. A description of the algorithm follows.

First, the column vector $\mathbf{a}_i$ $(i \leq N)$ of $\mathbf{A}$ producing maximum correlation with $\mathbf{y}$ is detected. This vector is swapped with the first column vector $\mathbf{a}_1$. The so arranged $\mathbf{A}$ is appended by $\mathbf{y}$ forming $\mathbf{X} = [\mathbf{A}|\mathbf{y}]$. The subsequent columns of $\mathbf{A}$ are pivoted as follows. Using the Gram–Schmidt orthogonalization concept [15], if $\mathbf{q}_1$ is the unit vector in the direction of $\mathbf{a}_1$, the portion of $\mathbf{a}_j$ $(j = 2$ to $N)$ and $\mathbf{y}$ in a direction orthogonal to $\mathbf{a}_1$ will be given by $(\mathbf{a}_j - \mathbf{q}_1^T \mathbf{a}_j \mathbf{q}_1)$ and $(\mathbf{y} - \mathbf{q}_1^T \mathbf{y} \mathbf{q}_1)$, respectively; these are referred to as the rotated vectors $\mathbf{a}_j^*$ and $\mathbf{y}^*$ with respect to $\mathbf{a}_1$.

At the ith stage of selection, the rotated variable vectors ($\mathbf{a}_j^*$) are given by

$$\mathbf{a}_j^* = \mathbf{a}_j - (\mathbf{q}_1^T \mathbf{a}_j \mathbf{q}_1 + \cdots + \mathbf{q}_{i-1}^T \mathbf{a}_j \mathbf{q}_{i-1})$$

for $i = 2$ to n, $j = i$ to N, and the rotated output ($\mathbf{y}^*$) vector is given by

$$\mathbf{y}^* = \mathbf{y} - (\mathbf{q}_1^T \mathbf{y} \mathbf{q}_1 + \cdots + \mathbf{q}_{i-1}^T \mathbf{y} \mathbf{q}_{i-1});$$

the ith selected vector is the one for which $\mathbf{a}_j^*$ will have maximum correlation with the rotated output vector $\mathbf{y}^*$. Here, each normalized vector $\mathbf{a}_i^*$ ($i = 2$ to n) is in a plane orthogonal to the subspace spanned by earlier ($i - 1$) selected vector spaces. The selection procedure is repeated until n regressors are selected. All the column swappings are recorded in the permutation matrix $\mathbf{P}$.

E. Remarks

1. Compared to QRcp factorization, m-QRcp factorization is more appropriate for causal representations as it takes the output vector into account. However, both methods take into consideration near collinearity (i.e., one regressor being a linear function of one or more other regressor vectors) in $\mathbf{A}$ by ascribing lower importance to nearly collinear vectors.
2. Implementation of QRcp factorization and m-QRcp factorization using Householder rotations is more robust than the Gram–Schmidt orthogonalization approach.
3. SVD can indicate the closeness to rank redundancy and hence the *number* of significant regressors required in a subset selection problem.
4. The present methods of subset selection are all numerically robust and computationally efficient. No explicit parameterizations are necessary for subset selection. Implementations of SVD and QRcp are available [22]. Alternative methods of subset selection are also possible [23, 24].

III. NETWORK-SIZE OPTIMIZATION THROUGH SUBSET SELECTION

A. Basic Principle

Consider the linear modeling problem

$$\mathbf{y} = \mathbf{A}\boldsymbol{\theta}, \tag{3}$$

where $\mathbf{A} = [\mathbf{a}_1, \ldots, \mathbf{a}_i, \ldots, \mathbf{a}_N]$ contains N m-regressor vectors $\mathbf{a}_i$, $\mathbf{y}$ is the output vector, and $\boldsymbol{\theta}$ is the N-parameter vector. The two prime aspects concerning optimal modeling are (i) collinearity (among the regressors within **A**) and (ii) orthogonality of the regressors with respect to **y**. The collinear regressor(s) in **A** are redundant. On the other hand, a regressor orthogonal to y may not be redundant (if $N > 2$), because the relationship between **y** and the regressors in **A** within (3) is a *group phenomenon*. This aspect has been further discussed in [5].

An optimal model has to be parsimonious [2]. Parsimony can be achieved through the elimination of redundancy in the model (a) by eliminating the collinear regressors in **A** and (b) by accommodating only those regressors in **A**, which *collectively* contain maximum information about the output in some appropriate statistical index sense, such as the minimization of the C_p statistic [25] discussed next.

The C_p statistic is given by

$$C_p = RSS_p/RSS_N - (m - 2p),$$

where m is the number of data sets, N is the maximum number of regressors, p is the number of regressors constituting the optimal model ($1 \leq p \leq N$), RSS_i is the residual sum of squares and error with i regressors. In linear modeling, the C_p statistic is often used for the assessment of the optimality of a parsimonious model [26].

The aforementioned concepts are applicable for linear models. In the present study, these are applied to optimize the size of a neural network by applying the methods in those sections of the network that are linear or can be configured to be linear.

The optimization is applied to determine (i) which of the candidate inputs to the neural network constitute the best set of inputs, and (ii) at the hidden layer(s) which links between the post–hidden layer stage and the subsequent stage should be retained for representative modeling.

B. Selection of Optimum Set of Input Nodes

Assume that there are n number of inputs, for each of which there are m sets of data points, together constituting an $m \times n$ matrix **A**. In a classification problem, m is the number of experiments performed or the number of subjects and n is the number of properties or features. In case of a multi-input, single- or multi-output process, n is the number of inputs and m is the number of data points available for each input. In case of a discrete-time causal expression or a time series, n is the number of appropriately time delayed regressors and m is the number of data points for each regressor. The objective is to determine which of the n inputs carry significant information.

The subset selection of **A** will identify the $m \times q$ subset $\mathbf{A}_1$ containing the prime information of **A**, as discussed in Section II. The subset selection can be performed by using QRcp factorization of **A** following SVD of **A**, to determine the number of dominant modes of **A**. Alternatively, m-QRcp factorization may be used on **X** (= [**A**|**y**]) and the number of inputs as well as the specific inputs are selected using the minimization of the C_p value.

C. Selection of Optimum Number of Hidden Nodes and Links

1. Optimization of Homogeneous Network

In case of a homogeneous network, all the hidden nodes are connected with all the input nodes and thus the structure is homogeneous. The optimization is performed as follows. An overparameterized network with a sufficient number of hidden nodes (say r) is considered. Following crude learning of the network, an $m \times r$ matrix **B** is formed at the post–hidden layer stage, where m is the length of the epoch or the number of iterations. The number of dominant singular values of **B** will indicate the number of hidden nodes to be retained. The reduced network is retrained to convergence.

2. Optimization of Nonhomogeneous Network

In case of a nonhomogeneous network, all possible combinations of connections between the input nodes and the hidden nodes are permitted. Because the links with the individual hidden nodes are different (unlike homogeneous networks), it is necessary to identify the specific nodes that are to be retained in the optimized structure. Two different approaches may be considered:

a. Using Singular Value Decomposition Followed by QR with Column Pivoting Factorization

Proceeding the same way as Section III.C.1, the $m \times r$ matrix **B** is formed, where r is the number of hidden nodes (including unity gain dummy nodes for direct links between an input node and the output node). The desired number of hidden nodes is ascertained using SVD of **B**. QRcp factorization is now performed on **B** to determine the significant columns of **B** and thus identify the specific nodes to be retained. The desired output **y** does not feature in this selection process.

b. Using Modified QR with Column Pivoting Factorization Coupled with C_p Statistic Assessment [17]

The reference output y is reverse nonlinearly transformed (with respect to the nonlinearity within the output node) to $\mathbf{y}'$. Following crude learning of the network, the candidate inputs to the output node, together with the transformed vector $\mathbf{y}'$, constitute a linear regression problem with $\mathbf{y}'$ being the response vector. The matrix $\mathbf{B}$ is formed the same way as before (Section III.C.1), and m-QRcp factorization is performed on $\mathbf{X}$ $(= [\mathbf{B}|\mathbf{y}'])$. The columns of $\mathbf{B}$ are successively selected from $\mathbf{X}$, and the corresponding C_p index is computed. The selected optimal subset is the one producing the minimum value of C_p, which indicates the desired specific set of links or nodes to be retained. The reduced network is retrained to convergence.

3. Remarks

1. For the same number of hidden nodes, the nonhomogeneous network is expected to incorporate larger degrees of nonlinearity within the network, compared to the homogeneous network. A nonhomogeneous network is a closer structural realization of the Kolmogorov–Gabor polynomial [27] through the neural network.
2. The optimization can be performed after crude training of the network. Experience shows that optimization can be performed quite early during the training. Other fast crude learning approaches (e.g., [28]) may also be used.
3. Although SVD is the most definitive method for the assessment of the rank of a matrix, QR factorization can also indicate the rank of a matrix [29]. So, to determine the desired number of hidden layer nodes, QR factorization may be used in place of SVD. Because, in a real-life situation, the distribution of the singular values (or the magnitudes of the diagonal elements of $\mathbf{R}$) of $\mathbf{B}$ may not show a significant jump, one should be rather conservative in deciding the desired number of hidden layer nodes.
4. The method of partial least squares (PLS) [30] offers a closely related approach to subset selection, where a set of mutually orthogonal vectors are determined to explain the output. Apparently, PLS is not as powerful as m-QRcp factorization coupled with C_p assessment [17]. No detailed comparison between PLS and other subset selection methods are available.

IV. INTRODUCTION TO ILLUSTRATIVE EXAMPLES

Three complex examples are studied to illustrate the application of the methods of optimization discussed in this chapter. The first two examples are time series depicting nonlinear dynamics of real-life processes, and the third one is on a complex input–output process. In all cases, feedforward neural network architectures are considered, with unity gain input nodes and sigmoidal nonlinearity between 0 and 1 for nonlinear nodes; the learning of the networks is performed using the back-propagation algorithm. Because a single hidden layer can adequately model most nonlinear systems [31], three-layer networks with a single hidden layer are considered. The study covers both homogeneous and nonhomogeneous networks. The optimization of the networks through SVD, QRcp factorization coupled with SVD, and m-QRcp factorization coupled with the optimality assessment through the C_p statistic are studied.

V. EXAMPLE 1: MODELING OF THE MACKEY–GLASS SERIES

The Mackey–Glass (MG) equation [18], which models the nonlinear oscillations occurring in physiological processes, is given by

$$x(k+1) - x(k) = \alpha x(k-\tau)/(1 + x^{\gamma}(k-\tau)) - \beta x(k),$$

with typically $\alpha = 0.2$, $\beta = 0.1$, and $\gamma = 10$. For $\tau = 17$ (Fig. 1), the attractor of the series has a fractal dimension of 1.95 [32].

The series can be modeled as

$$x(k+p) = f(x(k), x(k-T), x(k-2T), \ldots, x(k-(N-1)T)),$$

where p is the prediction or lead time and N can be typically between 4 and 8 [33]. Here, the values of $N = 6$, $p = 6$, and $T = 17$ have been used.

A homogeneous feedforward neural network having 6 input nodes, 11 hidden nodes, and 1 output node (i.e., a 6–11–1 network) is considered to model the MG

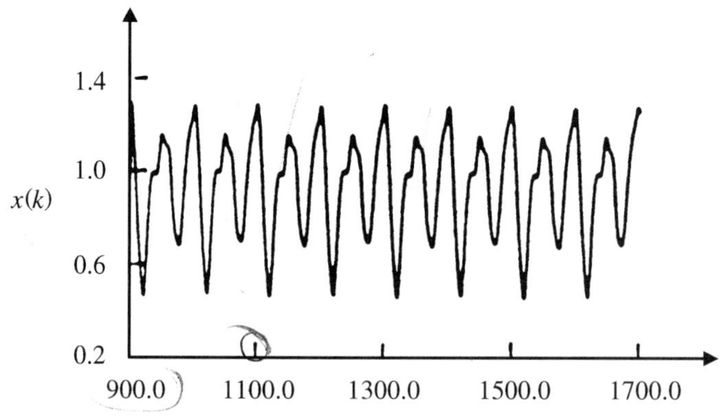

Figure 1 Mackey–Glass series ($\tau = 17$).

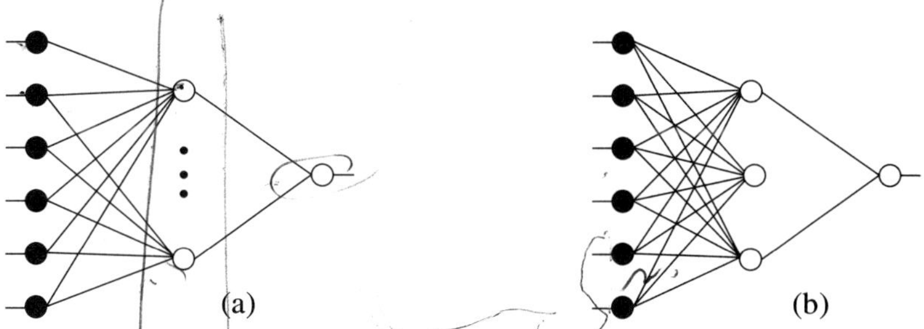

Figure 2 (a) Homogeneous 6–11–1 network modeling the MG series; (b) reduced 6–3–1 network (○ a node, • a node passing data as they are).

series. For all exercises a 300×6 data set is used for training, and the subsequent 200×6 data set is used for the validation test; the lead time p is considered to be 6.

The network used (Fig. 2a) has all 11 hidden nodes linked with all 6 input nodes and the output node. The training is performed with a 300×6 input data set. Throughout the training, SVD is performed on the 99×11 matrix **B**, a subset of the available 300×11 matrix (the size of **B** is not a limitation) at the post–hidden layer stage to determine the optimum number of hidden nodes. The results (Table I, Fig. 3) show three to four singular values being relatively dominant throughout. So three hidden nodes are considered necessary. Apparently, the selection is possible even at an early stage with crude convergence.

Both the 6–11–1 and the reduced 6–3–1 networks (Fig. 2b) are trained to convergence and the validation is tested. The validation root mean square er-

Table I
Selection of Optimum Number of Hidden Nodes Using Singular Value Decomposition

Number of epochs	Singular values of 99×11 matrix **B**	Number of nodes selected
10	20.1, 2.7, 0.8, 0.3, 0.2, 0.104, . . . , 0.003	3
1000	20.0, 2.4, 1.2, 0.8, 0.4, 0.328, . . . , 0.006	3

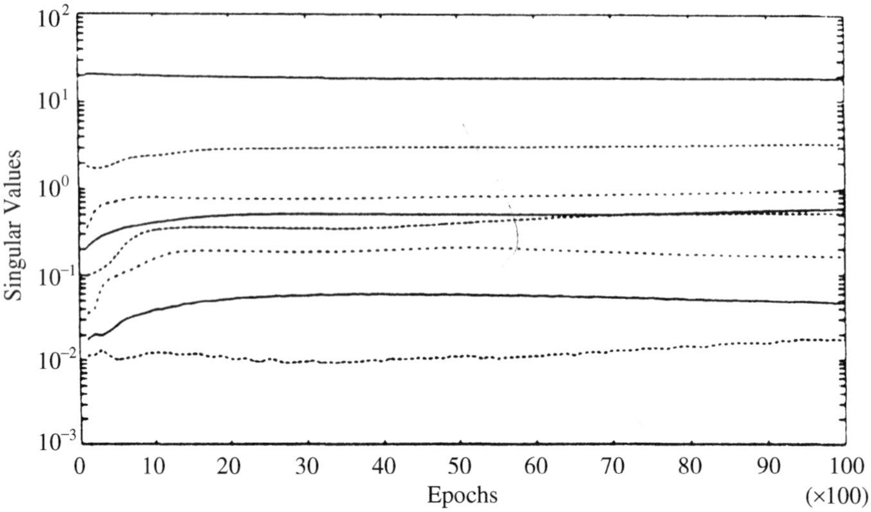

Figure 3 Distribution of the singular values of **B** during the training of the 6–11–1 network.

ror (RMSE) for the two networks work out to be 0.137 and 0.092, respectively. See Fig. 4.

Remark. The estimation and validation performances are also quite close, which validates the reduction in the size of the network. Further results on this series appear in [34].

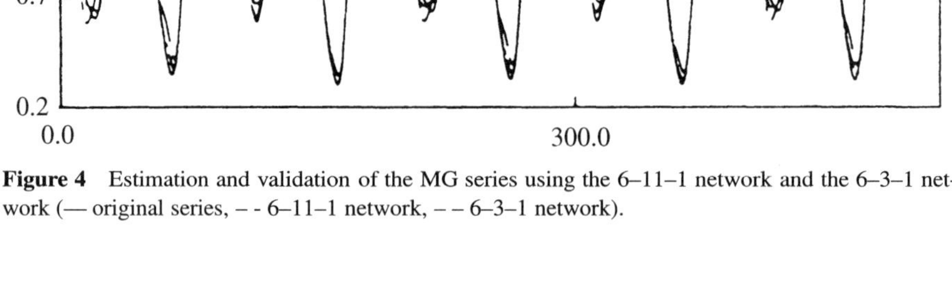

Figure 4 Estimation and validation of the MG series using the 6–11–1 network and the 6–3–1 network (— original series, - - 6–11–1 network, – – 6–3–1 network).

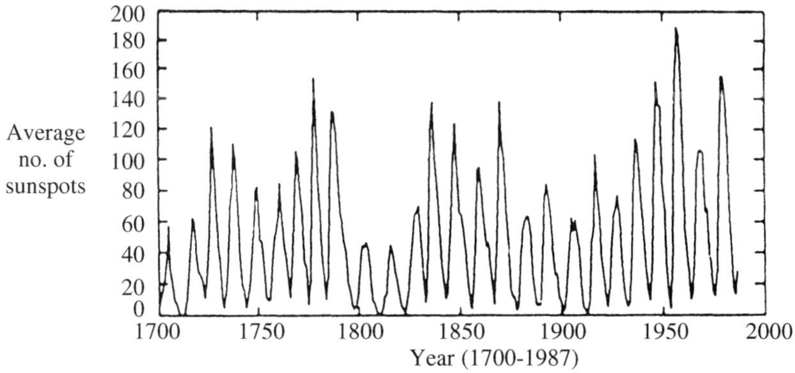

Figure 5 Yearly averaged series of sunspot numbers (from 1700 to 1987).

VI. EXAMPLE 2: MODELING OF THE SUNSPOT SERIES

The series of yearly averaged sunspot numbers (obtained from the daily observations from more than 50 observatories) have been of great interest to researchers and analysts [19, 35]. Data from the year 1700 are available (see Fig. 5). In the present study, the first 221 data points (over 1700 to 1920) are used for modeling, and data over the next 33 years are used for the validation study.

A. PRINCIPLE OF MODELING A QUASIPERIODIC SERIES

The three basic attributes of a nearly periodic series are the periodicity, the pattern over the periodic segments, and the scaling factor associated with each periodic segment. In case of a quasiperiodic series, all three features may vary. There are different approaches for modeling a quasiperiodic series like the sunspot series [2, 5, 35]. In the present study, the most dominant periodicity (N) is detected by using the singular value ratio (SVR) spectrum or the periodicity spectrum [5] (see Appendix B). The successive nearly periodic segments of the sunspot series are compressed or expanded to length N as follows.

Let the objective be to replace $y(1), \ldots, y(N^*)$ by the set $x(1), \ldots, x(N)$, where

$$\begin{aligned} x(j) &= y(j^*) + [y(j^*+1) - y(j^*)](r_j - j^*), \\ r_j &= (j-1)(N^*-1)/(N-1) + 1, \end{aligned} \tag{4}$$

and j^* is the integer part of r_j. The transformed pseudo-segments are arranged in consecutive rows of the data matrix **X**. The modeling proceeds as follows.

An $m \times N$ data window $\mathbf{A}(K)$ is assumed moving over **X**, thus tracking the dynamic variations in the data. $\mathbf{A}(K)$ is singular value decomposed. If $\sigma_1^2 \gg \sigma_2^2$, most of the information energy will be contained in the most dominant mode $\mathbf{u}_1\sigma_1\mathbf{v}_1^T = \mathbf{z}(K)\mathbf{v}_1^T$, where $\mathbf{z} = [z_1, \ldots, z_i, \ldots, z_m]^T$. So a sensible approach is to model the sequence of elements within **z** and to use the model to produce one-step-ahead prediction $z_{(m+1|m)}$, which can lead to one-(pseudo)period-ahead prediction $z_{(m+1|m)}\mathbf{v}_1^T$, the assumption being that the pattern $\mathbf{v}_1^T$ remains unaltered over the predicted segment. Similarly, p-period-ahead prediction will be given by $z_{(m+p)|m}\mathbf{v}_1^T$. In the present case, both homogeneous and nonhomogeneous neural networks are used for the modeling of the $\{z_i(K)\}$ series.

B. Sunspot Series Model

The occurrence of peaks at row lengths of 11 and its multiples in the SVR spectrum (Fig. 6) of the sunspot series shows that the prime periodicity is 11 (years). The data series shows 19 apparent periodic segments over the first 221 data points. So the data set is transformed into 19 periodic segments each of length 11 using (4). The transformed data are arranged into a 19×11 matrix **X**, where successive periodic segments occupy the successive rows of the matrix. Here a 4×11 data

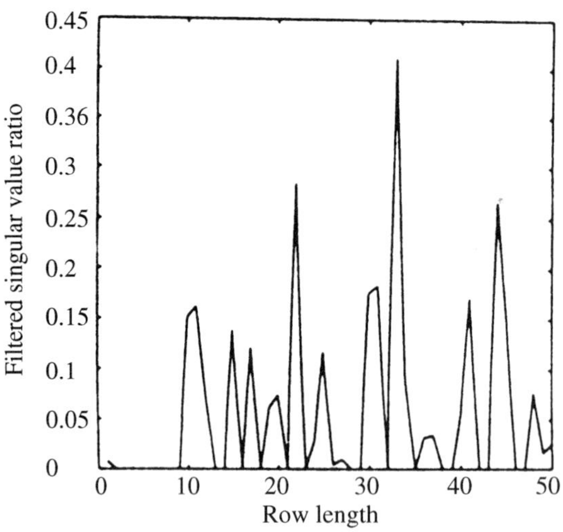

Figure 6 SVR spectrum or the periodicity spectrum of the sunspot series.

window $\mathbf{A}(K)$ is considered moving over $\mathbf{X}$, where $K = 4, \ldots, 19$. For each K, SVD is performed on $\mathbf{A}(K)$, and the vector $\mathbf{z}(K)(= [z_1\ z_2\ z_3\ z_4]^T)$ is obtained. The most dominant decomposition component is found to be sufficiently strong (with $\sigma_1^2/\sigma_2^2 > 16$) to justify approximation of $\mathbf{A}(K)$ by the most dominant mode $\mathbf{z}\mathbf{v}_1^T$. For neural network modeling, $z_1(K)$, $z_2(K)$, and $z_3(K)$ are used as the inputs and $z_4(K)$ is used as the output. The modeling exercises are detailed next.

1. Homogeneous Network

Because the number of training pattern sets is 16 (considering one pattern for each value of K), initially a 3–15–1 homogeneous network is considered as shown in Fig. 7a. During the course of the training, SVD of the data matrix $\mathbf{B} \in \mathbb{R}^{16\times 15}$ at the post–hidden layer stage is performed; because five singular values work out to be relatively dominant, five hidden nodes appear to be necessary. The normalized values of the dominant singular values of $\mathbf{B}$ are shown in Table II. The reduced 3–5–1 network (Fig. 7b) is retrained to convergence, and one-period-ahead prediction is computed for three successive periods using both 3–15–1 and 3–5–1 networks. The learning curves for the two networks shown in Fig. 8 and

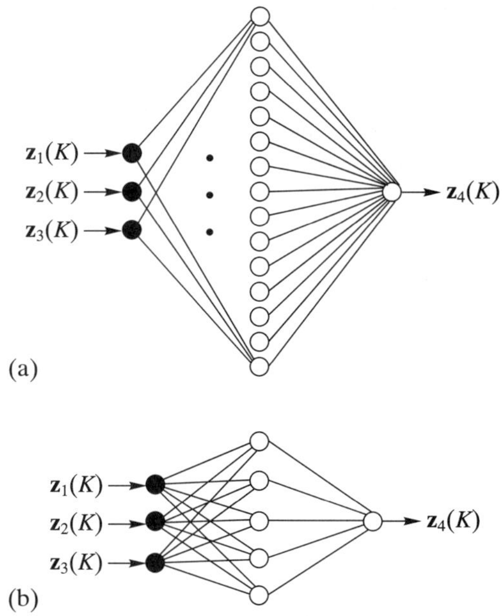

Figure 7 (a) 3–15–1 homogeneous network modeling $\{z_1\}$ in the sunspot series model ($\mathbf{z}\mathbf{v}_1^T$); (b) reduced 3–15–1 network.

Table II
Normalized Singular Values of B for 3–15–1 Homogeneous Network (Sunspot Series)

Iterations	Normalized singular values					
500	1, 0.17,	$4.85E-2$,	$3.23E-3$,	$9.68E-4$,	$3.02E-4$, ...	
5,000	1, 0.17,	$8.81E-2$,	$2.93E-2$,	$1.15E-2$,	$9.66E-4$, ...	
10,000	1, 0.21,	0.16,	$8.53E-2$,	$1.30E-2$,	$4.66E-3$, ...	
20,000	1, 0.43,	0.25,	0.19,	0.10,	$2.83E-2$, ...	
30,000	1, 0.25,	0.17,	0.13,	$8.6E-2$,	$4.97E-2$, ...	

the prediction performances shown in Fig. 9 display close conformity between the 3–15–1 network and the reduced 3–5–1 network.

2. Nonhomogenous Network

A nonhomogeneous 3–10–1 network is used, where all possible combinations of links between the input and the output layer through the hidden layer are considered. The network structure is shown in Fig. 10a; let the hidden nodes be numbered sequentially from 1 (at the top) to 10. Nodes 8, 9, and 10 connect input nodes directly with the output node. At different stages during the training, SVD followed by QRcp factorization–based subset selection is performed on the matrix $\mathbf{B} \in \mathbb{R}^{16\times 10}$. Five singular values being significant, five (specific) hidden nodes are selected; the selection is seen to be fairly consistent as shown in Table III. The

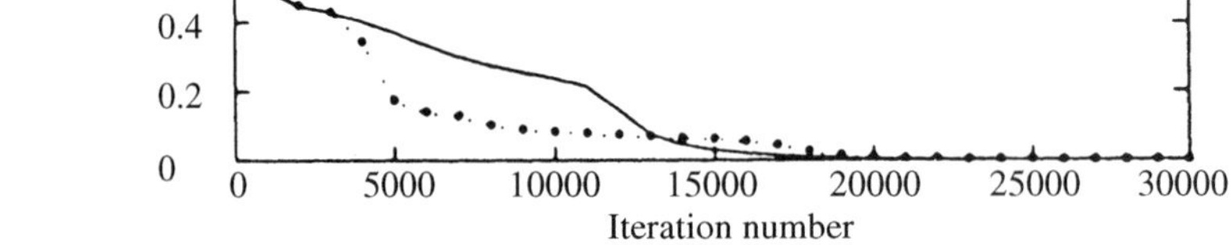

Figure 8 Learning curves for the 3–15–1 and the reduced 3–5–1 networks.

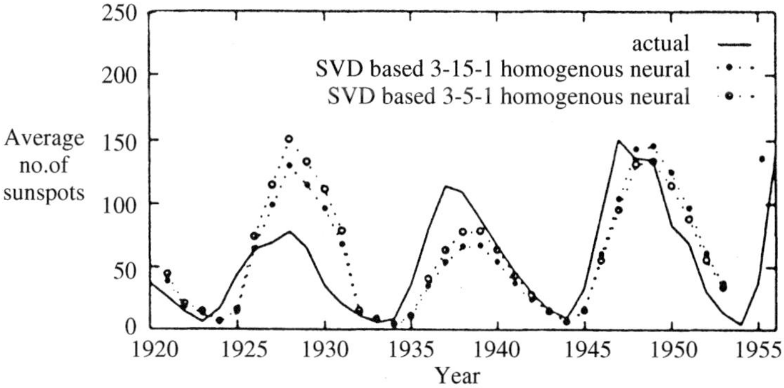

Figure 9 1- to 11-year-ahead prediction of the sunspot series over 1921 to 1953 using homogeneous networks.

reduced network is retrained and the one-period-ahead prediction computed over three successive pseudo-periods (without retraining of the network) are produced.

The prediction error (in terms of the mean square error per sample) for the nonhomogeneous and the reduced homogeneous (3–5–1) networks is 85.96% and

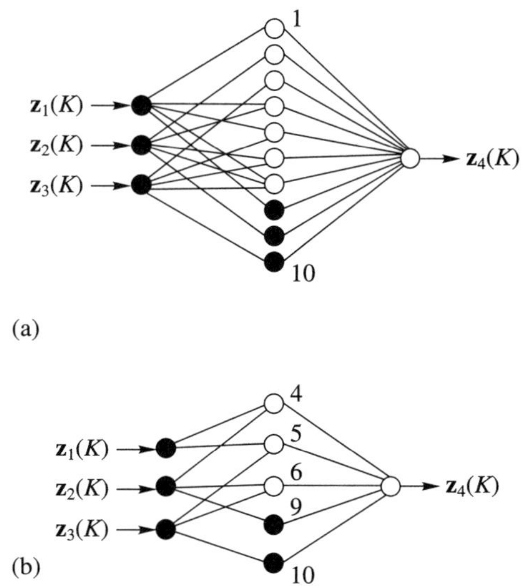

Figure 10 (a) Nonhomogeneous 3–10–1 network modeling of the sunspot series; (b) reduced 3–5–1 network.

Table III

Selection of Hidden Nodes of 3–10–1 Nonhomogeneous Network (Sunspot Series)

Iterations	Nodes selected (Fig. 10a)
10,000	9, 4, 10, 8, 6
20,000	9, 10, 4, 4, 6
30,000	9, 10, 5, 4, 6
40,000	9, 5, 10, 4, 6
50,000	9, 5, 4, 10, 6

119.33% of that obtained for the homogeneous 3–15–1 network. Thus, the nonhomogeneous network appears to offer the best modeling strategy; the performance of the homogeneous 3–5–1 network is also comparable to the much larger 3–15–1 homogeneous network.

3. Remarks

1. Even though the underlying assumption relating to $\mathbf{v}_1(K)$ being the same for the predicted period is only approximately true for the quasiperiodic sunspot series, the *periodic* prediction performance is reasonably good. The main reasons are (a) the capability of SVD to extract the prime repetitive feature, when the data matrix is suitably configured to accommodate the repetitive structure in the signal, and (b) the strength of neural network modeling. Note that relatively longer steps-ahead prediction has been possible through the present method compared to alternative methods [2, 10, 35], hence the ability to recognize a greater degree of determinism in the series.

2. Here, the neural network operates with orthogonally transformed data, enabling substantial reduction in the size of the network irrespective of the complex nature of the series.

3. From a numerical point of view, SVD is one of the most robust orthogonal transformations. Hence, the use of SVD with or without QRcp factorization is expected to be much more robust compared to eigenvalue-based approaches [16].

4. The present method of modeling [36] through the determination of the prime periodic component in a quasiperiodic series is worth noting. This, together with the fact that rank-one approximation is used in modeling in terms of $\mathbf{u}_1\sigma_1\mathbf{v}_1^T$, makes the method relatively immune to noise contamination. Further, the complete left and right singular vector matrices need not be stored, which adds to the computational advantage.

Table IVa

Selection Based on m-QRcp Factorization and C_p Statistic in Rocket Engine Testing Problem

Epoch	m-QRcp selection	C_p values	Number of nodes selected
4000	**10, 12, 11,** 15, 14, 5 . . .	34.5, 4.6, **1.5,** 2.7, 3.7, 4.6 . . .	3 ($C_p = 1.5$)
6000	**10, 12, 11,** 15, 5, 7 . . .	36.0, 6.3, **1.6,** 2.9, 4.7, 7.4 . . .	3 ($C_p = 1.6$)

VII. EXAMPLE 3: MODELING OF THE ROCKET ENGINE TESTING PROBLEM

This is a widely studied problem [20, p. 380, 37]. Here, the chamber pressure (y) is the output, which can be expressed as

$$y = f(x_1, x_2, x_3, x_4),$$

where x_1 is the temperature of the cycle, x_2 is the vibration, x_3 is the drop (shock), x_4 is the static fire, and y is the chamber pressure. Altogether 24 sets of data are available, out of which the first 19 sets are used for modeling and the rest are left for validation.

The problem is modeled using a 4–15–1 network with exhaustive choice of links between the layers of the network as shown in Fig. 11a. The network is trained with a 19×4 data set, and at different stages of training m-QRcp factorization coupled with C_p statistic–based subset selection is performed on the

Table IVb

Selection Based on QRcp Factorization and SVD in Rocket Engine Testing Problem

Epoch	QRcp selection	Singular values	Number of nodes selected
4000	**4, 11, 14, 2,** 5, 6, 7, 8 . . .	8.89, 0.66, 0.29, 0.20, 0.05, 0.05, 0.01, 0.009 . . .	4
6000	**4, 11, 2, 14,** 5, 6, 7, 8 . . .	8.9, 0.66, 0.34, 0.21, 0.07, 0.06, 0.01, 0.003 . . .	4

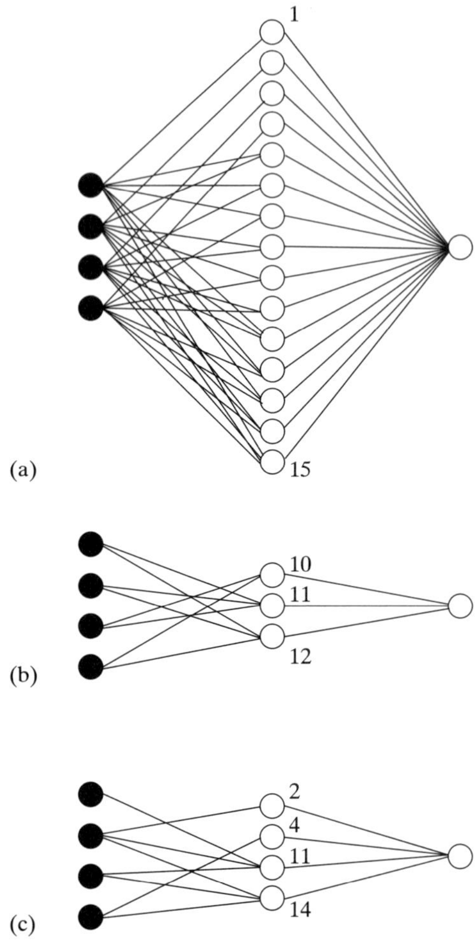

Figure 11 (a) Overparameterized 4–15–1 nonhomogeneous network; (b) reduced 4–3–1 network obtained through m-QRcp and C_p statistic; (c) reduced 4–4–1 network obtained through SVD and QRcp.

matrix $\mathbf{X} = [\mathbf{B}|\mathbf{y}']$, where $\mathbf{B} \in \mathbb{R}^{19\times 15}$ is formed from the data at the post–hidden layer stage. The selection of modes and the C_p values for two different stages during training are shown in Table IVa. The C_p statistic shows a distinct minimum for three *successively* selected hidden nodes marked 10, 12, and 11; the reduced 4–3–1 network (having three nonlinear nodes with three inputs) is shown in Fig. 11b. The training and the validation performance are shown in Fig. 12.

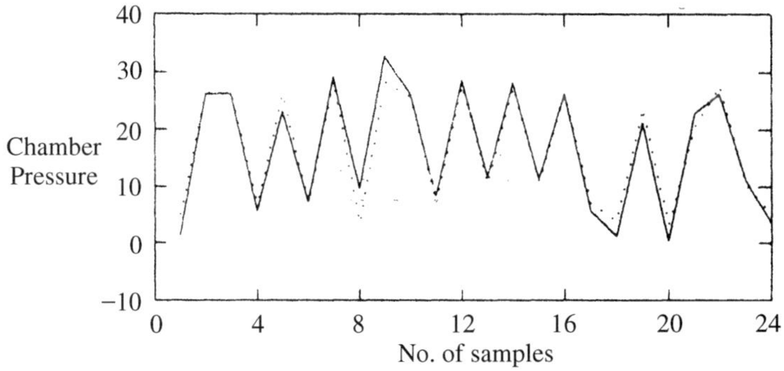

Figure 12 Estimation and validation performance of the chamber pressure in the rocket engine problem (— actual data, ··· estimation/validation).

The exercise of optimizing the network is repeated with QRcp factorization and SVD performed on the matrix **B** during the training of the network. The distribution of the singular values and the respective selections of the hidden layer nodes are shown in Table IVb. Because four singular values are relatively dominant, four hidden nodes are required for the reduced network shown in Fig. 11c. The performances of the original and the reduced networks are shown in Table V.

Remarks.

1. The detection of optimality through minimization of the C_p statistic coupled with m-QRcp factorization is seen to be conclusively distinct (see Fig. 13); further, knowledge of the output is inherently taken into consideration in m-QRcp factorization, hence the relative superiority of this approach. The distribution of the singular values may not show decisive jumps and so it may be somewhat difficult to decide the number of nodes to be selected for QRcp factorization.

Table V
Comparative Validation Performance for the Rocket Engine Problem

	Exhaustive non-homogeneous model	QRcp-SVD-based modeling	m-QRcp-C_p-based modeling
Network	4–15–1 network (16 nodes, 47 links)	4–4–1 network (5 nodes, 12 links)	4–3–1 network (4 nodes, 11 links)
RMSE	2.996	2.113	1.944

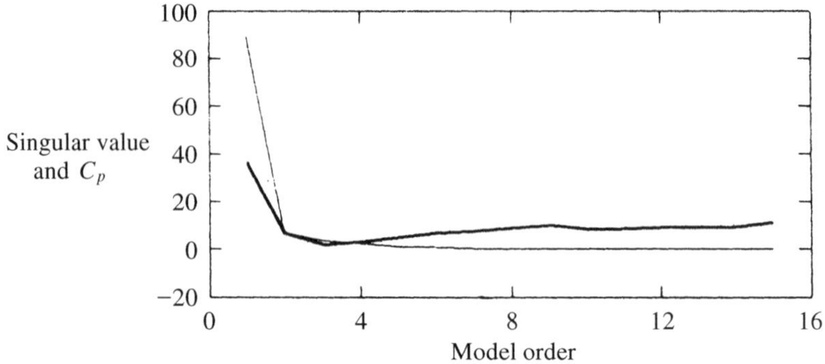

Figure 13 Profile of the C_p statistic (—) and the singular values (—) of **X**.

2. In the exhaustive nonhomogeneous network (Fig. 11a), direct links between the inputs and the output have not been considered because of the limited amount of data.

VIII. ASSESSMENT OF CONVERGENCE IN TRAINING USING SINGULAR VALUE DECOMPOSITION

The convergence in the training of a neural network is usually assessed in terms of the output error remaining almost unchanged at a low value. If an $m \times n$ input data set is used for training, the output error for m different sets of input has to be studied. SVD offers an alternative method for the convergence assessment through the rank-oneness assessment of the output matrix over several epochs (or iterations).

The training through one m-long epoch implies m number of network-weight updates, which generates an m-output vector, and g epochs will produce an $m \times g$ matrix $\mathbf{Y}_g$ at the output. At true convergence, all the columns of $\mathbf{Y}_g$ should be identical to $\mathbf{y}_R$, the corresponding reference output vector. On the other hand, as long as the training is not complete, the columns of $\mathbf{Y}_g$ will keep changing. So, the degree of convergence can be assessed from the closeness of $\mathbf{Y}_g$ to rank-oneness.

Let the SVD of $\mathbf{Y}_g$ be performed. The ratio of the energy contained in the most dominant decomposed mode $\mathbf{u}_1\sigma_1\mathbf{v}_1^T$ and the total reference output energy is given by

$$c = \sigma_1^2 / Q_{Rg},$$

where

$$Q_{Rg} = g\mathbf{y}_R^T\mathbf{y}_R.$$

Ideally, at convergence $c = 1$, so the percentage of residual energy at convergence can be defined as $\kappa = 1 - c$.

Remarks.

1. κ will be insensitive to a local minimum if g is large enough to cover any such minima.
2. The output of the network and the reference data sets are mean extracted before computing κ to make σ_1 insensitive to the mean value for nonzero-mean data.

EXAMPLE (Convergence in training for the Mackey–Glass series). Consider the 6–3–1 homogeneous neural network model of the Mackey–Glass series (Section V). At different stages of training, the output matrix $\mathbf{Y}_g$ is formed with $g = 200$, the epoch length being 209. As shown in Fig. 14, the progression of learning depicted by the profile conforms to that shown by the output error profile.

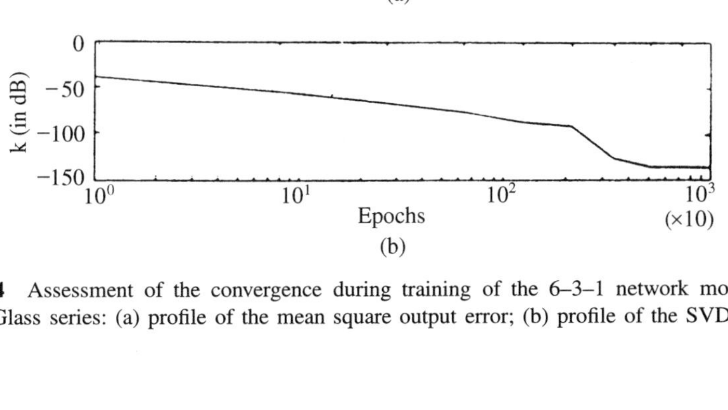

Figure 14 Assessment of the convergence during training of the 6–3–1 network modeling the Mackey–Glass series: (a) profile of the mean square output error; (b) profile of the SVD-based index κ.

IX. CONCLUSIONS

It has been shown that orthogonal transformation through singular value decomposition and various forms of QR with column pivoting factorization can offer robust and efficient methods for the optimization of feedforward neural networks. The optimization of homogeneous networks is much simpler than that of nonhomogeneous networks, although the latter attracts more interest as they can have larger density of nonlinearity.

Here, the sensible objective will be to produce meaningful optimization of the network such that the possibility of learning representative information about the underlying process through the available data is enhanced. It is not necessary to expect a unique solution in terms of the optimized structure, because the neural network is inherently nonlinear and hence many solutions may produce close results.

Orthogonal transformation can lead to meaningful optimization of neural networks with relatively less computational effort irrespective of the problems of collinearity within the data, noise associated with the data, or the uncertainty concerning the available knowledge of the process.

APPENDIX A: CONFIGURATION OF A SERIES WITH NEARLY REPEATING PERIODICITY FOR SINGULAR VALUE DECOMPOSITION-BASED ANALYSIS

Consider a process or series: $\{x(\cdot)\} = \{x(1), x(2), \ldots\}$. The successive n-long segments of the series can be arranged in a matrix $\mathbf{X}$ such that the successive segments occupy successive rows of the matrix as follows:

$$\mathbf{X} = \begin{vmatrix} x(1) & x(2) & \cdots & x(n) \\ x(n+1) & x(n+2) & \cdots & x(2n) \\ \vdots & \vdots & & \vdots \\ x((m-1)n+1) & x((m-1)n+2) & \cdots & x(mn) \end{vmatrix}.$$

SVD of the $m \times n$ matrix $\mathbf{X}$ is given by $\mathbf{X} = \mathbf{U}\boldsymbol{\Sigma}\mathbf{V}^T = \mathbf{Z}\mathbf{V}$.

If the series is strictly or nearly periodic with fixed periodicity of n, and if the periodic segments have the same pattern, irrespective of the scaling over the successive segments, Rank($\mathbf{X}$) will be 1, and only σ_1, the first singular value of $\mathbf{X}$, will be nonzero, whereas all other singular values will be zero.

Consider two other possibilities:

(i) If the successive apparently periodic segments of the series have the same period length but almost, yet not exactly similar patterns over the successive segments, Rank (**X**) will be > 1, where the closeness of **X** to rank-oneness will be given by σ_1/σ_2.

(ii) If the series has an apparently repetitive pattern but the successive segments are of different period lengths, the series may be arranged in **X** as follows. First the prime periodicity (say n) in the data series is determined using the SVR spectrum (see Appendix B); next, the successive nearly repetitive segments are identified (say in terms of the relatively regularly occurring features like the peaks or the valleys), and these (pseudoperiodic) segments are compressed or expanded in time (using (4)) to the same period length (n) and arranged in successive rows of the matrix **X** having the row length of n.

If SVD of **X** shows one singular value to be significantly dominant, **X** can be expressed as in (2).

APPENDIX B: SINGULAR VALUE RATIO SPECTRUM

The singular value ratio (SVR) spectrum offers a unique way of detecting the presence and the periodicity of a dominant periodic component (which need not be sinusoidal) in any composite signal or data sequence $\{x(k)\}$. The concept of the SVR spectrum can be briefly stated as follows.

Let the series $\{x(k)\}$ be arranged into a matrix **X** having row length of n, as shown in Appendix A. If $\{x(k)\}$ is strictly periodic with period length N, σ_1/σ_2 of **X** will be infinity, if $n = iN$, where i is a positive integer. If i is a noninteger or if $\{x(k)\}$ deviates from periodicity, σ_1/σ_2 will decrease. For a random series σ_1/σ_2 can be as low as 1. Hence, if the data matrices $\mathbf{X}(n)$ are formed with varying row length n, the corresponding pattern of σ_1/σ_2 of $\mathbf{X}(n)$ will show peaks at the values of n for which there is a dominant periodic component of period length n or any of its multiples present in $\{x(k)\}$. The σ_1/σ_2 values may be filtered such that the peaks in the profile are pronounced. Further discussions on the SVR spectrum and its applications appear in [5, 38].

REFERENCES

[1] D. Sarkar. Randomness in generalization ability: a source to improve it. *IEEE Trans. Neural Networks* 7:676–685, 1996.

[2] E. P. Box and G. M. Jenkins. *Time Series Analysis, Forecasting and Control.* Holden-Day, San Francisco, 1976.

[3] G. W. Stewart. Collinearity and least squares regression. *Statist. Sci.* 2:68–100, 1987.
[4] D. A. Belssley, E. Kuh, and R. E. Welsch. *Regression Diagnostics, Identifying Influential Data and Sources of Collinearity.* Wiley, New York, 1980.
[5] P. P. Kanjilal. *Adaptive Prediction and Predictive Control. IEE Control Engrg. Ser.*, No. 52. Peter Peregrinus, Stevenage, 1995.
[6] Y. Le Cun, J. S. Denker, and S. A. Solla. Optimal brain damage. In *Advances in Neural Information Processing Systems* (D. S. Touretzky, Ed.), Vol. 2. Morgan Kaufmann, San Mateo, CA, 1990.
[7] R. Reed. Pruning algorithms — a survey. *IEEE Trans. Neural Networks* 4:740–747, 1993.
[8] A. Levin, T. K. Leen, and J. E. Moody. Fast pruning using principal components. In *Advances in Neural Information Processing Systems* (J. D. Cowan, G. Tesauro, and J. Alspector, Eds.), Vol. 6. Morgan Kaufmann, San Mateo, CA, 1994.
[9] N. Murata, S. Yoshizawa, and S. Amari. Network information criterion — determining the number of hidden units for an artificial neural network model. *IEEE Trans. Neural Networks* 5:865–871, 1994.
[10] M. Cottrell, B. Girard, Y. Girard, M. Mangeas, and C. Muller. Neural modelling for time series: a statistical stepwise method for weight elimination. *IEEE Trans. Neural Networks* 6:1355–1363, 1995.
[11] D. B. Fogel. An information criterion for optimal neural network selection. *IEEE Trans. Neural Network* 2:490–497, 1991.
[12] H. Akaike. A new look at statistical model identification. *IEEE Trans. Auto. Control* 6:716–723, 1974.
[13] E. J. Hannan and B. G. Quinn. The determination of the order of an autoregression. *J. Roy. Statist. Soc. Ser. B* 41:190–195, 1979.
[14] S. J. Hanson and L. Pratt. A comparison of different biases for minimal network construction with backpropagation. In *Advances in Neural Information Processing Systems* (D.S. Touetzky, Ed.), Vol. 1, pp. 177–185, 1989.
[15] G. H. Golub and C. F. Van Loan. *Matrix Computations.* Johns Hopkins University Press, Baltimore, 1989.
[16] A. J. Laub. Numerical linear algebra aspects of control design computations. *IEEE Trans. Automat. Control* 30:727–764, 1985.
[17] P. P. Kanjilal, G. Saha, and T. J. Koickal. On robust non linear modelling of a complex process with large number of inputs using m-QRcp factorization and C_p statistic. *IEEE Trans. Systems Man Cybernet.*, 1997, to appear.
[18] M. C. Mackey and L. Glass. Oscillations and chaos in physiological control systems. *Science* 197:287–289, 1977.
[19] N. O. Weiss. Periodicity and aperiodicity in solar magnetic activity. *Philos. Trans. Roy. Soc. London Ser. A* 330:617–625, 1990.
[20] N. Draper and H. Smith. *Applied Regression Analysis*, 2nd ed. Wiley, New York, 1981.
[21] F. Deprettere, Ed. *SVD and Signal Processing, Algorithms, Applications and Architectures.* North-Holland, Amsterdam, 1988.
[22] MATLAB matrix software. The Math Works, Inc., Sherborn, MA.
[23] S. Chen, S. A. Billings, and W. Luo. Orthogonal least squares methods and their application to non-linear system identification. *Internat. J. Control* 50:1873–1896, 1989.
[24] S. V. Huffel and J. Vandewalle. *The Total Least Squares Problem: Computational Aspects and Analysis.* SIAM, Philadelphia, 1991.
[25] C. L. Mallows. Some comments on C_p. *Technometrics* 15:661–675, 1973.
[26] C. Daniel and F. S. Wood. *Fitting Equations to Data*, 2nd ed. Wiley, New York, 1980.
[27] A. G. Ivakhnenko. Past, present, and future of GMDH. In *Self-organizing Methods in Modelling.* (S. J. Farlow, Ed.), pp. 105–119. Marcel Dekker, New York, 1984.

[28] F. Biegler-König and F. Bärmann. A learning algorithm for multilayer neural networks based on linear least squares problems. *Neural Networks* 6:127–131, 1993.
[29] T. F. Chan. Rank revealing QR factorization. *Linear Algebra Appl.* 88/89:67–82, 1987.
[30] T. R. Holcomb and M. Morari. PLS/neural networks. *Comput. Chem. Engrg.* 16:393–411, 1992.
[31] G. Cybenko. Aproximation by superposition of a sigmoidal function. *Math. Control Signals Systems* 2:303–314, 1989.
[32] J. D. Farmer. Chaotic attractors of an infinite-dimensional dynamic system. *Physica D* 4:366–393, 1982.
[33] A. Lapedes and R. Farber. Nonlinear signal processing using neural networks. Technical Report LA-UR-2662, Los Alamos National Laboratory, 1987.
[34] P. P. Kanjilal and D. N. Banerjee. On the application of orthogonal transformation for the design and analysis of feedforward networks. *IEEE Trans. Neural Networks* 6:1061–1070, 1995.
[35] M. Casdagli. Chaos and deterministic versus stochastic nonlinear modelling. *J. Roy. Statist. Soc. Ser. B* 54:303–328, 1992.
[36] P. P. Kanjilal and S. Palit. Modelling and prediction of time series using singular value decomposition and neural networks. *Comput. Electric. Engrg.* 21:299–309, 1995.
[37] A. Desrochers and S. Mohseni. On determining the structure of a nonlinear system. *Internat. J. Control* 40:922–938, 1984.
[38] P. P. Kanjilal, S. Palit, and G. Saha. Fetal ECG extraction from single-channel maternal ECG using singular value decomposition. *IEEE Trans. Biomed. Engrg.* 44:51–59, 1997.

Sequential Constructive Techniques

Marco Muselli
Istituto per i Circuiti Elettronici
Consiglio Nazionale delle Ricerche
16149 Genoa, Italy

I. INTRODUCTION

The theoretical and practical problems deriving from the application of the back-propagation algorithm have led to the introduction of a new class of learning techniques, called *sequential constructive methods*, that allows the treatment of training sets containing several thousands of samples.

The computational cost of these algorithms is kept low by adopting two independent methodologies: first, the neural network is constructed in an incremental way by subsequently adding units to the hidden layer. With this approach the learning process does not require the contemporary updating of the whole weight matrix (as in the back-propagation algorithm), but only the modification of a small portion of the network.

Second, the size of the training set employed for the addition of a new hidden neuron decreases in the course of the algorithm, thus allowing a further increase of the convergence speed.

Unfortunately, these interesting features have not yielded a wide application of the sequential constructive methods to the solution of real-world problems. A rea-

Optimization Techniques

son for this could be the lack of a detailed description that presents the general approach of these algorithms along with the specific implementative procedures adopted.

Furthermore, for a correct analysis of the sequential constructive techniques it is necessary to perform a series of comparative tests showing both the properties of the resulting neural networks and the actual reduction of the computational cost.

This chapter may be a first step in that direction; it is subdivided into three distinct parts. In the first part the theoretical and practical problems involved in the application of the back-propagation algorithm are analyzed (Section II) and the solutions adopted by sequential constructive methods to overcome these obstacles are pointed out (Section III).

The general (Section IV) and specific (Sections V and VI) approaches employed by these algorithms form the subject of the second part of this chapter. The main theoretical results (along with the relative proofs) and the implementative aspects are described here in great detail.

The results obtained through the application of sequential constructive methods to several experimental tests are contained in the third part (Section VII); the comparison with the back-propagation algorithm allows an objective evaluation of the performances offered by these training techniques.

II. PROBLEMS IN TRAINING WITH BACK PROPAGATION

The most widely used method for supervised training of neural networks is certainly the back-propagation algorithm [1–5]. Its ease of implementation is one of the main reasons for this large diffusion and makes back propagation a flexible tool for the solution of many problems belonging to a variety of applicative fields [6–10].

Its ability to obtain an optimal or near-optimal configuration for a given training set has been further increased by the introduction of appropriate methodologies that accelerate the convergence of the procedure [11–16]. However, these improvements leave unchanged the basic kernel of the method which performs a minimization of the error made by the current neural network on the given training set.

In the original version, still widely employed, the back-propagation algorithm searches for the global minimum in the weight space by applying the method of steepest descent [17]. Although the implementation of this optimization technique is straightforward, its convergence properties are poor, because it can be prematurely stopped by a flat region or a local minimum of the cost function. Better results can be obtained by making some changes that improve the reliability of the method.

First of all, it is necessary to obtain a good initial point from which to begin the search. This is a crucial step because the algorithm of steepest descent is basically an optimization method that pursues the local minimum closest to the initial point. Procedures for approaching this problem can be found in the literature [18, 19] and have a great influence on the behavior of back propagation.

Also the updating rules for the current point in the search have been the subject of several modification proposals, not always supported by precise theoretical motivations. Most of them try to adapt the search trajectory to the behavior of the cost function so as to avoid getting stuck in an unsatisfactory local minimum [13, 14, 20, 21].

Finally, the expression of the cost function to minimize plays an important role in the determination of the convergence properties for the back-propagation algorithm. Usually it contains a measure of the error made by the current neural network on the given training set, together with other optional quantities that attempt to improve the generalization ability of the final configuration.

An important contribution in this direction is offered by the regularization theory [22, 23] which has been successfully applied to the training of connectionist models. In particular, the method of weight decay [15, 24] has allowed the achievement of interesting results in the treatment of real-world problems.

These and other techniques, which we omit for the sake of brevity, have led to refined versions of the back-propagation algorithm that are especially suited for application to specific fields. As an example, consider the problem of handwritten character recognition: the introduction of appropriate methodologies has produced neural networks with increasing generalization ability, very close to that presented by the human brain [8, 25, 26].

In spite of these promising results, however, there are some difficulties, both theoretical and practical, that thwart the employment of the back-propagation algorithm, particularly when the dimension of the input space or the size of the training set increases. In the next sections we shall analyze in detail the following two important problems:

- The network architecture must be fixed *a priori*.
- Optimal solutions cannot be obtained in polynomial time.

A. Network Architecture Must Be Fixed *A Priori*

The back-propagation algorithm provides the weight matrix for a feedforward neural network with fixed architecture: the number of hidden layers and the number of neurons in each layer must therefore be chosen beforehand. Let us denote by g: $\mathcal{R}^n \rightarrow \mathcal{R}^m$ the input–output transformation performed by the final multilayer perceptron when the training process is completed. The integers m and n then correspond to the dimension of the input and the output space, respectively. It can be shown that the network architecture determines the complexity of the function g that can be realized.

In fact, there are important theorems that assert the general validity of connectionist models: every (Borel) measurable function can be approximated to within an arbitrary precision by a feedforward neural network with a single hidden layer containing a sufficient number of units [27–29]. Unfortunately, the proofs of these theorems are not constructive and the choice of the network architecture is a matter of trial and error in most practical cases.

Let us denote by f the unknown function that has generated (eventually in the presence of noise) the samples contained in the given training set. It should be pointed out that the number of weights plays a fundamental role in the determination of the generalization ability that the final neural network exhibits in a specific application. If this number is too small, then many input–output relations contained in the given training set will not be satisfied; thus, the corresponding transformation g is a poor approximation of the unknown function f [30].

On the other hand, if the number of weights in the neural network is too high, an overfitting of the available data will occur with great probability; consequently, the generalization ability of our connectionist model will be low even if the error on the given training set is close to zero. In general, this means that the resulting neural network has memorized the available samples without extracting sufficient information on the underlying input–output function f.

A quantitative analysis of this phenomenon has been the subject of an important series of papers in the fields of mathematical statistics and machine learning [31–35]. In particular, a proper quantity, called the *Vapnik–Chervonenkis (VC) dimension*, has been defined, which measures the complexity of the trainable model. Unfortunately, a direct determination of the VC dimension for a given neural network is very difficult even when the number n of inputs is small [36].

Furthermore, although such an analysis has great theoretical relevance, the resulting relations give unusable values in most practical situations, because they refer to a worst case study. For this reason, an estimate of the VC dimension obtained by applying simplified hypotheses [34] does not allow for an efficient forecast of the number of hidden units to employ.

Other approaches have been proposed to achieve, in a theoretical way, alternative measures of the complexity of the connectionist model [37, 38]. Nevertheless, at present the optimal neural network architecture for a given real-world problem is mainly obtained through the application of heuristic rules and the execution of subsequent trials following a cross-validation procedure [39, 40]. This generally requires a high computing time, which increases rapidly with the dimension of the input space or the size of the training set.

B. Optimal Solutions Cannot Be Obtained in Polynomial Time

The choice of the configuration for the neural network to be trained is not the only problem inherent in the application of back propagation. In fact, there are some basic theoretical drawbacks that arise even when the architecture considered is very simple. In particular, it has been shown that the task of deciding if a given training set can be entirely satisfied by a given multilayer feedforward neural network is NP-complete [41, 42]. This result prevents us from obtaining optimal solutions in a reasonable time even for small values of the number n of inputs.

This limitation is closely related to the definition of *learnability* in the field of machine learning [43]. In short, a problem is called *learnable* if there is an algorithm, having polynomial computing time in its fundamental variables (number of inputs, complexity of the function f, etc.), which is able to find a satisfying approximation g to the unknown function f. Because the task of training neural networks with fixed architecture is NP-complete, we cannot use back propagation to establish the learnability of a practical problem.

This theoretical drawback is emphasized by the technical difficulties encountered in the application of the method. As previously noted, the search for the optimal weight matrix involves the minimization of a proper cost function often containing many flat areas and local minima, which can create problems for many optimization methods. Thus, it can be convenient to study different training algorithms for neural networks that try to avoid these theoretical and practical problems. A proposal in this direction is offered by the class of constructive methods, which forms the subject of the following section.

III. CONSTRUCTIVE TRAINING METHODS

The theoretical limitations involved in the application of the back-propagation algorithm have given rise to several alternative proposals, which can be subdivided into two classes: pruning methods and constructive techniques.

The former have the aim of achieving the neural network with minimal complexity for the solution of a given problem, rather than accelerating the training process. In fact, a multilayer perceptron containing a smaller number of weights generally has a lower VC dimension and consequently presents a better generalization ability for a given training set.

To this end, pruning methods implement the following approach: at first a larger network containing a higher number of hidden units than necessary is trained (by using some learning algorithm). Then the application of proper techniques [44–46] allows the location and removal of the connections (and eventually the neurons) that have a negligible influence on the behavior of the input–output transformation g.

It should be pointed out that these methods are often able to find possible inputs that are not relevant in the determination of the outputs. This is an important result for both the modeling of physical systems and the automatic control of processes.

However, as it follows from this short description, pruning methods by themselves cannot overcome the drawbacks involved in the application of the back-propagation algorithm. In fact, it is still necessary to know an upper bound on the number of hidden units needed for a neural network that approximates the unknown function f.

Furthermore, an optimal or near-optimal weight matrix for the redundant multilayer perceptron must be obtained within a reasonable execution time. As pointed out in Section II, the back-propagation algorithm cannot achieve this result, particularly when the number of hidden neurons is too high.

Nevertheless, the practical importance of the class of pruning methods should be emphasized: their employment allows us to obtain interesting information on the relevance of every connection and neuron contained in the multilayer perceptrons obtained through the application of a training algorithm.

A symmetrically opposite approach is followed by constructive methods [6]: after the training process, they provide both the configuration of the resulting neural network and the weight values for the relative connections. The learning is typically performed by subsequently adding hidden units to the network architecture until all the input–output relations in the given training set are satisfied. In general, the topology of the connections among the neurons is fixed beforehand and the addition of a new neuron simply implies the redetermination of a (small) portion of the global weight matrix.

This approach leads to learning techniques that present a high convergence speed in the construction of the multilayer perceptron and consequently allow the treatment of complex training sets. Nevertheless, because the updating of the weight matrix involves only a restricted number of connections (in most cases those associated with the neuron to be added), some kinds of regularities in the given training set could be missed. This may reduce the generalization ability of neural networks trained by a constructive method. A technique for dealing with this drawback is proposed in Section VI for the case of classification problems with binary inputs.

In the following two sections we shall describe how constructive methods try to overcome the limitations involved in the employment of the back-propagation algorithm. It should be pointed out, however, that some training methods [47, 48] provide the number of hidden neurons for the resulting neural networks together with the corresponding weight matrix without executing an incremental construction of its configuration. These techniques can still be inserted in the class of constructive methods because they do not work on a fixed architecture. However, some considerations contained in the following two sections may not be applied in this particular case.

A. Dynamic Adaptation to the Problem

At first glance, the lack of a fixed value for the number of hidden neurons could seem the cause of an increase in the training complexity, as it introduces an additional unknown quantity in the learning process. On the contrary, the possibility of adapting the network architecture to the given problem is one of the advantages of constructive techniques.

In fact, there is no need to find an estimate of the complexity of the resulting multilayer perceptron; it is automatically determined during the training on the grounds of the samples contained in the given training set. This is surely the main advantage of constructive algorithms; each of them tries to obtain the minimal neural network that satisfies all the input–output relations in the given training set by using proper heuristic methods.

Unfortunately, only in two particular cases [49, 50] is a theoretical support provided which asserts the optimality (in some sense) of the multilayer perceptron generated by a constructive method. For most algorithms it is only shown that the learning process converges (at least asymptotically) to a configuration that provides correct outputs for all the given samples. However, this result can be achieved if the training set is *consistent*, that is, if it does not contain an input pattern with two different outputs.

Although such a convergence theorem ensures the stability of the method employed, its practical relevance is moderated because all the real-world problems are affected by the presence of noise. Because of this, even when the given training set is consistent the fulfillment of all its samples can lead to a neural network with low generalization ability [32].

In fact, the presence of noise can increase the number of hidden neurons (and consequently the complexity of the multilayer perceptron) so as to take into account disturbed patterns that do not follow the behavior of the function f to be approximated. A general approach to the solution of this problem is not yet available.

B. High Training Speed

The possibility of adapting the neural network architecture to the current problem also has important effects on the convergence speed of the training process. In most constructive methods, the addition of a new hidden unit implies the updating of a small portion of weights, generally only those regarding the neuron to be added.

Hence, it is possible to employ training algorithms for a single neuron that present good convergence properties and allow us to obtain (at least asymptotically) an optimal set of weights [51, 52]. In most cases, the learning process does not involve a search for the global minimum of a proper cost function, as for the back-propagation algorithm, but is based on simpler procedures that lead to a higher convergence speed.

Some constructive methods do not even require the training of the output layer because the associated convergence theorems give suitable values for the corresponding weights. Furthermore, there are some techniques, such as the class of sequential constructive algorithms, in which only a portion of the training set is considered for the addition of a new hidden unit. In these cases, the aforementioned stability properties are maintained and a relevant saving of computation time can be achieved.

However, besides the practical interest of constructive methods, is there any deeper theoretical motivation to prefer this approach with respect to back propagation? Unfortunately, at present it is not possible to give a definitive answer to this question.

Baum in a review [53] has pointed out that the possibility of choosing in an adaptive way the architecture of the neural network can allow us to avoid the NP-completeness result found by Judd [41]. In fact, in the loading problem, the user has no control over the structure of the multilayer perceptron, which is fixed *a priori*.

This conjecture could be proved by generating constructive methods that are able to obtain an optimal configuration for some basic problems (like the intersection of halfspaces) in a polynomial execution time. In this case, we should conclude the superiority of the incremental approach with respect to back propagation. Unfortunately, only when the training uses examples and queries has it been possible to achieve a result of this kind [50]. However, this additional information (queries) makes the task of constructing the final neural network easier, for which no light is shed on the original question.

IV. SEQUENTIAL CONSTRUCTIVE METHODS: GENERAL STRUCTURE

Among the variety of constructive techniques for the supervised training of feedforward neural networks, we can determine a class of algorithms, denoted as *sequential*, which are characterized by a common methodology. To this class of methods the rest of this chapter will be dedicated and the implementative choices made by each algorithm will be described and discussed. Particular attention will be paid to the theoretical analysis of the convergence properties and the experimental evaluation of the properties of the multilayer perceptrons obtained.

Let S be a training set containing a finite number s of input–output relations $(\mathbf{x}_j, \mathbf{y}_j)$, $j = 1, \ldots, s$, which characterize the problem to be solved. Suppose that S is *consistent*; that is, there are no samples $(\mathbf{x}_{j_1}, \mathbf{y}_{j_1})$ and $(\mathbf{x}_{j_2}, \mathbf{y}_{j_2})$ having the same input pattern $\mathbf{x} = \mathbf{x}_{j_1} = \mathbf{x}_{j_2}$ and different output vector $\mathbf{y}_{j_1} \neq \mathbf{y}_{j_2}$. If this is not the case, we can adopt proper techniques to remove every ambiguity (e.g., simple statistical methods, such as the nearest neighbor algorithm [54], can be applied in the treatment of classification problems).

Denote again by f the unknown function that has generated the samples $(\mathbf{x}_j, \mathbf{y}_j)$, eventually in the presence of noise. The domain D of this function depends on the problem considered and the range of values that can be assumed by an input pattern $\mathbf{x}$. If n is the dimension of the input space, the most frequent choices are surely $D \subset \mathcal{R}^n$ (real inputs) and $D = B^n$ (binary inputs), where B is a set containing two values employed for the coding of Boolean data. In the following we set $B = \{-1, +1\}$, although all the results are still valid, with minor changes, for other definitions of the set B (in particular, $B = \{0, 1\}$).

The class of sequential constructive methods has been expressly developed for the solution of classification problems where the range of the function f is given by $\{-1, +1\}^m$, m being the number of components of the output patterns $\mathbf{y}_j$. The general technique employed for the construction of a neural network that approximates f can be easily described by introducing a generalization of the concept of decision lists defined in [55].

A. Sequential Decision Lists for Two-Class Problems

Consider at first the case $m = 1$ where the output patterns y_j of the given training set are single-valued. In this situation the function f subdivides its domain D into two disjoint subsets D_{+1} and D_{-1} given by

$$D_{+1} = \{\mathbf{x} \in D \mid f(\mathbf{x}) = +1\}, \qquad D_{-1} = \{\mathbf{x} \in D \mid f(\mathbf{x}) = -1\}.$$

Because $D_{+1} \cup D_{-1} = D$ this separation can be viewed as the result of a classification process of the input patterns based on the output of the function f. Then let us introduce the following:

Definition 1. An ordered sequence of pairs (L_j, d_j), $j = 1, \ldots, h+1$, where $L_j \subset D$ and $d_j \in \{-1, +1\}$, will be called a *sequential decision list for a two-class problem*, or simply a 1-*SDL*, if $L_{h+1} = D$.

The first element L_j of every pair will be called the *choice set*, whereas the latter d_j will be called the *pertaining class*.

```
if x ∈ L_1 then

    g(x) = d_1

else if x ∈ L_2 then

    g(x) = d_2

    .

    .

    .

else

    g(x) = d_{h+1}

end if
```

Figure 1 Procedure implementing the function g associated with the sequential decision list (L_j, d_j), $j = 1, \ldots, h+1$, for a two-class problem.

This definition includes as special cases the decision list presented in [55] and the neural decision list introduced in [56], where the choice sets L_j are halfspaces in the domain $\mathcal{R}^n$.

Every 1-SDL is associated with a function g: $D \longrightarrow \{-1, +1\}$ given by $g(\mathbf{x}) = d_j$, j being the first (least) index for which $\mathbf{x} \in L_j$. Because $L_{h+1} = D$ the function g is defined on the whole domain D. By following the interpretation given in [55], the value of $g(\mathbf{x})$ for any $\mathbf{x} \in D$ can be obtained through the application of a sequence of nested statements if-then-else as shown in Fig. 1.

Now consider a threshold neuron whose output y is defined as follows:

$$y = \operatorname{sgn}\left(\sum_{j=0}^{k} u_j z_j\right) = \begin{cases} +1, & \text{if } \sum_{j=0}^{k} u_j z_j \geqslant 0, \\ -1, & \text{otherwise,} \end{cases} \tag{1}$$

where $u_1, \ldots, u_k$ are the weights corresponding to the inputs $z_1, \ldots, z_k$, respectively. The bias u_0 is included in the summation by adding a new component $z_0 = +1$. The relevance of sequential decision lists for the incremental construction of neural networks is pointed out by the following basic result [57]:

THEOREM 1. *The function g associated with a given 1-SDL* (L_j, d_j), $j = 1, \ldots, h+1$, *can always be realized by a perceptron containing a single hidden layer and an output threshold neuron.*

Proof. The assertion of the theorem can easily be shown by providing the weights and the activation functions for the desired neural network. First of all, we put in the hidden layer h neurons whose output z_j, $j = 1, \ldots, h$, is given by

$$z_j = \varphi_j(\mathbf{x}) = \begin{cases} +1, & \text{if } \mathbf{x} \in L_j, \\ -1, & \text{otherwise.} \end{cases} \tag{2}$$

Then we verify that the following set of weights for the final threshold neuron leads to the realization of the given function g:

$$u_0 = \sum_{j=1}^{h} u_j + d_{h+1}, \qquad u_j = d_j 2^{h-j} \qquad \text{for } j = 1, \ldots, h. \tag{3}$$

To this end take a generic input pattern $\mathbf{x} \in D$ and denote by j^* the first index of the 1-SDL for which $\mathbf{x} \in L_j$. When $j^* \leqslant h$ Eq. (2) gives

$$z_{j^*} = +1, \qquad z_j = -1 \qquad \text{for every } j = 1, \ldots, j^* - 1$$

from which we obtain

$$\begin{aligned} \sum_{j=0}^{h} u_j z_j &= \sum_{j=1}^{h} (1 + z_j) u_j + d_{h+1} \\ &= d_{j^*} 2^{h-j^*+1} + \sum_{j=j^*+1}^{h} (1 + z_j) d_j 2^{h-j} + d_{h+1}. \end{aligned}$$

However,

$$\begin{aligned} \left| \sum_{j=j^*+1}^{h} (1 + z_j) d_j 2^{h-j} + d_{h+1} \right| &\leqslant \sum_{j=j^*+1}^{h} 2^{h-j+1} + 1 \\ &= \sum_{j=0}^{h-j^*} 2^j = 2^{h-j^*+1} - 1 < \left| d_{j^*} 2^{h-j^*+1} \right| \end{aligned}$$

for which the corresponding output of the final threshold neuron will be

$$y = \operatorname{sgn}\left(\sum_{j=0}^{h} u_j z_j \right) = \operatorname{sgn}\left(d_{j^*} 2^{h-j^*+1} \right) = d_{j^*}.$$

In the complementary case $j^* = h + 1$, we have from Eq. (2) $z_j = -1$ for every $j = 1, \ldots, h$, and obtain the following output y:

$$y = \operatorname{sgn}\left(\sum_{j=0}^{h} u_j z_j\right) = \operatorname{sgn}\left(\sum_{j=1}^{h} (1 + z_j) u_j + d_{h+1}\right) = d_{h+1}. \qquad (4)$$

■

Unfortunately, the choice (3) for the weights u_j of the output neuron provides integer values that increase exponentially with the size h of the hidden layer (given by the length of the 1-SDL). This makes extremely difficult or impossible both the simulation on a conventional computer and the implementation on a physical support, even for moderate values of h.

An alternative set of weights u_j that may not present this drawback can be obtained by taking the groups of consecutive pairs (L_j, d_j) in the given 1-SDL having the same pertaining class d_j. Let l be the number of these groups and h_i, $i = 1, \ldots, l$, the index of the last pair (L_{h_i}, d_{h_i}) belonging to the ith group. Then we have

$$d_{j+1} = d_j \qquad \text{for every } j = h_{i-1} + 1, \ldots, h_i,\ i = 1, \ldots, l,$$

where $h_0 = 0$ and $h_l = h$ by definition.

The construction followed by the proof of Theorem 1 is still valid if we employ the following values for the weights u_j of the final threshold neuron [58]:

$$u_0 = \sum_{j=1}^{h} u_j + d_{h+1},$$

$$u_j = d_j \left(\sum_{k=h_i+1}^{h} |u_k| + 1 \right) \qquad \text{for } j = h_{i-1} + 1, \ldots, h_i,\ i = 1, \ldots, l. \qquad (5)$$

To verify this assertion, we take again a generic input pattern $\mathbf{x} \in D$ and denote by j^* the first index of the 1-SDL for which $\mathbf{x} \in L_{j^*}$. Moreover, let i^* be the index of the group containing the pair (L_{j^*}, d_{j^*}). When $j^* \leqslant h$ Eq. (2) gives

$$z_{j^*} = +1, \qquad z_j = -1 \qquad \text{for every } j = 1, \ldots, j^* - 1$$

for which we can write

$$\begin{aligned} \sum_{j=0}^{h} u_j z_j &= \sum_{j=1}^{h} (1 + z_j) u_j + d_{h+1} \\ &= \sum_{j=j^*}^{h_{i^*}} (1 + z_j) u_j + \sum_{j=h_{i^*}+1}^{h} (1 + z_j) u_j + d_{h+1} \end{aligned}$$

$$= \left(2 + \sum_{j=j^*+1}^{h_{i^*}} (1 + z_j)\right) d_{j^*} \left(\sum_{j=h_{i^*}+1}^{h} |u_j| + 1 \right)$$
$$+ \sum_{j=h_{i^*}+1}^{h} (1 + z_j) u_j + d_{h+1}$$

because all the weights u_j belonging to the same group are equal.

However,

$$\left| \sum_{j=h_{i^*}+1}^{h} (1 + z_j) u_j + d_{h+1} \right|$$
$$\leqslant 2 \sum_{j=h_{i^*}+1}^{h} |u_j| + 1 < 2\left(\sum_{j=h_{i^*}+1}^{h} |u_j| + 1 \right)$$
$$\leqslant \left| \left(2 + \sum_{j=j^*+1}^{h_{i^*}} (1 + z_j)\right) d_{j^*} \left(\sum_{j=h_{i^*}+1}^{h} |u_j| + 1 \right) \right|.$$

Thus, the corresponding output of the final threshold neuron will be given by

$$y = \operatorname{sgn}\left(\sum_{j=0}^{h} u_j z_j \right)$$
$$= \operatorname{sgn}\left(\left(2 + \sum_{j=j^*+1}^{h_{i^*}} (1 + z_j)\right) d_{j^*} \left(\sum_{j=h_{i^*}+1}^{h} |u_j| + 1 \right) \right) = d_{j^*}.$$

In the complementary case $j^* = h + 1$, the choice (5) for the weights u_j leads again to Eq. (4) for which the verification is completed.

If the given 1-SDL is formed by a single group of consecutive pairs (L_j, d_j), all the pertaining classes d_j for $j = 1, \ldots, h$ are equal and Eq. (5) becomes

$$u_0 = h d_1 + d_{h+1}, \qquad u_j = d_j \qquad \text{for } j = 1, \ldots, h. \tag{6}$$

Consequently, the weights u_j associated with the inputs of the final neuron are binary, whereas the bias u_0 increases linearly with the number h of hidden units. Unfortunately, in the opposite case, when every pair (L_j, d_j) belongs to a separate group, we have $d_{j+1} = -d_j$ for every $j = 1, \ldots, h - 1$ and Eq. (5) provides the same exponentially increasing values for the weights u_j as Eq. (3).

The pertaining classes d_j of the 1-SDL can be freely chosen only in the case $D \subset \{-1, +1\}^n$, when the inputs are binary. In this situation it could be convenient to generate pairs (L_j, d_j) belonging to the same group because, in general,

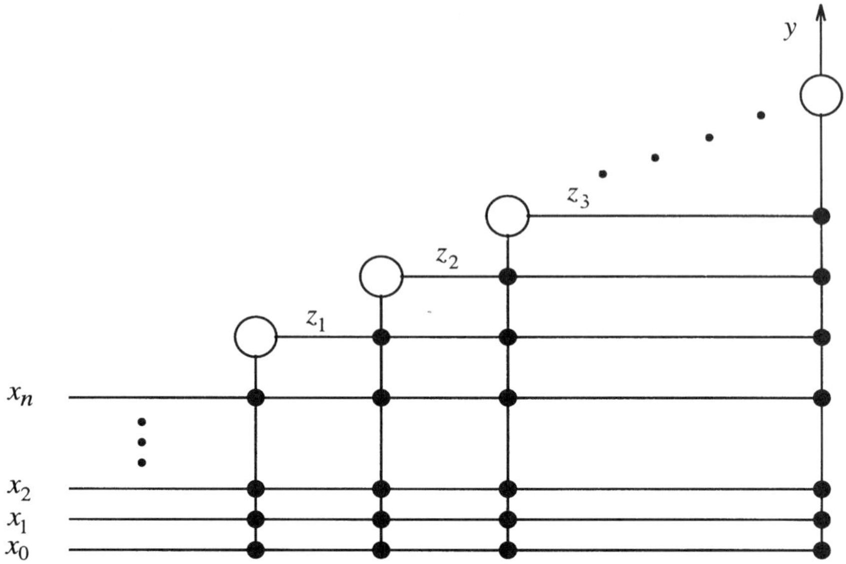

Figure 2 General architecture of a cascade net.

this choice leads to neural networks containing fewer connections. On the contrary, when the inputs are real numbers there are training sets for which the value of the pertaining class d_j, for some j, is fixed and consequently no decision is allowed.

In practical applications, sequential constructive methods select the pertaining classes by following proper criteria that take account of the samples contained in the given training set. It could therefore be useful to choose freely the values d_j without obtaining a multilayer perceptron with intractable weights. The employment of a different architecture, called a *cascade net* [56], allows this requirement to be satisfied.

In the cascade net (Fig. 2), every hidden layer contains a single unit that is fed by all the outputs of previous neurons together with the input pattern considered. As in the two-layer perceptron case, the final neuron is again a threshold unit whose activation function is given by Eq. (1).

We can state the following theorem [56]:

THEOREM 2. *The function g associated with a given 1-SDL* (L_j, d_j), $j = 1, \ldots, h + 1$, *can always be realized by a cascade net containing h hidden neurons.*

Proof. The assertion will be verified by providing the weights and the activation functions for the desired cascade net. The h hidden layers contain as many units whose output z_j, $j = 1, \ldots, h$, is given by

$$z_j = \psi_j(\mathbf{x}, z_1, \ldots, z_{j-1}) = \begin{cases} +1, & \text{if } \mathbf{x} \in L_j \text{ and } z_i = -1 \text{ for every } i = 1, \ldots, j-1, \\ -1, & \text{otherwise.} \end{cases} \tag{7}$$

With this choice an input pattern $\mathbf{x} \in D$ can activate at most one of the hidden neurons by setting its output to the value $+1$. Then we verify that there is a weight vector $\mathbf{u}$ for the output threshold unit that yields the value $g(\mathbf{x})$ generated by the given 1-SDL. It is sufficient to take

$$u_0 = \sum_{j=1}^{h+1} d_j, \qquad u_j = d_j \qquad \text{for } j = 1, \ldots, h. \tag{8}$$

As a matter of fact, denote by j^* the first index for which $\mathbf{x} \in L_{j^*}$. In the case $j^* \leqslant h$, we obtain

$$z_{j^*} = +1, \qquad z_j = -1 \qquad \text{for every } j \neq j^*.$$

The corresponding output of the cascade net is then given by Eq. (1)

$$y = \operatorname{sgn}\left(\sum_{j=0}^{h} u_j z_j\right) = \operatorname{sgn}\left(\sum_{j=1}^{h} (1 + z_j) d_j + d_{h+1}\right) = \operatorname{sgn}(2d_{j^*} + d_{h+1}) = d_{j^*}. \tag{9}$$

In the complementary case $j^* = h + 1$, we have $z_j = -1$ for every $j = 1, \ldots, h$, and the application of the activation function (1) leads to

$$y = \operatorname{sgn}\left(\sum_{j=0}^{h} u_j z_j\right) = \operatorname{sgn}\left(\sum_{j=1}^{h} (1 + z_j) d_j + d_{h+1}\right) = d_{h+1}. \qquad \blacksquare$$

When all the first h pairs (L_j, d_j) of the given 1-SDL have the same pertaining class d_j, the assignments (6) and (8) for the two-layer perceptron and the cascade net, respectively, are equivalent. Thus, in this special case the lateral connections among the hidden neurons have no effect on the final output.

This consideration can be further extended to the situation where we have l groups of consecutive pairs (L_j, d_j) having the same pertaining class. In this case only the lateral connections among the neurons associated with different groups are necessary. The resulting neural network contains therefore l hidden layers, each of which is fed by the outputs of previous layers and the input pattern; this architecture can be called a *generalized cascade net*. To avoid excessive notational complications, we omit the detailed study of this configuration.

B. Sequential Decision Lists for Multiclass Problems

So far we have considered two-class problems where the output patterns are given by single binary values ($m = 1$). Now we wish to extend the results of the previous section to the general case $m > 1$, in which the output vectors $\mathbf{y}_j$ of the training set S have m binary components. We will use the term *multiclass problem* to denote this situation, because a different class can be associated with a particular sequence of outputs. The definition of 1-SDL can be directly extended in the following way.

Definition 2. An ordered sequence of pairs $(L_j, \mathbf{d}_j)$, $j = 1, \ldots, h+1$, where $L_j \subset D$ and $\mathbf{d}_j$ is a nonnull vector with m components in $\{-1, 0, +1\}$, will be called a *sequential decision list for a multiclass problem* or *m-SDL* if $L_{h+1} = D$ and $\mathbf{d}_{h+1}$ does not contain null components.

With this definition the vector $\mathbf{d}_j$ may not correspond to the pertaining class of the patterns $\mathbf{x} \in L_j$ because it leaves the outputs corresponding to its null components undetermined. However, it should be noted that in the case $m = 1$, Definitions 1 and 2 coincide.

Furthermore, the association of a function $g\colon D \longrightarrow \{-1, +1\}^m$ with any m-SDL is straightforward: for every input pattern $\mathbf{x} \in D$ the kth component $g_k(\mathbf{x})$ of the corresponding output $g(\mathbf{x})$ is given by $g_k(\mathbf{x}) = d_{jk}$, j being the first (least) index for which $\mathbf{x} \in L_j$ and $d_{jk} \neq 0$. Again, the constraints $L_{h+1} = D$ and $d_{h+1,k} \neq 0$ for every $k = 1, \ldots, m$ ensure the correct definition of the function $g(\mathbf{x})$ in the whole domain D.

The value of this function can still be obtained through the application of a sequence of nested statements if-then-else as shown in Fig. 3.

From Definitions 1 and 2 it follows that an m-SDL $(L_j, \mathbf{d}_j)$ is equivalent to m 1-SDLs (L_j, d_{jk}), $k = 1, \ldots, m$, each of which contains only the pairs with $d_{jk} \neq 0$. Then Theorems 1 and 2 can be extended to the multiclass case leading to the following results.

Theorem 3. *The function g associated with a given m-SDL $(L_j, \mathbf{d}_j)$, $j = 1, \ldots, h+1$, can always be realized by a perceptron containing a single hidden layer and m threshold neurons in the output layer.*

Proof. The activation functions $\varphi_j(\mathbf{x})$ for the h hidden units are given again by Eq. (2), whereas the weights for the output neurons can be obtained by ap-

```
for k = 1,...,m do

    if x ∈ L_1 and d_1k ≠ 0 then

        g_k(x) = d_1k

    else if x ∈ L_2 and d_2k ≠ 0 then

        g_k(x) = d_2k

        .

        .

        .

    else

        g_k(x) = d_h+1,k

    end if

end do
```

Figure 3 Procedure implementing the function g associated with the sequential decision list $(L_j, \mathbf{d}_j)$, $j = 1, \ldots, h+1$, for a multiclass problem.

plying Eq. (3) to each of the m 1-SDLs obtained by breaking the original m-SDL. ■

THEOREM 4. *The function g associated with a given m-SDL* $(L_j, \mathbf{d}_j)$, $j = 1, \ldots, h+1$, *can always be realized by a cascade net containing h hidden units and m threshold neurons in the output layer.*

Proof. It is sufficient to proceed as in the proof of Theorem 3 by applying Eqs. (7) and (8). ■

Furthermore, it is possible to subdivide each of the m 1-SDLs deriving from the original m-SDL into groups of consecutive pairs (L_j, d_{jk}) having the same pertaining class d_{jk}; assignments (5) for the weights in the output layer can therefore be used. In general, this choice leads to a better range of variability for the two-layer perceptron or a lower number of connections for the resulting cascade net. For the sake of simplicity, we omit an exhaustive description of this generalization referring to [58] for some implementative details.

C. General Procedure for Two-Class Problems

Sequential constructive methods build the neural network associated with a classification problem by generating a sequential decision list starting from the given training set S. The kernel of these techniques is the learning algorithm for the hidden units, which has the aim of constructing neurons with activation functions given by Eq. (2) or (7) for some choice sets L_j. The following definition will be used:

DEFINITION 3. Let Q^+ and Q^- be two subsets of the input space; a neuron will be called a *partial classifier* if it provides output $+1$ for at least one pattern in Q^+ and output -1 for all the elements of Q^-.

As one can note, a partial classifier forms a choice set L_j containing all the patterns for which it provides output $+1$ in the input space. The general structure of this set is determined by the activation function of the neuron; for example, if threshold units are employed, the sets L_j are given by halfspaces in the domain $\mathcal{R}^n$.

Furthermore, in all the practical cases it is possible to change from Eq. (2) for the two-layer perceptron to Eq. (7) for the cascade net by modifying directly the weights of the input connections. In Section V, such a passage will be explained in detail for every choice of the sets L_j used in the construction of sequential decision lists.

Nevertheless, the general procedure for obtaining the multilayer perceptron associated with a finite training set S is common to every sequential constructive method. In the case of two-class problems ($m = 1$), its outline is reported in Fig. 4.

To obtain the convergence of this algorithm, it is necessary to suppose that the training set S is consistent; in this case the sets Q^+ and Q^- formed at step 2 are always disjoint ($Q^+ \cap Q^- = \emptyset$). It is, however, possible, as previously noted, to allow the treatment of ambiguous training sets by using proper tricks, such as the elimination of critical samples or the employment of simple statistical methods, like the nearest neighbor algorithm [54], for the assignment of most probable outputs to ambiguous input patterns.

In all the sequential constructive methods present in the literature, the activation functions employed for the hidden neurons always allow us to obtain, within a finite execution time, a partial classifier with a nonempty associated set R for at least one value of the pertaining class d_h. Thus, the elimination performed at step 4 reduces the size of the current training set, leading to the construction of a

GENERAL SEQUENTIAL CONSTRUCTIVE METHOD

(Procedure for two-class problems)

1. Set $h = 1$. (number of hidden neurons)

2. Choose a value d_h for the pertaining class of the hth pair (L_h, d_h) of the 1-SDL. Let Q^+ and Q^- contain the input patterns of the current training set having corresponding output d_h and $-d_h$ respectively.

3. Train the hth hidden neuron so as to obtain a partial classifier for Q^+ and Q^-. Let R be the subset of Q^+ containing the input patterns for which the hth hidden unit provides output $+1$.

4. Remove from the training set all the samples corresponding to the elements of R.

5. If there are two samples in the current training set having different output, then set $h = h + 1$ and go to Step 2.

6. Set $d_{h+1} = d_h$ and construct the resulting neural network by following the proof of the Theorem 1 or 2.

Figure 4 General procedure followed by sequential constructive methods for the solution of two-class problems.

neural network that satisfies all the given samples, after a finite number of iterations. This simple reasoning shows the assertion of the following theorem, which is widely employed to ensure the convergence of sequential constructive methods.

THEOREM 5. *Every sequential constructive method provides, in a finite execution time, a neural network that satisfies all the samples contained in a given finite consistent training set S if the learning algorithm for the hidden neurons* *allows us to obtain within a finite computing time a partial classifier, whichever pair of sets Q^+ and Q^- is considered.*

The hypothesis of Theorem 5, regarding the availability of a partial classifier, will be verified in Section V for every activation function used in the literature.

However, it should be pointed out that a better generalization ability for the resulting neural network can be attained by maximizing the size of the set R obtained at step 3. In general, this leads to a lower number of hidden neurons and consequently to a reduction in the network complexity. To achieve this goal, every sequential constructive technique employs proper heuristic methods.

At the beginning of the construction of the multilayer perceptron, the training of the output threshold neuron is performed so as to check the linear separability of the given training set S. In fact, in this particular situation, there is no need to apply the procedure in Fig. 4. In any case, when threshold activation functions are also used for the hidden layer, we can set $h = 2$ at step 1 because a single hidden unit can always be removed by modifying the weights of the output neuron.

An important role in the general procedure described in Fig. 4 is played by the choice of the pertaining class d_h (step 2) for the hth pair of the 1-SDL (associated with the hth hidden neuron). A widely used approach [57] is to compute the number of patterns in the sets R obtained for the two assignments $d_h = +1$ and $d_h = -1$. Let r^+ and r^- be these values; the choice of the pertaining class is made by considering the maximum of the ratios r^+/q^+ and r^-/q^-, where q^+ and q^- are the sizes of the sets Q^+ and Q^-, respectively.

Such an approach will be used in our experimental tests and is preferred to the simple comparison of the integers r^+ and r^-, because it takes into account the relative importance of the two classes. In particular, if $r^+ = q^+ < r^-$ we should miss the choice $d_h = +1$ that leads to the correct classification of the whole training set. Nevertheless, some sequential constructive methods [59] radically solve the problem by adding two hidden units with opposite pertaining classes at the same time.

D. General Procedure for Multiclass Problems

The natural extension of the procedure in Fig. 4 to the solution of multiclass problems ($m > 1$) is shown in Fig. 5. In this case, we can define m pairs of sets $(D_{k,+1}, D_{k,-1})$, with $k = 1, \dots, m$, containing the input patterns of the given training set S whose kth output is $+1$ and -1, respectively:

$$D_{k,+1} = \{\mathbf{x}_i \mid y_{ik} = +1, (\mathbf{x}_i, \mathbf{y}_i) \in S\}, \quad D_{k,-1} = \{\mathbf{x}_i \mid y_{ik} = -1, (\mathbf{x}_i, \mathbf{y}_i) \in S\}.$$

These subsets of the domain D will contain at any iteration the current training set for each output; the sets Q^+ and Q^- for the learning of the hidden neurons will

GENERAL SEQUENTIAL CONSTRUCTIVE METHOD

(Natural extension to multiclass problems)

1. Set $h = 1$. (number of hidden neurons)

2. Choose a vector $\mathbf{d}_h$ in $\{-1, 0, +1\}^m$ containing the pertaining classes for some of the m outputs ($d_{hk} = 0$ if the hth hidden neuron does not affect the kth output).

3. Set

$$Q^+ = \bigcap_{k=1}^{m} Q_k^+ \quad , \quad Q^- = \bigcup_{k=1}^{m} Q_k^-$$

where

$$Q_k^+ = \begin{cases} D & \text{if } d_{hk} = 0 \\ D_{k,d_{hk}} & \text{otherwise} \end{cases} \quad , \quad Q_k^- = \begin{cases} \emptyset & \text{if } d_{hk} = 0 \\ D_{k,-d_{hk}} & \text{otherwise} \end{cases}$$

4. Train the hth hidden neuron so as to obtain a partial classifier for Q^+ and Q^-. Let R be the subset of Q^+ containing the input patterns for which the hth hidden unit provides output $+1$.

5. For every k such that $d_{hk} \neq 0$, set $D_{k,d_{hk}} = D_{k,d_{hk}} \setminus R$.

6. If $D_{k,+1} \neq \emptyset$ and $D_{k,-1} \neq \emptyset$ for any output k then set $h = h + 1$ and go to Step 2.

7. Define the vector $\mathbf{d}_{h+1}$ in the following way

$$d_{h+1,k} = \begin{cases} +1 & \text{if } D_{k,-1} = \emptyset \\ -1 & \text{otherwise} \end{cases} \quad , \quad \text{for } k = 1, \ldots, m$$

and construct the resulting neural network by following the proof of the Theorem 3 or 4.

Figure 5 Natural extension of sequential constructive methods to the solution of multiclass problems.

be extracted from the pairs $(D_{k,+1}, D_{k,-1})$ by following the method described at step 3. It can easily be seen that, in this way, an m-SDL $(L_j, \mathbf{d}_j)$, $j = 1, \ldots, h+1$, is constructed in which every choice set L_j contains some input patterns of the training set S. Its form depends again on the activation function employed for the hidden units.

As one can note, in the multiclass case the choice of the vector $\mathbf{d}_h$ at step 2 is not so straightforward as in the solution of two-class problems. The variety of possible assignments increases the probability of constructing a neural network with low generalization ability. Nevertheless, it is always possible to consider the outputs one at a time by performing m executions of the procedure in Fig. 4; this choice has been adopted in our experimental tests.

A heuristic approach that allows us to construct multilayer perceptrons with low complexity has been proposed in [58]. It chooses initially for every output a pertaining class d_k (e.g., that containing the minimum number of samples in the training set S) and a lower threshold τ for the size of the set Q^+ at step 2. Then, when a new hidden neuron has to be added, a set of outputs K is selected by applying a greedy technique so as to obtain $|Q^+| \geqslant \tau$, where

$$Q^+ = \bigcap_{k \in K} D_{k,dk}.$$

The vector $\mathbf{d}_h$ is thus defined in the following way:

$$d_{hk} = \begin{cases} d_k, & \text{if } k \in K, \\ 0, & \text{otherwise.} \end{cases}$$

In general, the resulting networks contain a lower number of hidden neurons and therefore have a high probability of showing a good generalization ability. If the inputs are real, such an approach may need some minor corrections to take account of particular situations where there are no partial classifiers with nonnull R.

The convergence properties of sequential constructive methods ensured by Theorem 5 for two-class problems can easily be extended to the multiclass case. It is sufficient to subdivide the m-SDL obtained by applying the technique in Fig. 5 into m 1-SDLs, one for each output, according to the procedure described in Section IV.B. The application of Theorem 5 to each of these 1-SDLs determines a corresponding finite execution time t_k necessary for the construction of the portion of the neural network associated with the kth output. Because the number m of outputs is also finite, we obtain the following general result.

THEOREM 6. *Under the hypotheses of Theorem 5, the natural extension of a sequential constructive method to multiclass problems is able to provide, within a finite execution time, a neural network that satisfies all the samples contained in a given training set S.*

The choice of the vectors $\mathbf{d}_h$ can also be avoided by modifying the procedure in Fig. 5 and employing a proper algorithm for the training of the neurons in the output layer [58]. The resulting method is reported in Fig. 6; in this case the number h of hidden units is initially set to zero, emphasizing the preliminary learning of the output weights (also performed in the procedures described in Figs. 4 and 5).

If the training of the output layer does not lead to a multilayer perceptron that satisfies all the samples of the training set S, a new partial classifier is added to the hidden layer (step 5). The sets Q^+ and Q^- employed for the training of this neuron are generated at step 3 by taking at first the output k^+ that scores the highest number of errors on the input patterns of V_{k,d_k}, where d_k is the pertaining class chosen beforehand for the kth output (step 1).

The size of the auxiliary sets $V_{k,+1}$ and $V_{k,-1}$, initially equal to $D_{k,+1}$ and $D_{k,-1}$, respectively, decreases with the number of iterations (step 6). This prevents us from generating two hidden neurons having the aim of correcting the same output error. If all the input patterns belonging to the sets V_{k,d_k} are correctly classified (for every output k), the procedure in Fig. 6 tries to correct the errors associated with the pertaining classes $-d_k$. In particular, it considers the output k^- corresponding to the set $U_{k,-d_k}$ with the largest size.

This method can always construct a neural network that satisfies all the samples contained in a given (finite and consistent) training set S if the learning algorithm for the neurons in the output layer is able to obtain an optimal configuration, that is, a weight matrix that makes the minimum number of errors on the sets $D_{k,+1}$ and $D_{k,-1}$.

A method of this kind is the pocket algorithm [51, 60]; in fact, it can be shown that by applying this technique, the probability of obtaining an optimal weight vector for a single threshold neuron approaches unity when the number of iterations increases. Furthermore, the version with ratchet, very interesting from an applicative point of view, provides an optimal configuration within a finite execution time [52].

This property will be used to show the following result:

THEOREM 7. *Under the hypotheses of Theorem 5, the extension with output training of a sequential constructive method to multiclass problems is able to provide, within a finite execution time, a neural network that satisfies all the samples contained in a given training set S.*

Proof. In the procedure of Fig. 6, we can note that the repeated execution of steps 2–7 causes the addition of a (eventually null) number of partial classifiers for every output. Denote by I_k the set of hidden units generated when, at step 3, we take $Q^+ = U_{k,d_k}$ or $Q^+ = U_{k,-d_k}$. It can easily be seen that the sets R_h associated with these neurons (step 5) are all disjoint among them; thus, in the

GENERAL SEQUENTIAL CONSTRUCTIVE METHOD

(Extension with output training to multiclass problems)

1. For each output $k = 1, \ldots, m$ choose a pertaining class d_k. Set

$$V_{k,+1} = D_{k,+1} \quad , \quad V_{k,-1} = D_{k,-1} \quad , \quad h = 0$$

2. Train the m threshold neurons in the output layer. Let $U_{k,+1}$ ($U_{k,-1}$) be the set of input patterns in $V_{k,+1}$ ($V_{k,-1}$) that are not correctly classified by the current neural network. Denote with k^+ (k^-) the index of the output associated with the set U_{k,d_k} ($U_{k,-d_k}$) having maximum size.

3. If $U_{k^+,d_{k^+}} \neq \emptyset$ then set $Q^+ = U_{k^+,d_{k^+}}$, $Q^- = D_{k^+,-d_{k^+}}$, otherwise if $U_{k^-,-d_{k^-}} \neq \emptyset$ then set $Q^+ = U_{k^-,-d_{k^-}}$, $Q^- = D_{k^-,d_{k^-}}$, otherwise the current neural network satisfies all the samples in the training set S and the construction is complete.

4. Set $h = h + 1$.

5. Train the hth hidden neuron so as to obtain a partial classifier for Q^+ and Q^-. Let R_h be the subset of Q^+ containing the input patterns for which the hth hidden unit provides output $+1$.

6. If $U_{k^+,d_{k^+}} \neq \emptyset$, then set $V_{k^+,d_{k^+}} = V_{k^+,d_{k^+}} \setminus R_h$ otherwise set $V_{k^-,-d_{k^-}} = V_{k^-,-d_{k^-}} \setminus R_h$.

7. Go to Step 2.

Figure 6 Extension with output training of sequential constructive methods to the solution of multi-class problems.

worst case, the procedure will yield one of the following two situations:

$$D_{k,+1} \subset \bigcup_{h \in I_k} R_h \qquad \text{or} \qquad D_{k,-1} \subset \bigcup_{h \in I_k} R_h.$$

In both of these cases the application of Theorem 1 ensures the existence of a weight vector for the kth output neuron, which leads to the correct classification of the whole training set S. Because this configuration is surely optimal, it will be found by the pocket algorithm with ratchet in a finite number of iterations. ■

It is important to emphasize that the convergence theorems (Theorems 5–7), for sequential constructive methods, require the employment of a training algorithm, for the hidden neurons, which is able to generate a partial classifier within a finite execution time. In practice, this condition is always satisfied because all the activation functions used in the applications allow us to separate with relative simplicity at least a pattern of a given class from all the patterns belonging to the opposite one.

Unfortunately, such a separation leads to networks that memorize the samples of the given training set without performing any generalization. A way to avoid this undesired effect is to employ a learning algorithm for the hidden neurons that gives output $+1$ for most input patterns contained in Q^+. The techniques used to achieve this goal change from one sequential constructive method to another and form the subject of the following section.

V. SEQUENTIAL CONSTRUCTIVE METHODS: SPECIFIC APPROACHES

Consider two subsets Q^+ and Q^- of an n-dimensional domain D, containing a finite number of patterns q^+ and q^-, respectively. In particular, we shall analyze the cases $D \subset \mathcal{R}^n$ and $D \subset \{-1, +1\}^n$ because they are commonly used in the treatment of sequential constructive methods.

Let φ: $D \longrightarrow \{-1, +1\}$ be the activation function of a given neuron, which, in general, depends on a vector of parameters $\mathbf{w} \in \mathcal{R}^{n+1}$ (n input weights plus the bias w_0). We shall examine in detail training algorithms that generate a set of weights $\mathbf{w}$ for a neuron of this kind which provides output $+1$ for the greatest number of input patterns in Q^+ and output -1 for all the elements of Q^-. In the following we assume that the two sets Q^+ and Q^- are disjoint, that is, $Q^+ \cap Q^- = \emptyset$ (consistent training set).

Because the procedure followed by these training algorithms depends on the form of the activation function φ, we shall subdivide the available methods by considering the geometrical aspect of the choice set L associated with the partial classifier

$$L = \{\mathbf{x} \in D\colon \varphi(\mathbf{x}) = +1\}. \tag{10}$$

A. Halfspace Choice Set

Most sequential constructive methods employ threshold neurons for the construction of the hidden layer; their activation function is given by Eq. (1), here rewritten for convenience,

$$\varphi(\mathbf{x}) = \operatorname{sgn}\left(\sum_{i=0}^{n} w_i x_i\right) = \begin{cases} +1, & \text{if } \sum_{i=0}^{n} w_i x_i \geqslant 0, \\ -1, & \text{otherwise.} \end{cases} \tag{11}$$

As usual, we have added a component $x_0 = +1$ to the input pattern $\mathbf{x}$ so as to include the bias in the weights of the neuron. The value $\varphi(0) = +1$ has been arbitrarily chosen; all the following considerations are still valid if we set $\varphi(0) = -1$.

It can be argued from Eqs. (10) and (11) that the choice set L for threshold neurons is a halfspace in $\mathcal{R}^n$, whose position is univocally determined by the weight vector $\mathbf{w}$. The equation for the associated boundary hyperplane is then

$$\sum_{i=0}^{n} w_i x_i = 0.$$

When $D \subset \{-1, +1\}^n$ the input patterns are binary vectors with n components and correspond to the vertices of an n-dimensional hypercube. In this case, it is always possible to find a hyperplane that separates a pattern $\mathbf{x} \in D$ from all the other points of the set $\{-1, +1\}^n$; a possible choice for the weight vector is the following:

$$w_0 = 1 - n, \qquad w_i = x_i \qquad \text{for every } i = 1, \ldots, n. \tag{12}$$

A threshold neuron that performs this separation is called a *grandmother cell* and is used to show that a two-layer perceptron containing only threshold units can realize any Boolean function. It is sufficient to put at most $s/2$ grandmother cells in the hidden layer (where s is the size of the given training set) and a final threshold neuron that performs the logic operation OR among the outputs of the hidden units [6]. Nevertheless, such a configuration is interesting only from a theoretical point of view because it simply memorizes the samples of the training set without making any kind of generalization.

In the case $D \subset \mathcal{R}^n$, a similar construction has not been found; however, it can be shown that, given two sets of patterns U and V, there is always a partial classifier with threshold activation function, that is, a threshold neuron that separates some patterns of one set from all the elements of the other [61, 62]. This result can be obtained directly by considering the convex hulls formed by the points of U and V. However, a direct approach that gives the weight vector of such a unit is not available.

As previously pointed out, our aim is to generate a partial classifier that provides the correct output for the greatest number of patterns of Q^+, so as to minimize the number of connections in the final neural network. Unfortunately, the achievement of this result can lead, for some choices of the sets Q^+ and Q^-, to the solution of an NP-complete problem called the *densest-hemisphere problem* [63]. Consequently, to lower the total execution time, the learning techniques employed by sequential constructive methods only try to obtain a near-optimal partial classifier that approaches the desired configuration.

In the following sections we shall analyze in detail the main training algorithms used to achieve this result. For each of them we shall describe the procedure followed to obtain the weight vector $\mathbf{w}$ of the hidden unit (2) to be employed in the construction of the two-layer perceptron according to the proof of Theorem 1. From this configuration, the weight vector $\mathbf{v}$ for the threshold neuron (7) to be inserted in the corresponding cascade net can be obtained directly (under the same choice set L_j).

As a matter of fact, consider the addition of the jth hidden unit: the $n + j$ components of its weight vector $\mathbf{v}$ can be subdivided in the following way:

$$\begin{aligned} v_0 &= \text{neuron bias}, \\ v_1, \ldots, v_n &= \text{weights associated with the network inputs } x_1, \ldots, x_n, \\ v_{n+1}, \ldots, v_{n+j-1} &= \text{weights associated with the outputs } z_1, \ldots, z_{j-1} \text{ of the} \\ & \text{first } j-1 \text{ hidden neurons.} \end{aligned}$$

If the algorithm employed for the training of the (threshold) partial classifier has generated a weight vector $\mathbf{w}$ starting from the sets Q^+ and Q^-, a possible choice for the components of $\mathbf{v}$ is

$$\begin{aligned} v_0 &= w_0 + \sum_{i=n+1}^{n+j-1} v_i, \qquad v_i = w_i \qquad \text{for } i = 1, \ldots, n, \\ v_i &= -\tfrac{1}{2}\left(|w_0| + \sum_{k=1}^{n} |w_k| \cdot \max |x_k| + 1\right) \\ &\qquad\qquad\qquad\qquad \text{for } i = n+1, \ldots, n+j-1, \end{aligned} \tag{13}$$

$\max|x_k|$ being the maximum absolute value that can be assumed by the kth network input x_k. When the inputs are binary, we obtain directly $\max|x_k| = 1$ for every $k = 1, \ldots, n$, from which it follows that [56]

$$v_i = -\tfrac{1}{2}\left(\sum_{k=0}^{n}|w_k| + 1\right) \qquad \text{for } i = n+1, \ldots, n+j-1. \tag{14}$$

Now let us verify that Eq. (13) leads to the activation function (7), where L_j is the halfspace associated with the partial classifier having weight vector $\mathbf{w}$. At first, suppose that there is an index l such that $z_l = +1$ and $z_i = -1$ for $i \neq l$; then Eq. (13) gives

$$\begin{aligned}\sum_{i=0}^{n} v_i x_i + \sum_{i=n+1}^{n+j-1} v_i z_{i-n} &= \sum_{i=0}^{n} w_i x_i + \sum_{i=n+1}^{n+j-1} v_i(1 + z_{i-n}) \\ &= \sum_{i=0}^{n} w_i x_i - |w_0| - \sum_{k=1}^{n}|w_k| \cdot \max|x_k| - 1 < 0 \end{aligned}\tag{15}$$

for every value of the network inputs $x_1, \ldots, x_n$. In the complementary case, if $z_i = -1$ for every $i = 1, \ldots, j-1$ we obtain

$$\sum_{i=0}^{n} v_i x_i + \sum_{i=n+1}^{n+j-1} v_i z_{i-n} = \sum_{i=0}^{n} w_i x_i + \sum_{i=n+1}^{n+j-1} v_i(1 + z_{i-n}) = \sum_{i=0}^{n} w_i x_i. \tag{16}$$

Thus, the output of the threshold neuron is equal to that of the partial classifier having weight vector $\mathbf{w}$.

1. Irregular Partitioning Algorithm

The first sequential constructive method has probably been proposed by Ruján and Marchand [49] and is usually called the *regular partitioning algorithm*. This technique can only be applied when the training set has binary inputs; it is centered on the definition of regular partitioning for the hypercube whose vertices belong to the set $\{-1, +1\}^n$. Although the general approach followed is very similar to that described in Section IV, the training of the hidden units must satisfy a greater number of constraints, making more complex the construction of the resulting neural network.

In a subsequent paper the same authors, together with Golea [57], removed the regularity requirement for the partitioning of the hypercube and laid the bases for the sequential constructive methods as described in this chapter. The technique proposed in [57] is sometimes called the *irregular partitioning algorithm* (IPA) and is still devoted to the treatment of training sets with binary inputs.

A recent paper [56] extends this method to the general case $D \subset \mathcal{R}^n$, thus allowing for a direct application to many real-world problems. However, the training algorithm for the hidden neurons is basically unchanged and still employs a greedy technique to find a near-optimal partial classifier.

The procedure followed is shown in Fig. 7. Here we have denoted by $|A|$ the number of elements of a finite set A.

Its computational kernel makes use of a proper method (steps 4 and 8) that verifies the existence of a hyperplane separating the two sets of patterns R_i and Q^-. Several algorithms in the literature can perform this task; in particular, the perceptron algorithm [64], used in [57], and the linear programming, employed in [56].

In the first case, a weight vector is initially chosen and is subsequently updated so as to correct eventual classification errors on the patterns of the training set. At every iteration, a sample $(\mathbf{x}, y)$ is randomly extracted and the output $\varphi(\mathbf{x})$ of the threshold neuron associated with the current weight vector $\mathbf{w}$ is computed by applying Eq. (11). If $y \neq \varphi(\mathbf{x})$ the vector $\mathbf{w}$ is modified according to the following rule:

$$w_i = w_i + \eta y x_i, \tag{17}$$

η being a real constant called the *learning rate* (generally $0 < \eta \leqslant 1$).

It can be shown [65] that the perceptron algorithm converges in a finite number of iterations if the training set is linearly separable; in the opposite case, the method cycles indefinitely through feasible weight vectors. Unfortunately, there is no estimate for the number of iterations needed to establish if a given problem can be solved by a threshold neuron. Thus, it is difficult to assign a maximum duration n_t (to be used at steps 4 and 8) for the perceptron algorithm after we conclude a separating hyperplane for the sets R_i and Q^- does not exist.

In fact, if n_t is too low, we can achieve a wrong result even when treating a linearly separable training set; in the opposite case, for great values of n_t, the computational cost of this learning algorithm can become too expensive.

An alternative approach, employed in [56], is given by linear programming, which converges in polynomial time [66] to the separating neuron (if one exists). It tries to maximize the distance between the current hyperplane and the samples of the training set while ensuring the correct classification of all the input patterns. Other techniques can be used, such as the Minover algorithm [67], the Adatron method [68], etc., but a detailed analysis of their advantages and drawbacks goes beyond the scope of this chapter.

In the procedure of Fig. 7, three auxiliary sets are employed: W contains the patterns of Q^+ to be considered for the initialization of a new unit, R_i holds the portion of Q^+ correctly classified by the current threshold neuron (associated with the weight vector $\mathbf{w}_i$), and V contains the elements of Q^+ that can be included in R_i to increase its size.

IRREGULAR PARTITIONING ALGORITHM

(Method for generating a partial classifier)

1. Set $i = 1, W = Q^+, \mathbf{w} = (-1, 0, \ldots, 0)^t, R = \emptyset$.

2. If $W = \emptyset$ then output the threshold neuron associated with the weight vector $\mathbf{w}$.

3. Choose at random an input pattern $\mathbf{x} \in W$ and set

$$W = W \setminus \{\mathbf{x}\} \quad , \quad R_i = \{\mathbf{x}\} \quad , \quad V = Q^+ \setminus \{\mathbf{x}\}$$

4. Use a proper method to check if the sets R_i and Q^- are linearly separable.

5. If R_1 and Q^- are not linearly separable, then go to Step 3 otherwise let $\mathbf{w}_i$ be the vector associated with the separating hyperplane.

6. If $V = \emptyset$ go to Step 11.

7. Choose at random an input pattern $\mathbf{x} \in V$ and set

$$R_i = R_i \cup \{\mathbf{x}\} \quad , \quad V = V \setminus \{\mathbf{x}\}$$

8. Use a proper method to check if the sets R_i and Q^- are linearly separable.

9. If R_i and Q^- are not linearly separable, then set $R_i = R_i \setminus \{\mathbf{x}\}$, otherwise let $\mathbf{w}_i$ be the vector associated with the separating hyperplane.

10. Go to Step 6.

11. Set $W = W \setminus R_i$. If $|R_i| > |R|$ set $R = R_i, \mathbf{w} = \mathbf{w}_i$.

12. Go to Step 2.

Figure 7 Procedure employed by the irregular partitioning algorithm to generate partial classifiers.

As one can note, through the repeated application of steps 2–12, the algorithm generates a pool of candidate units among which the best partial classifier is chosen. The decision is based on the size of the set R_i associated with every neuron. In particular, the set R and the vector $\mathbf{w}$ are used to memorize the partial classifier that separates the greatest number of elements of Q^+ from the patterns of Q^-.

A drawback of this approach is the employment of a greedy technique that requires a high number of checks for linear separability at steps 4 and 8. This number can increase exponentially with the dimension n of the input space leading to an excessive execution time even for (relatively) small values of n.

Finally, we pointed out that step 4 can be omitted when the inputs are binary, because the grandmother cell solution (12) allows us to obtain directly the weight vector $\mathbf{w}_i$ for any pattern $\mathbf{x}$.

2. Carve Algorithm

A different approach to obtain the construction of a partial classifier for two sets of input patterns Q^+ and Q^- is to consider the convex hull generated by the elements of Q^-. It could be pointed out that only the points of Q^+ outside this convex hull can be separated from the elements of Q^- through a hyperplane. This is the basic idea of the *carve algorithm* (CA).

The original version [61] uses the beneath–beyond method [69] directly for the construction of convex hulls and leads to a total execution time that is exponential in the number n of inputs. In fact, by employing this technique, it is possible to generate an optimal partial classifier that correctly classifies the maximum number of patterns in Q^+ and Q^- (an NP-complete task).

A near-optimal result can be obtained much faster by considering only a restricted number of faces of the convex hull; this can be done by applying a proper algorithm, which derives from the gift-wrapping method [70] and has polynomial complexity. The procedure followed by this technique can be described by introducing the following definition, widely used in the theory of convex sets.

DEFINITION 4. Let A be a subset of $\mathcal{R}^n$. A hyperplane that passes through (at least) a point $\mathbf{x} \in A$ and has the remainder of this set in only one halfspace is called a *supporting hyperplane* of A. The elements of A that belong to (at least) a supporting hyperplane are called *boundary points*.

If the number of elements in A is finite, it can easily be seen that, given any direction in $\mathcal{R}^n$, we can always find at least one supporting hyperplane of A which is orthogonal to that direction. Moreover, apart from some critical cases, there are infinitely many supporting hyperplanes that pass through a boundary point.

With these premises, the procedure employed by the CA to find a partial classifier for the sets Q^+ and Q^- can be outlined as in Fig. 8. To obtain a pool of candidate neurons among which the best partial classifier is chosen, steps 2 and

CARVE ALGORITHM

(Method for generating a partial classifier)

1. Choose at random a direction in $\mathcal{R}^n$ and find a supporting hyperplane of Q^- which is orthogonal to this direction. The set of points of Q^- that belong to this initial hyperplane is called *boundary set*.
2. Choose at random another direction in $\mathcal{R}^n$ and find a supporting hyperplane of Q^- which is orthogonal to this direction and passes through (at least) a point of the boundary set.
3. Rotate the initial hyperplane around its intersection with the second hyperplane until it touches another point of Q^-. Output the threshold neuron associated with this final hyperplane.

Figure 8 Procedure employed by the carve algorithm to generate partial classifiers.

3 are repeatedly executed with the same initial hyperplane and the whole procedure is iterated for a fixed number of times. Nevertheless, it can be shown that the execution time for this procedure remains polynomial [62].

Sometimes, at step 2, a direction is randomly chosen that does not allow the existence of a supporting hyperplane containing a point of the boundary set. In this case, new different directions are selected at random until a supporting hyperplane satisfying the required conditions is found or a maximum number of iterations is exceeded.

3. Target Switch Algorithm

A different technique for the construction of partial classifiers was initially proposed by Zollner *et al.* [71] to treat training sets with binary inputs and subsequently extended to the general case $D \subset \mathcal{R}^n$ by Campbell and Perez Vicente [59]. To better explain the procedure followed by the target switch algorithm (TSA), let us denote by $\mathbf{x}_j^+$, $j = 1, \ldots, q^+$, the patterns contained in the set Q^+ and by $\mathbf{x}_j^-$, $j = 1, \ldots, q^-$, the elements of Q^-. If $\mathbf{w}$ is the weight vector of a threshold neuron, we can assign to every pattern $\mathbf{x}_j^+$ of the set Q^+ a real value μ_j^+ given by

$$\mu_j^+ = \sum_{i=1}^{n} w_i x_{ji}^+. \tag{18}$$

In the same way we can obtain a corresponding value μ_j^- for every $\mathbf{x}_j^- \in Q^-$.

By using this notation, we can outline the procedure followed by the TSA as shown in Fig. 9.

As in the irregular partitioning algorithm (Fig. 7), the TSA employs a set R and a weight vector $\mathbf{w}$ to hold the best partial classifier found during the repeated execution of steps 2–10. At step 3, a near-optimal threshold neuron can be obtained by applying the pocket algorithm with ratchet [51]; in fact, as pointed out in Section IV.D, it converges in a finite number of iterations to the optimal configuration. Other similar procedures can be constructed by substituting the perceptron learning method, which forms the kernel of the pocket algorithm, with alternative training techniques for threshold neurons, such as the Minover algorithm [67].

The auxiliary sets W^+ and W^-, initially equal to Q^+ and Q^-, respectively (step 2), are subsequently modified through the process of target switching, so as to achieve a linearly separable training set. In particular, at step 4 a suitable bias v_0 is determined that allows the correct classification of all the patterns in W^+. It should be noted, however, that the assignment $v_0 = -\mu_h^+$ leads to a hyperplane containing some elements of W^+; to avoid this problem, a small positive quantity δ is subtracted by v_0.

If the resulting threshold neuron does not correctly classify some patterns of W^- ($\mu_h^+ > \mu_k^-$), the element $\mathbf{x}_k^- \in W^-$ that leads to the maximum value μ_k^- is found. Then the vector $\mathbf{x}_l^+$, with $\mu_l^+ \leqslant \mu_k^-$, that presents the maximum overlapping ν_l with $\mathbf{x}_k^-$ is moved from W^+ to W^-. Such a modification of the set W^+ leads to a partial classifier (if there is one) for the given training set in a finite number of iterations.

Finally, step 10 allows us to obtain a pool of candidate weight vectors $\mathbf{v}$ among which that providing the correct output for the greatest number of patterns in Q^+ is selected.

4. Oil Spot Algorithm

If the input patterns have binary components, we can employ some basic properties to obtain the hyperplane associated with a partial classifier for the sets Q^+ and Q^-. As previously noted, in the case $D \subset \{-1, +1\}^n$ the binary vectors are placed at the vertices of an n-dimensional hypercube centered at the origin of the space $\mathcal{R}^n$. Thus, two patterns $\mathbf{x}_j$ and $\mathbf{x}_k$ will be called *contiguous* if they are connected by a single edge of the hypercube; because they differ by only one

TARGET SWITCH ALGORITHM

(Method for generating a partial classifier)

1. Set $\mathbf{w} = (-1, 0, \ldots, 0)^t$, $R = \emptyset$.

2. Set $W^+ = Q^+$, $W^- = Q^-$.

3. Use a proper method to find the weight vector $\mathbf{v}$ of a near-optimal threshold neuron for the training set formed by the sets W^+ e W^-.

4. Find the minimum value μ_h^+ by applying Eq. (18) to the patterns of the set W^+ ($\mu_h^+ \leq \mu_j^+$ for every $\mathbf{x}_j^+ \in W^+$) and set $v_0 = -\mu_h^+$.

5. Find the maximum value μ_k^- by applying Eq. (18) to the patterns of the set W^- ($\mu_k^- \geq \mu_j^-$ for every $\mathbf{x}_j^- \in W^-$).

6. If $\mu_h^+ > \mu_k^-$ go to Step 9 (a partial classifier is found).

7. Consider the subset $V^+ = \{\mathbf{x}_j^+ \in W^+ \mid \mu_j^+ \leq \mu_k^-\}$ and find a pattern $\mathbf{x}_l^+ \in V^+$ such that $\nu_l \geq \nu_j$ for every $\mathbf{x}_j^+ \in V^+$ being

$$\nu_j = \sum_{i=1}^{n} x_{ji}^+ x_{ki}^- \qquad (\mathbf{x}_k^- \text{ is the pattern of } W^- \text{ associated with } \mu_k^-)$$

8. Remove the pattern $\mathbf{x}_l^+$ from W^+ and add it to W^- (target switching). Go to Step 3.

9. If $|W^+| > |R|$ then set $\mathbf{w} = \mathbf{v}$ and $R = W^+$.

10. Set $Q^+ = Q^+ \setminus W^+$. If $Q^+ = \emptyset$, then output the threshold neuron having weight vector $\mathbf{w}$, otherwise go to Step 2.

Figure 9 Procedure employed by the target switch algorithm to generate partial classifiers.

component, it is possible to define the *orientation* of the linking edge through the quantity [72]

$$\omega_{jk} = \sum_{i=1}^{n} (x_{ki} - x_{ji}),$$

which can be either positive (+2) or negative (−2).

It can easily be seen that two edges are *parallel* if the corresponding pairs of vertices differ by the same component; moreover, two parallel edges are *congruent* if their orientation coincides. Finally, according to the classification induced by the sets Q^+ and Q^-, the edges that join two vertices belonging to opposite classes will be called *critical*.

By using these definitions, it is possible to show that the sets Q^+ and Q^- can be separated by a threshold neuron if the following two conditions are satisfied [73]:

1. The patterns of Q^+ must be the nodes of a connected subgraph of the hypercube $\{-1, +1\}^n$.
2. Two parallel critical edges must be congruent.

This property is used by the oil spot algorithm (OSA) [72] to construct partial classifiers; its implementation is shown in Fig. 10.

Steps 3–7 have the aim of determining a subset $R \subset Q^+$ of vertices of the hypercube which can be separated from the patterns of Q^- through a threshold neuron. The method followed is to construct, step by step, a connected subgraph of vertices of the hypercube checking at every iteration (step 7) that condition 2 mentioned previously is satisfied. The choice of the weight vector for the partial classifier (step 8) follows from geometrical considerations [72]. The value for the bias has been modified to prevent the separating hyperplane from containing either patterns of Q^+ or Q^-.

To increase the generalization ability of the resulting neural network, the following smoothing rule is used [72]: the set Q^+ is enlarged by adding to it all the contiguous vertices that do not belong to Q^-. In this way a precise output is assigned to some patterns not included in the training set by following an approach similar to that used by the nearest neighbor method [54].

Nevertheless, the choice of the initial vertex (step 3) is crucial for the resulting generalization ability, because it determines the portion of the hypercube that will be analyzed for the construction of the set R. Its influence can be reduced by iterating the procedure in Fig. 10, likewise the sequential constructive techniques previously described. Every time a set R is completed, its elements are removed from Q^+ to avoid the achievement of the same threshold neuron. When this reduction leads to $Q^+ = \emptyset$, the best partial classifier obtained (that associated with the set R having greatest size) is inserted in the neural network to be constructed.

OIL SPOT ALGORITHM

(Method for generating a partial classifier)

1. Set $\mathbf{w} = (-1, 0, \ldots, 0)^t$, $R = \emptyset$.

2. If $Q^+ = \emptyset$, then output the threshold neuron having weight vector $\mathbf{w}$.

3. Choose at random a pattern $\mathbf{x} \in Q^+$ and set $R = \{\mathbf{x}\}$, $Q^+ = Q^+ \setminus \{\mathbf{x}\}$.

4. Let V be the set of vertices which are contiguous with the patterns of R.

If $V \subset Q^-$ go to Step 7.

5. Choose at random a pattern $\mathbf{x} \in V \cap Q^+$. Let E_x (E_R) be the set of critical edges that have a vertex in $\mathbf{x}$ (in a pattern of R).

6. If there are two parallel edges of E_x and E_R which are not congruent,

then set $Q^- = Q^- \cup \{\mathbf{x}\}$ and go to Step 5,

otherwise set $R = R \cup \{\mathbf{x}\}$ and go to Step 4.

7. The weights $w_1, \ldots, w_n$ for a threshold neuron that separates R from Q^- are given by

$$w_i = \beta(\lambda_i(1) - \lambda_i(-1)) \quad \text{for } i = 1, \ldots, n$$

where β is an arbitrary positive real value and $\lambda_i(c)$ is the number of patterns in R that have the ith component equal to c. The following formula provides the value for the bias w_0

$$w_0 = \beta - \min_{\mathbf{x} \in R} \left(\sum_{i=1}^{n} w_i x_i \right)$$

Figure 10 Procedure employed by the oil spot algorithm to generate partial classifiers.

B. Hyperplane Choice Set

When the input patterns have binary components ($D \subset \{-1, +1\}^n$), a fast training algorithm for partial classifiers having the following activation function is available [58]:

$$\varphi(\mathbf{x}) = \begin{cases} +1, & \text{if } \left| \sum_{i=0}^{n} w_i x_i \right| \leqslant \delta, \\ -1, & \text{otherwise.} \end{cases} \tag{19}$$

Units of this kind are called *window neurons* because of the behavior shown by Eq. (19) (see Fig. 11).

The real quantity δ is called the *amplitude* and is meaningful only by an implementative point of view. In fact, the considerations of the present section are still valid in the particular case $\delta = 0$; however, due to the limited precision offered by a machine (either a computer or a dedicated support), the weighted sum of the inputs in Eq. (19) can move away from its theoretical value. The introduction of a small amplitude δ is therefore necessary to allow the practical use of the window neuron. It can easily be seen that in the case $\delta = 0$ the choice set L associated with a window unit is a hyperplane in the space $\mathcal{R}^n$, whose position is univocally determined by the weight vector $\mathbf{w}$.

Also with the activation function (19), generating a grandmother cell for any pattern $\mathbf{x}$ is straightforward: it is sufficient to set

$$w_0 = -n, \qquad w_i = x_i \qquad \text{for } i = 1, \ldots, n.$$

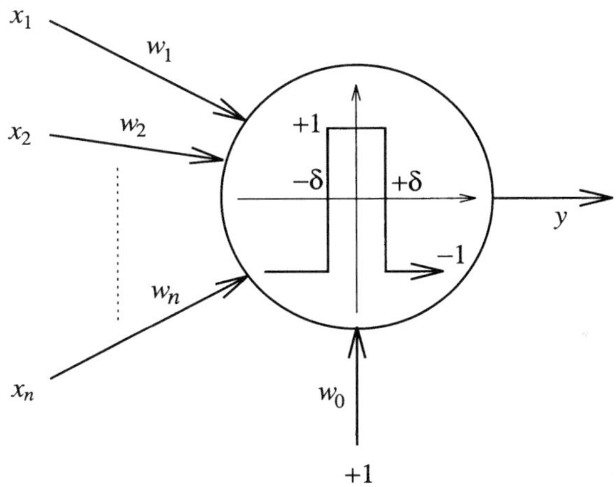

Figure 11 Behavior of the activation function of window neurons.

The construction described in Section V.A can thus be used to show that a two-layer perceptron containing only window neurons in the hidden layer is able to realize any Boolean function. From this theoretical result we can conclude a certain generality of the window neuron that motivates its employment in the solution of real-world problems.

Even in this case, the proposed training algorithm gives the weight vector $\mathbf{w}$ for the partial classifier (2) to be inserted in the resulting two-layer perceptron. From this unit we can obtain directly the corresponding vector $\mathbf{v}$ for the window neuron to be included in the cascade net architecture (under the same choice set L_j).

In particular, the choice (13) introduced for the threshold activation function is still valid in this situation. As a matter of fact, Eqs. (15) and (16) also hold for window neurons, thus proving the correctness of the weight vector $\mathbf{v}$ defined by Eq. (13). Furthermore, we can remember that it is possible to employ the simplified expression (14) for the components v_i, with $i = n+1, \ldots, n+j-1$, because the training set has binary inputs.

1. Sequential Window Learning Algorithm

The interest in window neurons is motivated by the existence of a fast and efficient learning algorithm that allows the treatment of training sets containing a high number of samples. Before describing this technique, let us introduce some theoretical results that prove the validity of the proposed approach. The following two theorems play a crucial role in the construction of the partial classifier.

THEOREM 8. *Given a set V containing q linearly independent input patterns $\mathbf{x}_j$, with $j = 1, \ldots, q$, in the space $\mathcal{R}^{n+1}$, it is always possible to construct a window neuron that provides a desired output y_j for each vector $\mathbf{x}_j \in V$.*

Proof. Consider the matrix A, having dimension $q \times (n+1)$, whose rows are given by the input patterns $\mathbf{x}_j$, $j = 1, \ldots, q$. Because these vectors are linearly independent among them, a nonsingular minor B with size q can be extracted from A. Suppose, without loss of generality, that B contains the first q columns of A ($i = 0, 1, \ldots, q-1$); thus, the following system of linear algebraic equations:

$$B\mathbf{w} = \mathbf{z}, \tag{20}$$

where the jth component of the vector $\mathbf{z}$ is given by

$$z_j = 1 - y_j - \sum_{i=q}^{n} w_i x_{ji} \qquad \text{for } j = 1, \ldots, q, \tag{21}$$

has a unique solution in $\mathcal{R}^q$.

If the values w_i, for $i = q, \ldots, n$, in (21) are arbitrarily chosen, by solving Eq. (20) we obtain the weight vector for the desired window neuron. As a matter of fact, we have from Eq. (21)

$$\sum_{i=0}^{n} w_i x_{ji} = \sum_{i=0}^{q-1} w_i x_{ji} + \sum_{i=q}^{n} w_i x_{ji} = 1 - y_j$$

and the application of the activation function (19) gives

$$\varphi\left(\sum_{i=0}^{n} w_i x_{ji}\right) = \varphi(1 - y_j) = y_j \qquad \text{for } j = 1, \ldots, q$$

if $0 \leqslant \delta < 2$. ■

THEOREM 9. *If the arbitrary weights w_i used in Eq.* (21) *for the generation of the vector* $\mathbf{z}$ *are linearly independent in $\mathcal{R}$ as a $\mathcal{Q}$-vector space, then the window neuron obtained by Eq.* (20) *when $\mathbf{x}_j \in Q^+$, for $j = 1, \ldots, q$, provides output* $+1$ *for all the vectors that are linearly dependent with $\mathbf{x}_1, \ldots, \mathbf{x}_q$ and output* -1 *for any other input pattern.*

Proof. Because $\mathbf{x}_j \in Q^+$, the weights w_i of the resulting window neuron must satisfy the following system of linear algebraic equations:

$$\sum_{i=0}^{n} w_i x_{ji} = 0 \qquad \text{for } j = 1, \ldots, q. \tag{22}$$

Suppose again that the nonsingular minor B in Eq. (20) contains the components $i = 0, 1, \ldots, q-1$ of the input patterns $\mathbf{x}_1, \ldots, \mathbf{x}_q$ and consider a vector $\mathbf{x}$ different from them.

The $q+1$ vectors

$$(x_0, x_1, \ldots, x_{q-1})^t,\ (x_{10}, x_{11}, \ldots, x_{1,q-1})^t, \ldots, (x_{q0}, x_{q1}, \ldots, x_{q,q-1})^t$$

are linearly dependent in $\mathcal{Q}^q$, where $\mathcal{Q}$ is the rational field (t denotes the transpose operation). Hence, there are constants $\lambda_1, \ldots, \lambda_q \in \mathcal{Q}$, some of which are nonnull, such that

$$x_i = \sum_{j=1}^{q} \lambda_j x_{ji} \qquad \text{for } i = 0, 1, \ldots, q-1 \tag{23}$$

and consequently by using Eq. (22) we can write

$$\sum_{i=0}^{n} w_i x_i = \sum_{i=0}^{n} w_i x_i - \sum_{j=1}^{q} \lambda_j \sum_{i=0}^{n} w_i x_{ji} = \sum_{i=q}^{n} w_i \left(x_i - \sum_{j=1}^{q} \lambda_j x_{ji}\right). \tag{24}$$

Now we have two different situations. If the pattern $\mathbf{x}$ is linearly dependent with $\mathbf{x}_1, \ldots, \mathbf{x}_q$, the right-hand side of Eq. (24) vanishes for some rational constants

$\lambda_j,\ j = 1, \ldots, q$, that satisfying Eq. (23). Hence, the corresponding output of the window neuron will be $+1$.

In the opposite case, if $\mathbf{x}, \mathbf{x}_1, \ldots, \mathbf{x}_q$ are linearly independent in $\mathcal{Q}^{n+1}$, then there is at least one index i such that

$$x_i \neq \sum_{j=1}^{q} \lambda_j x_{ji},$$

with $q \leqslant i \leqslant n$. However, the quantities $x_i - \sum_{j=1}^{q} \lambda_j x_{ji}$ are rational numbers; so, if the weights w_i, for $i = q, \ldots, n$, are linearly independent in $\mathcal{R}$ as a $\mathcal{Q}$-vector space, we obtain from Eq. (24)

$$\sum_{i=0}^{n} w_i x_i = \sum_{i=q}^{n} w_i \left(x_i - \sum_{j=1}^{q} \lambda_j x_{ji} \right) \neq 0.$$

Hence, the corresponding output of the window neuron will be -1, assuming that the amplitude δ is small enough. ■

A possible choice for the arbitrary weights w_i that fulfills the hypothesis of Theorem 9 is the following [74]:

$$w_i = \sqrt{\gamma_i} \qquad \text{for } i = 0, 1, 2, \ldots, \tag{25}$$

where the γ_i are positive square-free (not divisible by a perfect square) integers, sorted in increasing order (by definition, $\gamma_0 = 1$). The first 50 values of the constants γ_i are shown in Table I.

Table I

First 50 Values of the Integer Constants γ_i

i	γ_i	i	γ_i	i	γ_i	i	γ_i	i	γ_i
1	1	11	15	21	33	31	47	41	66
2	2	12	17	22	34	32	51	42	67
3	3	13	19	23	35	33	53	43	69
4	5	14	21	24	37	34	55	44	70
5	6	15	22	25	38	35	57	45	71
6	7	16	23	26	39	36	58	46	73
7	10	17	26	27	41	37	59	47	74
8	11	18	29	28	42	38	61	48	77
9	13	19	30	29	43	39	62	49	78
10	14	20	31	30	46	40	65	50	79

Now let us recall the definition of the VC dimension [32]:

DEFINITION 5. Let C be a class of Boolean functions and S a set of points. We say C *shatters* S if C induces all Boolean functions on S. The *VC dimension* of C is the size of the largest set C shatters.

Then the following important result is obtained directly from Theorems 8 and 9:

COROLLARY 1. *The VC dimension of the class of window neurons is equal to* $n + 1$.

Proof. It can be verified directly that the following group of $n + 1$ vertices of the n-dimensional hypercube:

$$\begin{aligned}
\mathbf{x}_1 &= (+1, +1, +1, +1, \ldots, +1, +1),\\
\mathbf{x}_2 &= (+1, -1, +1, +1, \ldots, +1, +1),\\
\mathbf{x}_3 &= (+1, +1, -1, +1, \ldots, +1, +1),\\
\vdots \;\; &\;\; \vdots \;\; \vdots\\
\mathbf{x}_n &= (+1, +1, +1, +1, \ldots, -1, +1),\\
\mathbf{x}_{n+1} &= (+1, +1, +1, +1, \ldots, +1, -1)
\end{aligned}$$

forms a set of $n + 1$ linearly independent vectors in $\mathcal{R}^{n+1}$. Thus, according to Theorem 8, there is a window neuron that realizes any classification of these $n+1$ input patterns. Hence, the required VC dimension will not be less than $n+1$. Now consider a set V containing $q > n+1$ input patterns $\mathbf{x}_j$. Because they are linearly dependent in the space $\mathcal{R}^{n+1}$, Theorem 9 ensures that the outputs provided by any window neuron cannot be freely chosen. Thus, the required VC dimension is exactly $n + 1$. ■

As it has been shown that the VC dimension of the class of threshold neurons is also equal to $n + 1$ [75], we can conclude that the window and threshold units present, at least in principle, the same complexity.

Theorems 8 and 9 allow us also to extend the concept of the grandmother cell in the case of window neurons.

COROLLARY 2. *It is always possible to find a window neuron that separates two patterns* $\mathbf{x}_1$ *and* $\mathbf{x}_2$ *from the remaining vertices of the hypercube* $\{-1, +1\}^n$.

Proof. If $\mathbf{x}_1$ and $\mathbf{x}_2$ are two distinct patterns, then there is at least one component i such that $x_{1i} \neq x_{2i}$. Hence, the determinant of the minor

$$B = \begin{pmatrix} x_{10} & x_{1i} \\ x_{20} & x_{2i} \end{pmatrix}$$

SEQUENTIAL WINDOW LEARNING ALGORITHM

(Method for generating a partial classifier)

1. Choose at random two patterns $\mathbf{x}_1, \mathbf{x}_2 \in Q^+$ and obtain the weight vector $\mathbf{w}$ for the corresponding grandmother cell. Set $R = \{\mathbf{x}_1, \mathbf{x}_2\}$, $Q^+ = Q^+ \setminus \{\mathbf{x}_1, \mathbf{x}_2\}$.

2. Through the application of (20) and (21) find all the patterns of Q^+ that lead to a partial classifier once added to R. Let V be the set of these patterns.

3. If $V = \emptyset$, go to Step 6.

4. Add to R the element $\mathbf{x} \in V$ that leads to the generation of the weight vector $\mathbf{w}$ for a window neuron that correctly classifies the greatest number of patterns in Q^+. Let W be the set of these patterns.

5. Set $Q^+ = Q^+ \setminus W$. If $Q^+ \neq \emptyset$, go to Step 2.

6. Output the weight vector $\mathbf{w}$ for the required window neuron.

Figure 12 Procedure employed by the sequential window learning algorithm to generate partial classifiers.

is nonnull; by employing the choice (25) and Eqs. (20) and (21), it is then always possible to obtain the weights w_i for the desired window neuron. ∎

Thus, we can always associate a grandmother cell having a window activation function with any pair of input patterns.

Finally, Theorem 9 allows us to formulate a greedy technique for the construction of a partial classifier having a window activation function; this method tries to maximize the number of patterns in Q^+ with correct output. The procedure employed is shown in Fig. 12 and forms the kernel of the sequential window learning (SWL) algorithm [58].

The set R contains the linearly independent patterns that form the minor B employed in Eq. (20) to obtain the weight vector for the desired window neuron.

The choice (25) is used at steps 1 and 2 to compute the right-hand-side $\mathbf{z}$ through Eq. (21).

At every iteration of the procedure, the set R is enlarged (step 4) by adding to it the vector $\mathbf{x} \in Q^+$ that maximizes the number of input patterns correctly classified by the current window neuron. In this way we try to obtain a weight vector $\mathbf{w}$ that satisfies the greatest number of samples in the given training set. The algorithm can stop either at step 3, if it is impossible to add other patterns to R, or at step 5, if a single window neuron is able to provide the correct output for all the patterns of Q^+ and Q^-.

Naturally, we can iterate the procedure in Fig. 12 starting from different initial patterns $\mathbf{x}_1, \mathbf{x}_2$, so as to perform a further maximization of the number of patterns correctly classified. The corresponding changes to the SWL can be made by following the approach employed in the irregular partitioning method (Fig. 7).

VI. HAMMING CLUSTERING PROCEDURE

As pointed out in Section III, the addition of hidden units to the resulting neural network through the execution of a sequential constructive method involves the training of a reduced portion of the weight matrix. In many cases, this prevents us from determining some important features of the function f that generated the given training set S; consequently, the generalization ability of the final connectionist model is reduced.

When the input patterns have binary components, a way to overcome this problem is provided by the procedure of Hamming clustering (HC) [58, 76]. With this approach we can find redundant input variables that do not affect the behavior of one or more outputs. The removal of the corresponding connections therefore allows a lowering of the number of weights in the neural network and consequently its complexity.

HC is based on the following observation [77]: simple statistical methods, such as the nearest neighbor algorithm [54], show a high generalization ability in many real situations. Hence, the analysis of the local properties of the training set can be used to accelerate and improve the learning process. If the input patterns have binary components, a simple way to implement this suggestion is to gather the samples belonging to the same class that are close to each other according to the Hamming distance.

This produces some clusters in the input space that determine the extension of each class, thus providing an output to the patterns that are not contained in the training set (generalization). Such an approach can also lead to a simplification in the sets Q^+ and Q^- (used for the generation of hidden units), which increases the convergence speed of the training algorithm.

HC recalls the methods employed in the synthesis of digital networks [78], but shows a low computational cost that allows the treatment of training sets containing tens of thousands of samples. To describe the algorithm employed, let us introduce the following definition widely used in the design of logical networks.

DEFINITION 6. Let $\mathbf{x}$ be a binary vector in $\{-1, +1\}^n$ and I a subset containing m of its components. The set $C_m(\mathbf{x}, I)$, obtained by assigning all the possible values to the m components in I, is called an *m-cube*.

For example, if we take $\mathbf{x} = (-1, +1, -1, -1, +1)$ and $I = \{2, 4\}$, we obtain the 2-cube

$$\begin{aligned} C_2(\mathbf{x}, I) = \{&(-1, -1, -1, -1, +1), (-1, -1, -1, +1, +1), \\ &(-1, +1, -1, -1, +1), (-1, +1, -1, +1, +1)\} \end{aligned}$$

by considering all the possible pairs of values that can be assumed by the second and the fourth components of the vector $\mathbf{x}$. Note that the remaining components maintain the same values in every element of the set $C_2(\mathbf{x}, I)$.

Now it is always possible to construct a threshold or window neuron that separates an m-cube $C_m(\mathbf{x}, I)$ from the other vertices of the hypercube $\{-1, +1\}^n$. It is sufficient to employ the following weights:

$$w_0 = m + 1 - n, \qquad w_i = x_i \text{ for } i \notin I, \qquad w_i = 0 \text{ for } i \in I$$

in the case of a threshold unit and

$$w_0 = m - n, \qquad w_i = x_i \text{ for } i \notin I, \qquad w_i = 0 \text{ for } i \in I,$$

when the activation function is given by Eq. (19). Hence, the connections associated with the components belonging to the set I can be removed, because they do not affect the output of the neuron.

HC implements this consideration to simplify the construction of the partial classifier for the sets Q^+ and Q^-. It generates a group of m-cubes $C_m(\mathbf{x}, I)$, with the same component set I, starting from the patterns $\mathbf{x} \in Q^+$; every m-cube must not contain elements of Q^- so as to realize a clusterization of the set Q^+. A possible algorithm that pursues this result is shown in Fig. 13.

The procedure followed is very simple: it adds to the set I the components that maximize (at each iteration) the number of m-cubes $C_m(\mathbf{x}, I)$ generated by the elements of Q^+ and not containing patterns of Q^-. The set W to which the vectors $\mathbf{x}$ belong decreases with the number of iterations and causes the termination of the clustering process when $|W|$ falls below a given threshold σ. In the simulations we have set $\sigma = (q^+/2, 2n)$, where q^+ is the number of elements in Q^+.

After the application of HC, the construction of the partial classifier to be inserted in the neural network follows these steps:

PROCEDURE OF HAMMING CLUSTERING

1. Set $m = 0$, $I = \emptyset$, $W = Q^+$.
2. For every $i \notin I$ let V_i be the subset of W containing all the input patterns $\mathbf{x}$ for which $C_{m+1}(\mathbf{x}, I \cup \{i\}) \cap Q^- = \emptyset$. Let i^* be the component associated with the set V_i having maximum size.
3. If $|V_{i^*}| \geq \sigma$ then set $I = I \cup \{i^*\}$, $W = V_{i^*}$, $m = m + 1$ and go to Step 2.
4. Remove the components belonging to I from the sets Q^+ and Q^- for the training of the partial classifier.

Figure 13 Algorithm employed by the procedure of Hamming clustering.

- Set $w_i = 0$, for every $i \in I$. In fact, any change in these components of the patterns belonging to W does not lead to a vector of Q^- and vice versa.
- Train the weights w_i, for $i \notin I$, on the grounds of the sets R^+ and R^- obtained by removing the components of I from Q^+ and Q^-, respectively.

In this way we achieve a reduction in the dimension of the input space, which improves the convergence speed of the training algorithm. Furthermore, no modifications are required to the procedures of the previous sections for the generation of partial classifiers. This leads to a perfect modularity in the construction of the multilayer perceptron.

Other choices are possible under the same computational cost; for example, we can consider, for the generation of an m-cube, only the patterns having Hamming distance from the set Q^+ not greater than that from the set Q^-. However, the efficiency of a given rule is measured by the generalization ability of the resulting neural network and depends on the problem to be solved.

VII. EXPERIMENTAL RESULTS

The sequential constructive techniques described in Section V have been extensively tested on several benchmark problems to evaluate their performance. In particular, three basic properties have been examined in detail:

- complexity of the resulting configuration,
- convergence speed of the learning,
- generalization ability of the trained neural network.

A reference value for each of these quantities has been obtained by the application of the back-propagation algorithm (BPA) to every set of trials (except those regarding random Boolean functions). The implementation adopted is an optimized version of the standard procedure and makes use of the acceleration suggested by Vogl *et al.* [14].

As we shall see, such an implementation presents a high convergence speed that is comparable with that shown by some sequential constructive methods. Nevertheless, the resulting execution times for the BPA do not take account of the preliminary runs needed for the determination of the network architecture and are consequently underestimated.

To obtain a better evaluation of the complexity and the generalization ability of the neural networks generated by each constructive algorithm, we have performed two different groups of tests. In the first group (exhaustive learning), complete training sets deriving from artificial Boolean functions are presented to every method and the average number of hidden neurons contained in the final two-layer perceptron is considered. No test phase takes place in this case.

The second group of trials (generalization tests) has the aim of determining the performance of sequential constructive methods in the solution of practical problems. In these benchmarks, the generalization ability of the resulting neural networks will be analyzed in detail through the employment of both artificial tests and data sets deriving from real-world applications.

Thirty runs have been performed for every trial and every learning algorithm so as to obtain a sufficient statistic of the measured quantities. Apart from the extensive training on the parity and the symmetry functions, every run uses different training and test sets for the construction of the two-layer perceptron.

The implementation of the sequential constructive methods strictly follows the procedures reported in Figs. 7–12; the values of the parameters for every technique have been kept fixed throughout all the trials. In particular, we have noted that in the irregular partitioning algorithm, the stopping criterion employed at step 2 of Fig. 7 ($W = \emptyset$) leads to huge execution times for training sets containing more than 100 samples. Thus, we have imposed an upper bound $n_w = 10$ to the number of iterations of steps 3–12; this change provides a good compromise between the convergence speed and generalization ability of the resulting configuration.

A similar approach has also been used in the other sequential constructive methods; the only exception is the carve algorithm, whose duration depends on the number n_r of iterations of steps 2–3 (Fig. 8) and on the number n_i of different initial hyperplanes chosen at step 1. For these quantities the values $n_i = 50$ and $n_r = 50$ have been chosen.

At step 3 of the target switch algorithm (Fig. 9), the weight vector **w** for a near-optimal threshold neuron is found by adopting the thermal perceptron learning rule (TPLR) proposed by Frean [79]. Although the optimality of this method is not theoretically ensured, it generally requires a lower execution time than the pocket algorithm with ratchet to reach an equivalent configuration.

The TPLR follows the same approach employed by the perceptron algorithm, but the updating of the weight vector **w** is performed by using the following rule instead of Eq. (17):

$$w_i = w_i + \eta y x_i e^{-|\phi|/T}, \tag{26}$$

where

$$\phi = \sum_{i=0}^{n} w_i x_i.$$

It can easily be seen that the introduction of the factor $e^{-|\phi|/T}$ allows us to treat samples of the training set which present different values of ϕ in a different way. In fact, in the standard perceptron algorithm, the change of the weight vector **w** deriving from an input pattern **x** with high $|\phi|$ often leads to an increase in the total number of misclassifications. To avoid this undesirable effect, the TPLR biases the weight changes toward correcting errors for which ϕ is close to zero.

The parameter T, called *temperature*, controls the behavior of the TPLR: if T is much higher than $|\phi|$, the rules (17) and (26) become equivalent and all the samples in the training set are treated in the same way. On the other hand, if T is very small, the weights are frozen at their current values.

In practice, an annealing of the temperature T is performed by gradually reducing its value from a high initial T_0 to zero. With this trick, the weight vector **w** can stabilize around near-optimal configurations for a given problem. A further improvement in the convergence of the TPLR can also be obtained by decreasing at the same time the learning rate η from 1 to 0.

In our simulations of the TSA, $n_t = 1000$ iterations of the TPLR have always been performed, starting from an initial temperature $T_0 = 5n$, where n is the number of inputs. Some tests can produce equivalent neural networks even with lower values of n_t, but, as previously noted, every parameter has been kept constant throughout all the trials. Consequently, the reported execution times for the TSA are often excessive for the treatment of the corresponding problems.

The TPLR has also been employed as an alternative to linear programming (LP) in the IPA to evaluate the properties of the two different implementations proposed in the literature. The same values $T_0 = 5n$ and $n_t = 1000$ have been used for the initial temperature and the number of iterations.

In the sequential window learning algorithm, the search for all the patterns in Q^+ that lead to a partial classifier once added to R (step 2 of Fig. 12) can lead to high computing time when training sets containing many patterns are considered. Hence, an upper bound $n_v = 20$ for the dimension of the set V has been introduced.

All the execution times refer to a DECStation 3000/600 with 64 MB RAM under operating system Digital UNIX 4.0.

A. Exhaustive Learning

The first three groups of trials are devoted to evaluating the architecture complexity, measured by the number of hidden neurons included in the two-layer perceptrons generated by various sequential constructive methods. The training sets used in these runs refer to artificial Boolean functions and contain all the possible ($s = 2^n$) samples that can be extracted by the corresponding truth tables. The Hamming clustering (HC) procedure will never be applied because no generalization is required.

1. Parity Function

The output of the parity function is $+1$ if and only if the number of components with value $+1$ in the input pattern is odd. This benchmark has been widely used in the literature to evaluate the performance of training algorithms for multilayer perceptrons because it cannot be performed by a single threshold neuron [65].

It has been shown that a two-layer neural network containing at least n (number of inputs) threshold units [5] or $\lceil (n+1)/2 \rceil$ window neurons [58] can realize the parity function, $\lceil x \rceil$ being the integer not less than x. Extensive simulations of sequential constructive algorithms have been performed to determine the average complexity of the resulting neural networks for $n = 2, 3, \ldots, 9$. The corresponding results are shown in Fig. 14, together with the average CPU time employed by each sequential constructive method. We have also included the computational cost needed by a typical BPA to train a minimal neural network (containing n hidden units).

As one can note, only the IPA with LP and the SWL are always able to obtain the optimal configuration when the number n of inputs increases. Nevertheless, whereas the SWL employs a lower CPU time than the BPA, the computational cost of the IPA with LP becomes excessive for $n = 9$ (almost 30 hours for a single trial). The OSA is the fastest method but leads to neural networks containing a high number of hidden neurons.

2. Symmetry Function

In the symmetry function, the output is $+1$ if and only if the binary string associated with the current input pattern is symmetric (with respect to its center).

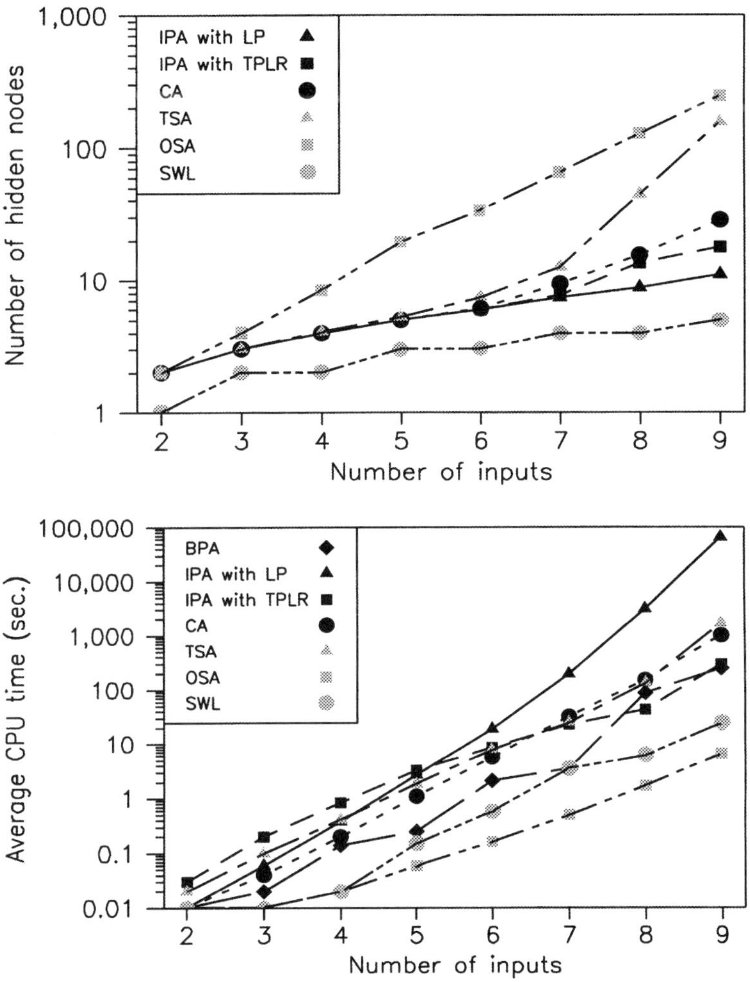

Figure 14 Number of hidden units and average CPU time (s) for the neural networks performing the parity function.

It has been shown that two hidden threshold neurons are sufficient to realize this Boolean function [5], whereas a single window unit can provide the correct target output [58]. Hence, both in this case and in the previous one, the minimal configuration containing window neurons presents a lower complexity.

The simulations performed for $n = 2, 3, \ldots, 9$ have given neural networks containing the average number of hidden neurons as in Fig. 15. The average execution times employed by each algorithm are also included.

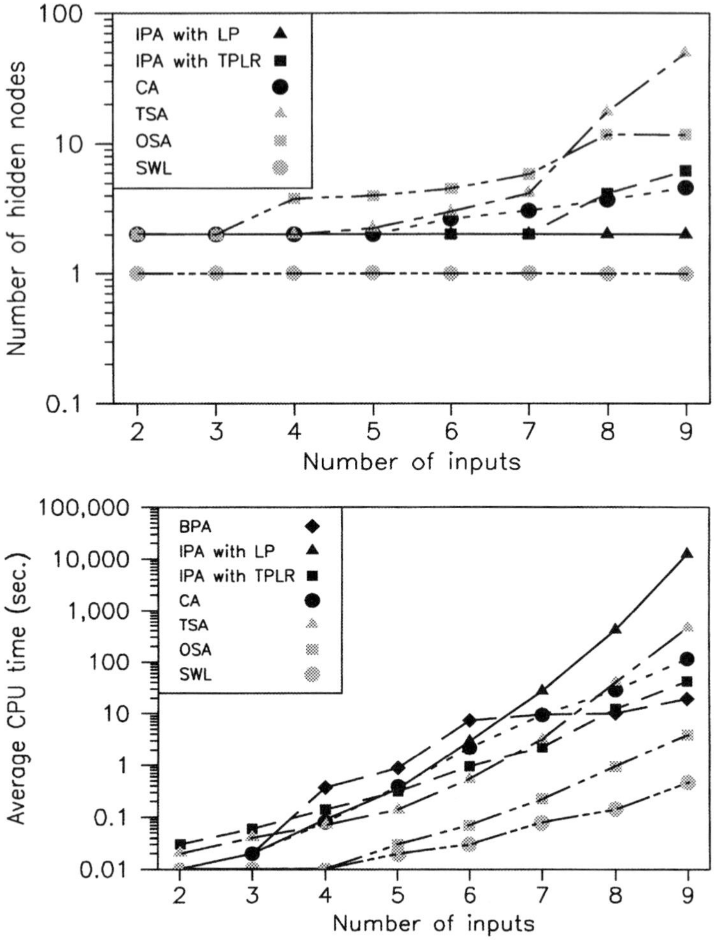

Figure 15 Number of hidden units and average CPU time (s) for the neural networks performing the symmetry function.

As in the previous test, only the IPA with LP and the SWL are always able to obtain the optimal configuration. The computational cost of the IPA with TPLR is comparable with that shown by the BPA. The SWL requires the lowest CPU time and generates the smallest two-layer perceptron for all the values of the number n of inputs. When threshold networks are considered, the CA represents a good compromise between training speed and efficiency.

3. Random Boolean Functions

A third group of trials has been devoted to determining the average complexity of a neural network performing a Boolean function whose output is randomly generated with uniform probability. This benchmark has been widely employed in the literature to compare different constructive training methods [80, 57, 81]. Unfortunately, a general upper bound on the number of hidden neurons contained by a minimal two-layer perceptron that realizes this kind of function is not known. Thus, a valid comparison with the computational cost of the BPA is not possible.

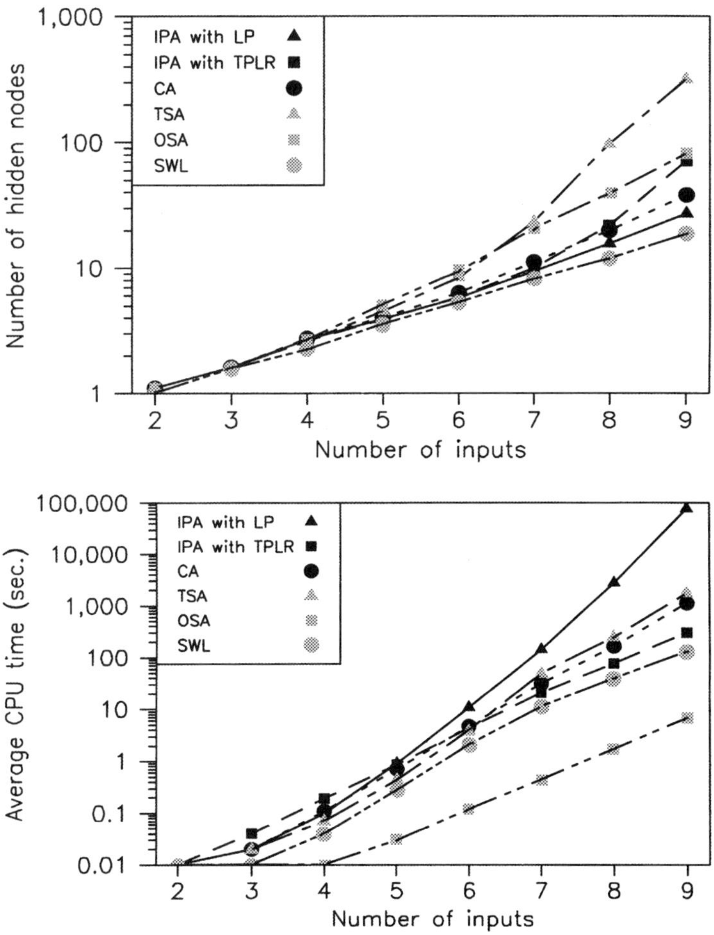

Figure 16 Number of hidden units and average CPU time (s) for the neural networks performing a random Boolean function.

Figure 16 shows the complexity of the neural networks generated by each sequential constructive algorithm for $n = 2, 3, \ldots, 9$, and the average CPU time needed for each training.

Also in this case the employment of window units allows us to minimize the number of hidden neurons contained in the resulting two-layer perceptrons. The computational cost of the SWL is again low; only the OSA employs a smaller CPU time for each training, but leads to neural networks with high complexity. The CA achieves again a good compromise between convergence speed and efficiency.

B. Generalization Tests

The generalization ability of neural networks generated by sequential constructive methods has been evaluated by considering six different applications. Two of them (symmetry and Monk's problems) have been artificially built and allow a direct comparison with other training algorithms, because several simulation results on these two benchmarks are available in the literature.

The other four groups of tests derive from a collection of data sets distributed by the machine learning group of the University of California at Irvine [82]. These real-world problems have been previously considered in [56] and hence offer an important validation for our implementation of the IPA. Moreover, the execution of the BPA again provides a valid reference measure for the results obtained.

In these tests, the influence of HC on the complexity and the generalization ability of the configurations generated by the OSA and SWL will also be analyzed. However, all the applications considered (except the symmetry problem) have real or discrete inputs; so HC and training algorithms that build binary neural networks, like the OSA and SWL, cannot be applied directly. The use of proper transformations allows the removal of this obstacle, but, as we shall see later, leads to a degradation of the resulting generalization ability.

The complexity of the resulting neural networks will be measured through the number of nonnull weights contained in each of them; in this way we can better evaluate the influence of the employment of HC.

1. Mirror Symmetry Detector

The first series of generalization tests refers to the symmetry function (Section VII.A.2) with $n = 30$. The training and test sets employed in each trial always contain samples that are equally subdivided between the two classes (symmetrical and nonsymmetrical patterns); hence, they allow an unbiased analysis of the properties of the resulting neural networks.

Table II

Number of Nonnull Weights for the Mirror Symmetry Problem

	Size s of the training set			
Algorithm	100	200	400	600
BPA	62.00	62.00	62.00	62.00
IPA with LP	62.00	62.00	62.00	62.00
IPA with TPLR	88.87	159.13	373.03	140.53
CA	177.73	272.80	415.40	512.53
TSA	77.50	116.77	706.80	125.03
OSA	647.43	1087.90	1675.80	2043.73
OSA with HC	95.57	126.57	135.57	141.87
SWL	30.00	30.00	30.00	30.00
SWL with HC	35.37	38.40	43.67	50.63

Every test set is formed by 4000 samples, whereas the number s of patterns in the training set assumes the values $s = 100, 200, 400, 600$. Tables II and III show, respectively, the average number of nonnull weights and the percentage of the test set correctly classified. The average CPU time of each training algorithm is reported in Table IV.

Here, the SWL achieves excellent results in a reduced computing time: it is always able to obtain the optimal window neuron performing the mirror symmetry function even when the training set contains only 100 samples. Obviously, the use of HC cannot improve this result, but allows the reduction of the number of

Table III

Generalization Ability for the Mirror Symmetry Problem

	Size s of the training set			
Algorithm	100	200	400	600
BPA	70.72%	86.10%	92.63%	94.60%
IPA with LP	80.25%	90.98%	96.12%	98.74%
IPA with TPLR	67.60%	86.55%	94.46%	96.12%
CA	59.16%	70.79%	81.21%	84.43%
TSA	71.09%	85.98%	90.63%	94.03%
OSA	69.55%	75.77%	82.07%	85.14%
OSA with HC	61.58%	73.75%	85.67%	89.30%
SWL	100.00%	100.00%	100.00%	100.00%
SWL with HC	67.82%	82.98%	90.36%	92.38%

Table IV
Average CPU Time (s) for Solving the Mirror Symmetry Problem

	Size s of the training set			
Algorithm	100	200	400	600
BPA	12.27	16.18	103.55	169.12
IPA with LP	98.91	1670.17	23908.64	97923.01
IPA with TPLR	10.45	46.20	209.98	545.72
CA	171.76	464.10	1286.53	2413.48
TSA	4.13	16.04	242.80	1258.31
OSA	4.45	5.98	9.64	13.65
OSA with HC	4.01	5.32	13.53	25.55
SWL	5.43	7.13	10.76	14.85
SWL with HC	11.18	12.24	19.52	32.01

connections in the networks generated by the OSA. In this case the generalization ability is also increased when $s = 400,\ 600$.

Among the sequential constructive algorithms for two-layer perceptrons with threshold units, the IPA achieves the best performances, but its computational burden is very high if LP is employed for the training of hidden neurons.

2. Monk's Problems

Three artificial classification problems have been proposed in [83] as benchmarks for training algorithms. They can be viewed as recognition tasks in a robot domain, where every instance is described by six attributes, which are listed in

Table V
Attributes and Possible Values for Monk's Problems

Attributes	Possible values
Head shape	Round, square, octagon
Body shape	Round, square, octagon
Is smiling	Yes, no
Holding	Sword, balloon, flag
Jacket color	Red, yellow, green, blue
Has tie	Yes, no

Table V, along with their possible values. With this characterization 432 different robots can be obtained.

The aim of each proposed benchmark is the determination of a general classification rule starting from a limited set of samples. The first two Monk's problems are noise-free, whereas in the third case the outputs of the training set can undergo small changes from their correct value. This last test can therefore provide a measure of the robustness of sequential constructive methods.

Because the inputs are discrete, a proper transformation must be applied to allow the employment of the SWL and OSA. The resulting binary training sets contain 15 inputs, each of which is associated with the presence of a particular attribute in the corresponding robot. The results obtained for the complexity and the generalization ability are shown in Tables VI and VII, respectively. The average CPU time for every trial is reported in Table VIII.

The employment of HC allows the achievement of good performances fot both the SWL and the OSA: the resulting neural networks show high generalization ability and discrete robustness (Monk's problem 3). Furthermore, the low computational cost of HC allows the reduction of the total CPU time in all three tests.

Interesting results are also offered by CA, which is again a good compromise between training speed and efficiency.

3. Real-World Problems

An interesting analysis of the properties of sequential constructive methods can be obtained through their application to real-world problems. Among the several benchmarks maintained at the UCI machine learning repository [82], we have considered the following four:

Table VI

Number of Nonnull Weights for Monk's Problems

	Monk's problem		
Algorithm	1	2	3
BPA	21.00	42.00	21.00
IPA with LP	25.20	59.07	27.80
IPA with TPLR	34.07	87.27	36.17
CA	38.20	77.93	37.33
TSA	70.93	147.70	75.60
OSA	373.37	1008.80	347.47
OSA with HC	13.17	74.20	41.43
SWL	38.13	16.00	67.57
SWL with HC	10.97	17.67	30.17

Table VII
Generalization Ability for Monk's Problems

Algorithm	Monk's problem 1	2	3
BPA	95.74%	84.31%	90.22%
IPA with LP	88.87%	83.67%	88.19%
IPA with TPLR	90.30%	83.08%	89.26%
CA	87.11%	82.31%	88.87%
TSA	84.73%	82.72%	83.42%
OSA	81.78%	74.70%	80.76%
OSA with HC	99.72%	91.42%	91.54%
SWL	95.66%	100.00%	83.36%
SWL with HC	100.00%	98.58%	92.18%

1. *Glass identification (GL)*: The correct identification of types of glass has great importance in criminological investigation. Starting from input patterns containing eight components measuring the concentration of a given element (Mg, Na, Al, ...) and a component associated with the refractive index, we want to determine if a given piece of glass is float-processed or not.

2. *Iris data set (IR)*: Three different types of iris plant, virginica, versicolor, and setosa, must be recognized by knowing the length and width of the sepal and petal. One class (setosa) is linearly separable from the other two; the latter are not

Table VIII
Average CPU Time (s) for Solving Monk's Problems

Algorithm	Monk's problem 1	2	3
BPA	8.30	36.33	4.15
IPA with LP	84.83	246.67	59.28
IPA with TPLR	6.03	16.89	5.74
CA	4.37	9.89	3.52
TSA	8.81	27.49	6.73
OSA	1.12	2.31	0.92
OSA with HC	0.46	0.81	0.51
SWL	17.02	1.62	26.53
SWL with HC	0.46	1.07	1.12

separable from each other. It is a benchmark widely used in the literature [84] to test pattern recognition methods.

3. *Chess endgame (CH)*: In this problem, the goal is to determine whether or not a given chess board configuration is a winning position for the white player [85]. White has a king and a rook, whereas black has a king and a pawn which is ready to be promoted to a queen. Every input pattern contains 35 binary components and one ternary component describing a board position of a chess endgame.

4. *United States congressional voting record database* (*V*0 *and V*1): This benchmark contains the results of 16 key votes for each member of the U.S. House of Representatives (V0). Every input can assume three values: "yea", "nay", or "?" where this last value is taken when a congressman did not vote, voted present, or voted present to avoid conflict of interest. This is a two-class problem because a congressman is either a Democrat or a Republican. The V1 data set is obtained from V0 by deleting the most informative input (vote on physician fee freeze), which makes the problem harder.

The set of samples available for each of these benchmarks has been subdivided in the following way: two-thirds of the patterns (randomly chosen in each of the 30 trials) form the training set, whereas the remaining one-third has been used to test the generalization ability of the resulting neural networks.

In all the groups of trials, the application of the OSA and SWL needs a proper transformation of the input patterns. In the case of GL and IR, the real attributes have been converted by employing a Gray code that holds the same resolution as that of the original data. The equivalent binary patterns contain 79 (GL) and 22 (IR) inputs, respectively.

For the data sets CH, V0, and V1, the same conversion used for Monk's problems has been adopted; the resulting binary samples have 37 (CH), 48 (V0), and 45 (V1) components. As one can note, these transformations increase highly the dimension of the input space, making the training process more difficult; consequently, the performances of the OSA and SWL will be poorer.

The average complexity of the resulting neural networks and the associated generalization ability is reported in Tables IX and X, respectively. The average CPU time for every trial is reported in Table XI. The BPA has been included again to provide an objective evaluation measure; it should be remembered, however, that the computational cost of the BPA does not take into account the preliminary runs needed for the determination of the number of hidden neurons and are consequently underestimated.

In all these tests, the TSA generates two-layer perceptrons with good generalization ability in reasonable training times. The version of the IPA with TPLR performs better than the IPA with LP, even if it leads to configurations with a greater number of connections. The generalization ability of the neural networks constructed by the OSA and SWL suffers from the data binarization, particularly in the GL and IR problems where the inputs are real.

Table IX
Number of Nonnull Weights for Five Real-World Problems

Algorithm	Real-world problem GL	IR	CH	V0	V1
BPA	40.00	10.00	148.00	34.00	64.00
IPA with LP	41.66	27.00	446.22	35.13	59.43
IPA with TPLR	103.33	45.83	657.89	78.20	128.00
CA	67.00	27.33	1794.50	88.97	137.07
TSA	150.33	39.33	557.96	62.33	144.53
OSA	1741.33	602.00	7329.40	1051.84	1841.53
OSA with HC	103.57	49.20	155.78	46.67	130.10
SWL	85.33	108.50	370.88	140.88	176.33
SWL with HC	47.03	43.43	102.40	31.90	55.57

Nevertheless, we can note the positive influence of HC, which generally leads to better configurations in a smaller training time. In particular, in the CH problem where the inputs are almost binary, the OSA and SWL achieve a better generalization ability than the other sequential constructive techniques considered.

Table X
Generalization Ability for Five Real-World Problems

Algorithm	Real-world problem GL	IR	CH	V0	V1
BPA	73.58%	95.44%	98.64%	94.55%	87.47%
IPA with LP	70.56%	91.67%	96.52%	92.71%	87.59%
IPA with TPLR	74.07%	92.53%	95.96%	95.66%	90.25%
CA	71.79%	91.67%	90.38%	92.76%	88.53%
TSA	74.88%	93.67%	96.81%	94.74%	87.78%
OSA	67.84%	74.13%	88.32%	91.63%	87.15%
OSA with HC	63.83%	81.07%	98.31%	93.33%	85.98%
SWL	50.43%	35.73%	93.05%	89.40%	82.64%
SWL with HC	55.30%	77.40%	98.03%	91.70%	82.71%

Table XI

Average CPU Time (s) for Solving Five Real-World Problems

	Real-world problem				
Algorithm	GL	IR	CH	V0	V1
BPA	142.64	92.81	798.87	24.23	141.18
IPA with LP	116.81	18.97	107232.43	2682.46	5734.09
IPA with TPLR	7.25	3.35	2116.79	16.49	32.39
CA	4.67	1.03	1178.17	9.93	15.98
TSA	7.14	0.53	2501.53	1.18	14.49
OSA	5.06	0.85	8945.01	3.32	5.35
OSA with HC	1.74	0.46	411.44	2.75	4.55
SWL	77.61	96.52	5146.10	713.52	995.61
SWL with HC	15.72	2.74	650.53	6.12	22.29

VIII. CONCLUSIONS

The class of sequential constructive methods has been analyzed in detail, from both the theoretical and the applicative point of view. The good convergence properties of these training algorithms have been validated by some basic theorems that show the common general procedure followed by them.

In particular, five different sequential constructive methods for the generation of the hidden layer have been considered and the specific implementative solutions adopted have been examined. Four of them, the irregular partitioning algorithm (IPA), the carve algorithm (CA), the target switch algorithm (TSA), and the oil spot algorithm (OSA), build neural networks containing threshold units. The fifth method, the sequential window learning (SWL) algorithm, employs window neurons for the construction of the hidden layer.

The aim of the experimental tests performed was to point out the properties of each sequential constructive technique; three main characteristics have been analyzed:

- the complexity of the resulting neural network (measured by the number of hidden units or by the number of nonnull weights),
- the generalization ability of this configuration, and
- the computing time needed for the learning.

Furthermore, the procedure of Hamming clustering (HC) has been introduced to improve the performances of training algorithms that build binary neural networks (OSA and SWL).

On the grounds of the experimental results and the comparison with the configurations generated by the back-propagation algorithm (BPA), we can draw the following conclusions:

• The IPA allows the achievement of neural networks having low complexity and generalization ability close to that obtained by the BPA. Unfortunately, the employment of linear programming (LP) to check the linear separability of a given training set leads to high execution times even for small values of the number of inputs. On the contrary, the version using the thermal perceptron learning rule (TPLR) requires a lower computational cost.

• The CA allows a good compromise between convergence speed and efficiency of the resulting configuration. However, the values achieved for the generalization ability are generally lower than those obtained by the BPA.

• The TSA and IPA with TPLR have similar performances. Nevertheless, the neural networks generated by the TSA are generally more complex and present a lower stability. Better results can be obtained by increasing the number n_t of iterations for the TPLR, but the computational cost can become too expensive.

• The OSA is the fastest sequential constructive method, but leads to configurations having great complexity and small generalization ability. The employment of HC allows a significant improvement of its performances and can even lower the total computing time.

• The SWL is able to build neural networks with few hidden units and high stability in a reasonable execution time. Unfortunately, the generalization ability of the resulting configurations is often inadequate. The application of HC allows us to overcome this drawback leading to interesting performances, particularly when the original problem to be solved has binary inputs.

Finally, the large diffusion of the benchmarks considered allows a direct comparison with other training algorithms so as to obtain an objective evaluation of the performances offered by sequential constructive methods.

REFERENCES

[1] P. J. Werbos. Beyond regression: new tools for prediction and analysis in the behavioral sciences. Ph.D. Thesis, Harvard University, 1974.

[2] Y. Le Cun. A learning procedure for asymmetric networks. In *Proceedings of Cognitiva*, Paris, pp. 599–604, 1985.

[3] D. B. Parker. Learning logic. Technical Report TR-47, MIT Center for Research in Computational Economics and Management Science, Cambridge, MA, 1985.

[4] D. E. Rumelhart, G. E. Hinton, and R. J. Williams. Learning representations by back-propagating errors. *Nature* 323:533–536, 1988.

[5] D. E. Rumelhart, G. E. Hinton, and R. J. Williams. Learning internal representations by error propagation. In *Parallel Distribute Processing* (D. E. Rumelhart and J. L. McClelland, Eds.), pp. 318–362, MIT Press, Cambridge, MA, 1986.

[6] J. Hertz, A. Krogh, and R. G. Palmer. *Introduction to the Theory of Neural Computation.* Addison-Wesley, Redwood City, CA, 1991.
[7] R. P. Lippmann. Review of neural networks for speech recognition. *Neural Comput.* 1:1–38, 1989.
[8] Y. Le Cun, B. Boser, J. S. Denker, D. Henderson, R. E. Howard, W. Hubbard, and L. D. Jackel. Backpropagation applied to handwritten zip code recognition. *Neural Comput.* 1:541–551, 1989.
[9] I. Guyon, V. Vapnik, B. Boser, L. Bottou, and S. A. Solla. Structural risk minimization for character recognition. In *Advances in Neural Information Processing Systems* (J. E. Moody, S. J. Hanson, and R. P. Lippmann, Eds.), Vol. 4, pp. 471–479. Morgan Kaufmann, San Mateo, CA, 1992.
[10] A. Lapedes and R. Farber. How neural nets work. In *Neural Information Processing Systems* (D. Z. Anderson, Ed.), pp. 442–456. American Institute of Physics, New York, 1987.
[11] B. S. Wittner and J. S. Denker. Strategies for teaching layered networks classification tasks. In *Neural Information Processing Systems* (D. Z. Anderson, Ed.), pp. 850–859. American Institute of Physics, New York, 1987.
[12] S. A. Solla, E. Levin, and M. Fleischer. Accelerated learning in layered neural networks. *Complex Systems* 2:625–639, 1988.
[13] D. Plaut, S. Nowlan, and G. Hinton. Experiments on learning by back propagation. Technical Report CMU-CS-86-126, Department of Computer Science, Carnegie Mellon University, Pittsburgh, 1986.
[14] T. P. Vogl, J. K. Mangis, A. K. Rigler, W. T. Zink, and D. L. Alkon. Accelerating the convergence of the back-propagation method. *Biol. Cybernet.* 59:257–263, 1988.
[15] A. H. Kramer and A. Sangiovanni-Vincentelli. Efficient parallel learning algorithms for neural networks. In *Advances in Neural Information Processing Systems* (D. S. Touretzky, Ed.), Vol. 1, pp. 40–48. Morgan Kaufmann, San Mateo, CA, 1989.
[16] S. Makram-Ebeid, J.-A. Sirat, and J.-R. Viala. A rationalized back-propagation learning algorithm. In *Proceedings of the International Joint Conference on Neural Networks*, Washington, DC, Vol. 2, pp. 373–380, 1989.
[17] D. G. Luenberger. *Introduction to Linear and Nonlinear Programming.* Addison-Wesley, Reading, MA, 1984.
[18] D. Nguyen and B. Widrow. Improving the learning speed of 2-layer neural networks by choosing initial values of adaptive weights. In *Proceedings of the International Joint Conference on Neural Networks*, Washington, DC, Vol. 3, pp. 21–26, 1989.
[19] G. P. Drago and S. Ridella. Statistically controlled activation weight initialization (SCAWI). *IEEE Trans. Neural Networks* 3:257–263, 1992.
[20] J. P. Cater. Successfully using peak learning rates of 10 (and greater) in back-propagation networks with the heuristic learning algorithm. In *Proceedings of the IEEE First International Conference on Neural Networks*, San Diego, Vol. 2, pp. 645–651, 1987.
[21] R. A. Jacobs. Increased rates of convergence through learning rate adaptation. *Neural Networks* 1:295–307, 1988.
[22] A. N. Tikhonov and V. Y. Arsenin. *Solutions of Ill-Posed Problems.* Winston, Washington, DC, 1977.
[23] T. Poggio, V. Torre, and C. Koch. Computational vision and regularization theory. *Nature* 317: 314–319, 1985.
[24] G. E. Hinton. Learning distributed representations of concepts. In *Proceedings of the Eighth Annual Conference of the Cognitive Science Society*, Amherst, MA, pp. 1–12, 1986.
[25] H. Drucker, R. Schapire, and P. Simard. Boosting performance in neural networks. *Internat. J. Pattern Recognition and Artificial Intelligence* 7:705–719, 1993.
[26] L. Bottou, C. Cortes, J. Denker, H. Drucker, I. Guyon, L. Jackel, Y. Le Cun, U. Miller, E. Säckinger, P. Simard, and V. Vapnik. Comparison of classifier methods: a case study in hand-

written digit recognition. In *Proceedings of the 12th IAPR International Conference on Pattern Recognition*, Los Alamos, CA, pp. 77–83, 1994.
[27] G. Cybenko. Approximation by superpositions of a sigmoidal function. *Math. Control Signals Systems* 2:303–314, 1989.
[28] K. Hornik, M. Stinchcombe, and H. White. Multilayer feedforward networks are universal approximators. *Neural Networks* 2:359–366, 1989.
[29] K. Funahashi. On the approximate realization of continuous mapping by neural networks. *Neural Networks* 2:183–192, 1989.
[30] C. M. Bishop. *Neural Networks for Pattern Recognition.* Oxford University Press, New York, 1995.
[31] V. N. Vapnik and A. Y. Chervonenkis. On the uniform convergence of relative frequencies of events to their probabilities. *Theory Probab. Appl.* 16:264–280, 1971.
[32] V. Vapnik. *Estimation of Dependencies Based on Empirical Data.* Springer-Verlag, New York, 1982.
[33] A. Blumer, A. Ehrenfeucht, D. Haussler, and M. K. Warmuth. Learnability and the Vapnik–Chervonenkis dimension. *J. Assoc. Comput. Mach.* 36:929–965, 1989.
[34] E. B. Baum and D. Haussler. What size net gives valid generalization? *Neural Comput.* 1:151–160, 1989.
[35] V. Vapnik and L. Bottou. Local algorithms for pattern recognition and dependencies estimation. *Neural Comput.* 5:893–909, 1993.
[36] V. Vapnik, E. Levin, and Y. Le Cun. Measuring the VC dimension of a learning machine. *Neural Comput.* 6:851–876, 1994.
[37] J. Denker, D. Schwartz, B. Wittner, S. Solla, R. Howard, L. Jackel, and J. Hopfield. Large automatic learning, rule extraction, and generalization. *Complex Systems* 1:877–922, 1987.
[38] D. Schwartz, V. K. Samalam, S. Solla, and J. S. Denker. Exhaustive learning. *Neural Comput.* 2:374–385, 1990.
[39] M. Stone. Cross-validatory choice and assessment of statistical predictions. *J. Roy. Statist. Soc. Ser. B* 36:111–147, 1974.
[40] G. Wahba and S. Wold. A completely automatic French curve: fitting spline functions by cross-validation. *Comm. Statist. Theory Methods* 4:1–17, 1975.
[41] J. S. Judd. Learning in networks is hard. In *Proceedings of the First International Conference on Neural Networks*, San Diego, Vol. 2, pp. 685–692, 1987.
[42] A. Blum and R. L. Rivest. Training a 3-node neural network is NP-complete. *Neural Networks* 5:117–127, 1992.
[43] L. G. Valiant. A theory of the learnable. *Comm. Assoc. Comput. Mach.* 27:1134–1142, 1984.
[44] J. Sietsma and R. J. F. Dow. Neural network pruning—why and how. In *Proceedings of the IEEE International Conference on Neural Networks*, San Diego, Vol. 1, pp. 325–333, 1988.
[45] Y. Le Cun, J. S. Denker, and S. A. Solla. Optimal brain damage. In *Advances in Neural Information Processing Systems* (D. S. Touretzky, Ed.), Vol. 2, pp. 598–605. Morgan Kaufmann, San Mateo, CA, 1990.
[46] B. Hassibi and D. G. Stork. Second order derivatives for network pruning: optimal brain surgeon. In *Advances in Neural Information Processing Systems* (S. J. Hanson, J. D. Cowan, and C. L. Giles, Eds.), Vol. 5, pp. 164–171. Morgan Kaufmann, San Mateo, CA, 1993.
[47] D. L. Gray and A. N. Michel. A training algorithm for binary feedforward neural networks. *IEEE Trans. Neural Networks* 3:176–194, 1992.
[48] J. V. Jaskolski. Construction of neural network classification expert systems using switching theory algorithms. In *Proceedings of the International Joint Conference on Neural Networks*, Baltimore, Vol. 1, pp. 1–6, 1992.
[49] P. Ruján and M. Marchand. Learning by minimizing resources in neural networks. *Complex Systems* 3:229–241, 1989.

[50] E. B. Baum. Neural net algorithms that learn in polynomial time from examples and queries. *IEEE Trans. Neural Networks* 2:5–19, 1991.

[51] S. I. Gallant. Perceptron-based learning algorithms. *IEEE Trans. Neural Networks* 1:179–191, 1990.

[52] M. Muselli. On convergence properties of pocket algorithm. *IEEE Trans. Neural Networks* 8:623–629, 1997.

[53] E. B. Baum. Review of neural network design and the complexity of learning, by S. Judd. *IEEE Trans. Neural Networks* 2:181–182, 1991.

[54] R. O. Duda and P. E. Hart. *Pattern Classification and Scene Analysis*. Wiley, 1973.

[55] R. L. Rivest. Learning decision lists. *Machine Learning* 2:229–246, 1987.

[56] M. Marchand and M. Golea. On learning simple neural concepts: from halfspace intersections to neural decision lists. *Network* 4:67–85, 1993.

[57] M. Marchand, M. Golea, and P. Ruján. A convergence theorem for sequential learning in two-layer perceptrons. *Europhys. Lett.* 11:487–492, 1990.

[58] M. Muselli. On sequential construction of binary neural networks. *IEEE Trans. Neural Networks* 6:678–690, 1995.

[59] C. Campbell and C. Perez Vicente. The target switch algorithm: a constructive learning procedure for feed-forward neural networks. *Neural Comput.* 7:1245–1264, 1995.

[60] S. I. Gallant. *Neural Networks Learning and Expert Systems*. MIT Press, Cambridge, MA, 1993.

[61] S. Young and T. Downs. CARVE — a constructive algorithm for real valued examples. In *Artificial Neural Networks—ICANN 94* (M. Marinaro and P. G. Morasso, Eds.), pp. 785–788. Springer-Verlag, Berlin, 1994.

[62] S. Young and T. Downs. Improvements and extensions to the constructive algorithm CARVE. In *Artificial Neural Networks—ICANN 96* (C. von der Malsburg, J. C. V. W. von Seelen, and B. Sendhoff, Eds.), pp. 513–518. Springer-Verlag, Berlin, 1996.

[63] D. S. Johnson and F. P. Preparata. The densest hemisphere problem. *Theoret. Comput. Sci.* 6:93–107, 1978.

[64] F. Rosenblatt. *Principles of Neurodynamics*. Spartan Press, Washington, DC, 1961.

[65] M. Minsky and S. Papert. *Perceptrons: An Introduction to Computational Geometry*. MIT Press, Cambridge, MA, 1969.

[66] N. Karmarkar. A new polynomial time algorithm for linear programming. *Combinatorica* 4:373–395, 1984.

[67] W. Krauth and M. Mezard. Learning algorithms with optimal stability in neural networks. *J. Phys. A* 20:745–752, 1987.

[68] J. K. Anlauf and M. Biehl. Properties of an adaptive perceptron algorithm. In *Parallel Processing in Neural Systems and Computers* (R. Eckmiller, G. Hartmann, and G. Hauske, Eds.), pp. 153–156. North-Holland, Amsterdam, 1990.

[69] H. Edelsbrunner. *Algorithms in Combinatorial Geometry*. Springer-Verlag, Berlin, 1987.

[70] R. A. Jarvis. On the identification of the convex hull of a finite set of points in the plane. *Inform. Process. Lett.* 2:18–21, 1973.

[71] R. Zollner, H. J. Schmitz, F. Wünsch, and U. Krey. Fast generating algorithm for a general three-layer perceptron. *Neural Networks* 5:771–777, 1992.

[72] F. M. Frattale Mascioli and G. Martinelli. A constructive algorithm for binary neural networks: the oil-spot algorithm. *IEEE Trans. Neural Networks* 6:794–797, 1995.

[73] M. R. Emamy-Khansary. On the cuts and cut number of the 4-cube," *J. Combin. Theory, Ser. A* 41:221–227, 1986.

[74] D. A. Marcus. *Number Fields*. Springer-Verlag, New York, 1977.

[75] R. S. Wenocur and R. M. Dudley. Some special Vapnik–Chervonenkis classes. *Discrete Math.* 33:313–318, 1981.

[76] M. Muselli. Hamming clustering: improving generalization in binary neural networks. In *Ar-*

tificial Neural Networks—ICANN 94 (M. Marinaro and P. G. Morasso, Eds.), pp. 1083–1086. Springer-Verlag, Berlin, 1994.

[77] L. Bottou and V. Vapnik. Local learning algorithms. *Neural Comput.* 4:888–900, 1992.

[78] H. W. Gschwind and E. J. McCluskey. *Design of Digital Computers*. Springer-Verlag, New York, 1975.

[79] M. Frean. A "thermal" perceptron learning rule. *Neural Comput.* 4:946–957, 1992.

[80] M. Mézard and J.-P. Nadal. Learning in feedforward layered networks: the tiling algorithm. *J. Phys. A* 22:2191–2203, 1989.

[81] M. Frean. The upstart algorithm: a method for constructing and training feedforward neural networks. *Neural Comput.* 2:198–209, 1990.

[82] C. J. Merz and P. M. Murphy. UCI repository of machine learning databases [http://www.ics.uci.edu/mlearn/mlrepository.html]. Department of Information and Computer Science, University of California, Irvine, CA, 1996.

[83] S. Thrun, J. Bala, E. Bloedorn, I. Bratko, B. Cestnik, K. De Jong, S. Džeroski, S. Fahlman, D. Fisher, R. Hamann, K. Kaufman, S. Keller, I. Kononenko, J. Kreuziger, R. Michalski, T. Mitchell, P. Pachowicz, Y. Reich, H. Vafaie, W. Van de Welde, W. Wenzel, J. Wnek, and J. Zhang. A performance comparison of different learning algorithms. Technical Report CMU-CS-91-197. Department of Computer Science, Carnegie Mellon University, Pittsburgh, 1991.

[84] R. A. Fischer. The use of multiple measurements in taxonomic problems. *Ann. Eugenics* 8:376–386, 1936.

[85] A. D. Shapiro. *Structured Induction and Expert Systems*. Addison-Wesley, Wokingham, UK, 1987.

Fast Backpropagation Training Using Optimal Learning Rate and Momentum

Xiao-Hu Yu
National Communications Research Laboratory
Department of Radio Engineering
Southeast University
Nanjing 210018, China

Li-Qun Xu
Intelligent Systems Research
Advanced Research and Technologies
BT Laboratories
Ipswich IP5 7RE, England

Yong Wang
National Communications Research Laboratory
Department of Radio Engineering
Southeast University
Nanjing 210018, China

I. INTRODUCTION

The *backpropagation algorithm* (BPA) has been playing a unique role in the application of neural networks to various domain problems. The standard BPA with a fixed learning rate and momentum usually suffers from extremely slow convergence. This is because the error (cost function) surface is far from a quadratic bowl in the weight space, exhibiting usually many plateaus and ravines arising from the nonlinearity of the sigmoid units used in a multilayer feedforward network [1]. Figure 1 shows two typical slices of such an error surface, which result from learning a six-parity check problem by varying two arbitrary weights of a converged neural network. The standard BPA, however, fails to account for these changes in error surface while pursuing its minimum. Therefore, the exploration of fast and robust learning algorithms is of particular interest.

To achieve faster convergence, the learning rate of an algorithm that defines the shift of the weight vector has to be dynamically varied in accordance with the

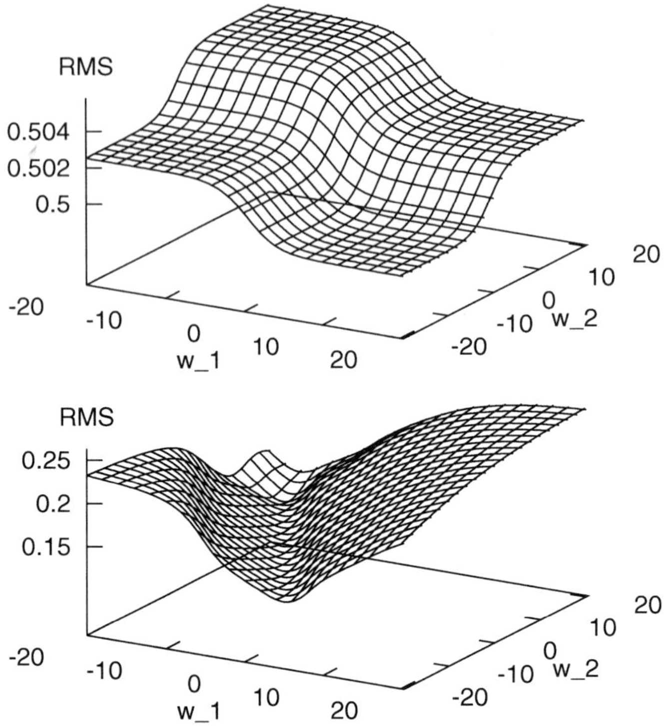

Figure 1 Illustration of typical error surfaces of a trained feedforward neural network for six-parity check problem.

region that the weight vector currently stands. In recent years, there has been a good deal of effort in developing efficient methods to speed up the convergence of the BPA by using a dynamic learning rate. The techniques can be largely divided into two categories: those based on weight-specific local variation information and those based on global knowledge of the state of the entire network.

The techniques of the first category include some *ad hoc* approaches [2, 3] and heuristic methods [4–8]. Jacobs [5], for example, presented a simple method for updating the learning parameters. He suggested dynamically increasing or decreasing the learning rate and momentum term by a fixed factor based on observations of the previous error signals. Variations of Jacobs' method were reported by Vogel [6] and compared in detail by Allred and Kelly [2]. Although these kinds of methods proved to work well for many cases, they may lead to overadjustment of the weights, resulting in dramatic divergence. In fact, the optimal learning rate varies almost randomly from iteration to iteration as discussed by Yu *et al.* [9]; it is not possible to obtain optimal learning parameters based on observations.

The second category of algorithms effectively exploits the second-order derivatives (Hessian matrix) information of the synaptic weights. They include Newton's method [10, 11], the Broyden–Fletcher–Goldgarb–Shanno (BFGS) and the Levenberg–Marquardt methods [12], and many others. These methods normally converge with a fewer number of iterations than that required by the standard BPA. The computational complexity at each iteration, however, increases quadratically with the number of weights, because of the Hessian or Jacobian matrix being very costly to compute and store. The extended Kalman type of algorithms such as those introduced by Singhal and Wu [13], Pushkorius and Feldkamp [14], and Kollias and Anastrassiou [15] also roughly belong to this category. A simplified version of these kinds of algorithms was recently reported by Monhandes *et al.* [16], where an estimate of the Hessian matrix norm was used to update the learning rate. Le Cun *et al.* [17] also proposed an interesting modification which can avoid directly computing the Hessian matrix. For a recent review of some of the preceding accelerated BPAs, readers are referred to the books of Haykin [18] and Bishop [19].

In this chapter, we explore a family of rather different approaches to accelerating backpropagation learning [9, 20, 21]. Instead of acquiring the Hessian matrix of the synaptic weights, we concern ourselves mainly with the derivative information with respect to the learning rate and momentum, which can be computed from an extended feedforward propagation procedure by way of a set of recursive formulas. Because the computational complexity of the set of formulas scales like that of the standard BPA, the estimation of the optimal dynamic learning rate can be achieved with a moderate overhead at each iteration. As a result, backpropagation learning can be accelerated considerably with a significant reduction in overall training time. In addition, because a near-optimal learning rate can be obtained at each iteration, the proposed technique is very robust to the initial setting of random weights in the sense that multiple trials of the same experiment for a task give consistently small standard deviation in performance, as opposed to the aforementioned *ad hoc* and heuristic methods.

This chapter is organized as follows. In Section II, the computational procedures are introduced, for recursively computing various derivatives of the cost function with respect to the learning parameters (learning rate and momentum), within the context of backpropagation learning of a multilayer feedforward neural network. Specifically, Section II.A is devoted to formulating a univariate cost function of the learning rate and computing its first four derivatives, whereas Section II.B considers a bivariate function of the learning rate and momentum, and computes their first two partial derivatives. In Section III, several methods are proposed for calculating the optimized dynamic learning rate for a BP algorithm.

First, an effective line search method employing only the first derivative is used to search for a valid optimal learning rate, with almost no extra computational and storage burden being introduced. Next, considering the fact that the univariate cost function of the learning rate can be well approximated by a parabola in the small neighborhood of its origin, a Newton-like method using the first two derivatives is presented. This method works very well unless the quadratic assumption becomes invalid. In this case, the higher-order (up to four) derivatives of the learning rate proved to be vital; hence, a higher-order derivatives method is proposed which is of more general application. Section IV discusses the strategy of simultaneously optimizing both the learning rate and the momentum. For this purpose, a second-order dual variable Taylor series expansion is used to approximate the cost function near the origin. The estimate of the optimal learning rate and momentum can then be pursued by simple algebraic manipulations. In Section V, we analyze the specification of the direction vector for the methods proposed. Possible choices include the negative gradient and the Gauss–Newton and Newton directions, though our attention is drawn to a modified conjugate gradient direction. In Section VI, we carry out analysis on the computational complexity of the preceding methods together with several well-studied classic methods for training feedforward neural networks, including the extended Kalman filter (EKF), the delta–bar–delta (DBD), and standard BP algorithm. Simulation results on three benchmark problems are presented. The advantages can be clearly seen of this family of algorithms over other methods in terms of the convergence rate, computational complexity, and robustness to network weights initialization.

II. COMPUTATION OF DERIVATIVES OF LEARNING PARAMETERS

We start by introducing backpropagation learning under the normal batch training mode. Given a set of P training pairs, $\{(X_1, D_1), (X_2, D_2), \ldots, (X_P, D_P)\}$, where X_s and D_s denote, respectively, an N_0-dimensional input pattern and an N_M-dimensional desired output pattern, backpropagation learning can be described as a process of minimizing the following cost function or the summed squared error in the weight space:

$$E(W) = \frac{1}{P \cdot N_M} \sum_{s=1}^{P} \|D_s - Y_s^M\|^2, \tag{1}$$

where W represents a vector consisting of all synaptic weights and biases in an M-layered network and Y_s^M is the actual output vector corresponding to the input X_s.

A. Derivatives of the Learning Rate

Let $W(k)$ be the current estimate of the optimal weight vector and $P(k)$ a descent direction at $W = W(k)$. The BPA is essentially a descent algorithm that adjusts the weight vector according to

$$W(k+1) = W(k) + \mu P(k), \tag{2}$$

where μ is a small constant, the so-called *learning rate*.

The optimal learning rate at $W = W(k)$, $\mu^*(k)$ say, should be such that

$$\mu^*(k) = \underset{\mu}{\operatorname{argmin}}\{h(\mu) \equiv E[W(k) + \mu P(k)] \mid \mu > 0\}. \tag{3}$$

Note that, given the current weight vector $W(k)$, when the direction vector $P(k)$ is specified, the cost function for the next iteration $E[W(k+1)]$ or $h(\mu)$ is a univariate function of μ. Once the derivatives of $h(\mu)$ w.r.t. μ are available, the estimation of $\mu^*(k)$ can be easily pursued.

In the following we give a set of formulas for efficient calculation of these derivatives. First, we consider the feedforward propagation in the mth layer (whose number of units is N_m) of an M-layered network. Given X_s, the sth input pattern of the network, the output of the unit i in the mth layer is given by

$$y^m_{s,i} = f\left([W^m_i(k) + \mu P^m_i(k)]^t Y^{m-1}_s\right), \tag{4}$$

where t denotes the transpose operation and $f(x)$ the unipolar sigmoid function $1/(1+e^{-x})$. $W^m_i(k)$ and $P^m_i(k)$ represent, respectively, the subset of weights and their corresponding directions associated with the connections to unit i of layer m; Y^{m-1}_s is the activation vector formed by all outputs from layer $m-1$ (including the bias):

$$Y^{m-1}_s = \begin{cases} [1 \quad y^{m-1}_{s,1} \cdots y^{m-1}_{s,N_m}]^t & \text{for } m > 1, \\ [1 \quad X^t_s\,]^t & \text{for } m = 1. \end{cases}$$

Next, we compute the first four derivatives of (4) w.r.t. the learning rate at its origin $\mu = 0$.

$$\frac{\partial y^m_{s,i}}{\partial \mu} = f'(a)b, \tag{5}$$

$$\frac{\partial^2 y^m_{s,i}}{\partial \mu^2} = f''(a)b^2 + f'(a)c, \tag{6}$$

$$\frac{\partial^3 y^m_{s,i}}{\partial \mu^3} = f^{(3)}(a)b^3 + 3f''(a)bc + f'(a)d, \tag{7}$$

$$\frac{\partial^4 y^m_{s,i}}{\partial \mu^4} = f^{(4)}(a)b^4 + 6f^{(3)}(a)b^2c + f''(a)[3c^2 + 4bd] + f'(a)e, \tag{8}$$

where the parameters a, b, c, d, e are specified, respectively, by

$$a = W_i^m(k)^t Y_s^{m-1}, \tag{9}$$

$$b = P_i^m(k)^t Y_s^{m-1} + W_i^m(k)^t \frac{\partial Y_s^{m-1}}{\partial \mu}, \tag{10}$$

$$c = 2P_i^m(k)^t \frac{\partial Y_s^{m-1}}{\partial \mu} + W_i^m(k)^t \frac{\partial^2 Y_s^{m-1}}{\partial \mu^2}, \tag{11}$$

$$d = 3P_i^m(k)^t \frac{\partial^2 Y_s^{m-1}}{\partial \mu^2} + W_i^m(k)^t \frac{\partial^3 Y_s^{m-1}}{\partial \mu^3}, \tag{12}$$

$$e = 4P_i^m(k)^t \frac{\partial^3 Y_s^{m-1}}{\partial \mu^3} + W_i^m(k)^t \frac{\partial^4 Y_s^{m-1}}{\partial \mu^4}, \tag{13}$$

and the first four derivatives of the unipolar sigmoid function $f(x)$ are

$$\begin{aligned} f'(x) &= f(x)[1 - f(x)], \\ f''(x) &= f'(x) - 2f(x)f'(x), \\ f^{(3)}(x) &= f''(x) - 2[f'(x)]^2 - 2f(x)f''(x), \\ f^{(4)}(x) &= f^{(3)}(x) - 6f''(x)f'(x) - 2f(x)f^{(3)}(x). \end{aligned}$$

Note that (5)–(8) can be recursively calculated for $m = 2, \ldots, M$, with the initial values ($m = 1$) taken as

$$\frac{\partial Y_s^0}{\partial \mu} = \frac{\partial^2 Y_s^0}{\partial \mu^2} = \frac{\partial^3 Y_s^0}{\partial \mu^3} = \frac{\partial^4 Y_s^0}{\partial \mu^4} = \mathbf{0}.$$

Finally, considering the cost function (1), the first four derivatives of $h(\mu)$ at $\mu = 0$ are immediately as follows:

$$h'(0) = F_c \sum_{s=1}^{P} E_s^t \frac{\partial Y_s^M}{\partial \mu}, \tag{14}$$

$$h''(0) = F_c \sum_{s=1}^{P} \left[E_s^t \frac{\partial^2 Y_s^M}{\partial \mu^2} + \left(\frac{\partial Y_s^M}{\partial \mu} \right)^t \frac{\partial Y_s^M}{\partial \mu} \right], \tag{15}$$

$$h^{(3)}(0) = F_c \sum_{s=1}^{P} \left[E_s^t \frac{\partial^3 Y_s^M}{\partial \mu^3} + 3 \left(\frac{\partial^2 Y_s^M}{\partial \mu^2} \right)^t \frac{\partial Y_s^M}{\partial \mu} \right], \tag{16}$$

$$\begin{aligned} h^{(4)}(0) = F_c \sum_{s=1}^{P} \bigg[& E_s^t \frac{\partial^4 Y_s^M}{\partial \mu^4} + 4 \left(\frac{\partial^3 Y_s^M}{\partial \mu^3} \right)^t \frac{\partial Y_s^M}{\partial \mu} \\ & + 3 \left(\frac{\partial^2 Y_s^M}{\partial \mu^2} \right)^t \frac{\partial^2 Y_s^M}{\partial \mu^2} \bigg], \end{aligned} \tag{17}$$

where we have defined the N_M-dimensional error vector $E_s = Y_s^M - D_s$ and the constant $F_c = 2/(P \cdot N_M)$.

It is particularly noted that the computational and storage demand required for calculating the higher-order derivatives of the learning rate has the same scale as that of the standard BPA, as opposed to the previous second-order methods [11, 12, 16], where the computational complexity increased quadratically with the scale of a network due to the need of acquiring the Hessian-like information of the synaptic weights.

From the expression of $h(\mu)$ in (3), one can also obtain $h'(\mu)$ and $h''(\mu)$ for arbitrary $\mu \geqslant 0$ as follows:

$$h'(\mu) = P(k)^t \nabla_w E\big[W(k) + \mu P(k)\big], \tag{18}$$

$$h''(\mu) = P(k)^t \nabla_w^2 E\big[W(k) + \mu P(k)\big] P(k), \tag{19}$$

where ∇_w and ∇_w^2 represent, respectively, the the gradient and the Hessian matrix of the cost function w.r.t. W. Note that (18) and (19) are especially useful when a second-order method, for example, Newton's method with $P(k) = -(\nabla_w E[W(k)])/(\nabla_w^2 E[W(k)])$, is used. In this case, the information on $\nabla_w|_{\mu=0}$ and $\nabla_w^2|_{\mu=0}$ is available, so the computation of $h'(0)$ and $h''(0)$, as is required by Methods 1–3 to be addressed in Section III, is straightforward.

B. Derivatives of the Learning Rate and Momentum

In order to improve the efficiency of the standard BPA, one simple technique is to add a *momentum* term to the gradient descent formula. This can effectively smooth out the fluctuations of weights adjustment in the course of learning. The modified formula of weights adjustment becomes

$$\Delta W(k) = \mu P(k) + \alpha \Delta W(k-1), \tag{20}$$

where $0 \leqslant \alpha < 1$.

We shall take a close look at the heuristic roles the momentum term plays. First, when $\Delta W(k-1)$ and $P(k)$ assume the same numerical sign, meaning that there may exist a large distance between the current weight vector position and the forthcoming minimum point, the weights adjustment is thus increased. So the inclusion of the momentum term here acts as an accelerating factor. Second, when $\Delta W(k-1)$ and $P(k)$ have opposite numerical sign, which means that an oscillation is currently taking place, the moving pace is decreased. The inclusion of the momentum term therefore stabilizes the motion of a learning process in the weight space.

We can write the cost function $E[W(k+1)]$ as a *bivariate* function of μ and α:

$$\begin{aligned} h(\mu,\alpha) &\equiv E\big[W(k)+\mu P(k)+\alpha\Delta W(k-1)\big] \\ &= \frac{1}{P\cdot N_M}\sum_{s=1}^{P}\|D_s - Y_s^M\|^2. \end{aligned} \tag{21}$$

To achieve the largest possible reduction in the cost function for the step k, the optimum learning rate and momentum term $(\mu^*(k), \alpha^*(k))$ should be such that

$$(\mu^*(k), \alpha^*(k)) = \operatorname*{argmin}_{\mu,\alpha}\big\{h(\mu,\alpha) \mid \mu > 0,\ \alpha \geqslant 0\big\}. \tag{22}$$

Note again that, given $W(k)$ and $\Delta W(k-1)$, when $P(k)$ is specified, $h(\mu,\alpha)$ is a bivariate function of μ and α. Once the partial derivatives of $h(\mu,\alpha)$ w.r.t. μ and α are available, the estimation of $\mu^*(k)$ and $\alpha^*(k)$ can be easily obtained.

Following the discussion in the previous subsection, we shall give a set of recursive formulas for calculating these derivatives. This time, given input X_s, the output associated with the unit i in the mth layer is given by

$$y_{s,i}^m = f\big(\big[W_i^m(k)+\mu P_i^m(k)+\alpha\Delta W_i^m(k-1)\big]^t Y_s^{m-1}\big), \tag{23}$$

where $\Delta W_i^m(k-1)$ represents the weights adjustment in the last step associated with the connections to unit i of layer m.

Next, we compute the first two partial derivatives of (23) with respect to (μ,α) at their origin (0, 0):

$$\frac{\partial y_{s,i}^m}{\partial\mu} = f'(\mathcal{A})\mathcal{B}, \tag{24}$$

$$\frac{\partial y_{s,i}^m}{\partial\alpha} = f'(\mathcal{A})\mathcal{C}, \tag{25}$$

$$\frac{\partial^2 y_{s,i}^m}{\partial\mu^2} = f''(\mathcal{A})\mathcal{B}^2 + f'(\mathcal{A})\mathcal{D}, \tag{26}$$

$$\frac{\partial^2 y_{s,i}^m}{\partial\alpha^2} = f''(\mathcal{A})\mathcal{C}^2 + f'(\mathcal{A})\mathcal{E}, \tag{27}$$

$$\frac{\partial^2 y_{s,i}^m}{\partial\mu\partial\alpha} = f'(\mathcal{A})\mathcal{B}\mathcal{C} + f'(\mathcal{A})\mathcal{R}, \tag{28}$$

where

$$\mathcal{A} = W_i^m(k)^t Y_s^{m-1},$$

$$\mathcal{B} = P_i^m(k)^t Y_s^{m-1} + W_i^m(k)^t \frac{\partial Y_s^{m-1}}{\partial\mu},$$

$$\mathcal{C} = \Delta W_i^m(k-1)^t Y_s^{m-1} + W_i^m(k)^t \frac{\partial Y_s^{m-1}}{\partial \alpha},$$

$$\mathcal{D} = 2P_i^m(k)^t \frac{\partial Y_s^{m-1}}{\partial \mu} + W_i^m(k)^t \frac{\partial^2 Y_s^{m-1}}{\partial \mu^2},$$

$$\mathcal{E} = 2\Delta W_i^m(k-1)^t \frac{\partial Y_s^{m-1}}{\partial \alpha} + W_i^m(k)^t \frac{\partial^2 Y_s^{m-1}}{\partial \alpha^2},$$

$$\mathcal{R} = P_i^m(k)^t \frac{\partial Y_s^{m-1}}{\partial \alpha} + \Delta W_i^m(k-1)^t \frac{\partial Y_s^{m-1}}{\partial \mu} + W_i^m(k)^t \frac{\partial^2 Y_s^{m-1}}{\partial \alpha \partial \mu}.$$

It is then readily shown from (21) that the first two partial derivatives of $h(\mu, \alpha)$ with respect to (μ, α) at $(0, 0)$ are as follows:

$$h'_{\mu}(0,0) = F_c \sum_{s=1}^{P} E_s^t \frac{\partial Y_s^M}{\partial \mu}, \tag{29}$$

$$h'_{\alpha}(0,0) = F_c \sum_{s=1}^{P} E_s^t \frac{\partial Y_s^M}{\partial \alpha}, \tag{30}$$

$$h''_{\mu^2}(0,0) = F_c \sum_{s=1}^{P} \left[E_s^t \frac{\partial^2 Y_s^M}{\partial \mu^2} + \left(\frac{\partial Y_s^M}{\partial \mu} \right)^t \frac{\partial Y_s^M}{\partial \mu} \right], \tag{31}$$

$$h''_{\alpha^2}(0,0) = F_c \sum_{s=1}^{P} \left[E_s^t \frac{\partial^2 Y_s^M}{\partial \alpha^2} + \left(\frac{\partial Y_s^M}{\partial \alpha} \right)^t \frac{\partial Y_s^M}{\partial \alpha} \right], \tag{32}$$

$$h''_{\mu\alpha}(0,0) = F_c \sum_{s=1}^{P} \left[E_s^t \frac{\partial^2 Y_s^M}{\partial \alpha \partial \mu} + \left(\frac{\partial Y_s^M}{\partial \mu} \right)^t \frac{\partial Y_s^M}{\partial \alpha} \right], \tag{33}$$

where, as before, we have defined the error vector $E_s = Y_s^M - D_s$, and the constant $F_c = 2/(P \cdot N_M)$.

Note that in the input layer $m = 1$ the following conditions:

$$\frac{\partial Y_s^0}{\partial \mu} = \frac{\partial Y_s^0}{\partial \alpha} = \frac{\partial^2 Y_s^0}{\partial \mu^2} = \frac{\partial^2 Y_s^0}{\partial \alpha^2} = \frac{\partial^2 Y_s^0}{\partial \mu \partial \alpha} = \mathbf{0} \tag{34}$$

hold, so that the desired derivatives on the right-hand sides of (29)–(33) can be recursively calculated for $m = 2, \ldots, M$.

III. OPTIMIZATION OF DYNAMIC LEARNING RATE

In this section, we consider the weights adjustment formula (2). We shall investigate several methods for computing the optimized dynamic learning rate $\mu^*(k)$ for each iteration based on the necessary derivative information of the cost function $h(\mu)$ at $\mu = 0$, the computation of which has been previously discussed in Section II.A.

A. Method 1: Learning Rate Search with an Acceptable δE

Using the first two derivatives of the learning rate, this method is based on a simple but effective line search algorithm due originally to Goldstein as can be found in Wolfe [22]. The method first tries to obtain a valid range for the learning rate based on information of $h'(0)$. Specifically, a learning rate μ is expected such that the descent value $\delta E = h(\mu) - h(0)$ is not beyond the area bounded between the two lines, both starting from the point $(0, h(0))$ but with $\tau_1 h'(0)$ and $\tau_2 h'(0)$ as their respective slopes:

$$\tau_2 h'(0) \leqslant \frac{h(\mu) - h(0)}{\mu} \leqslant \tau_1 h'(0)$$

or, equivalently,

$$h(0) + \tau_2 \mu h'(0) \leqslant h(\mu) \leqslant h(0) + \tau_1 \mu h'(0). \tag{35}$$

Considering $h'(0) < 0$, to make (35) meaningful, $0 < \tau_1 < \tau_2 < 1$ should be satisfied. As is clear, from the geometric point of view, the inequality on the left-hand side of (35) provides a lower bound that the descent δE needs to meet, whereas the one on the right-hand side tends to keep the valid μ away from zero.

As discussed in Yu *et al.* [9], in most cases $h(\mu)$ can be well approximated by a parabola in the small neighborhood of $\mu \geqslant 0$. So, a reasonable choice for τ_1 and τ_2 should lead to an optimal learning rate μ^* within the range of valid learning rates satisfying (35). In the case that $h(\mu)$ takes an exact quadratic form, that is,

$$h(\mu) = h(0) + h'(0)\mu + \tfrac{1}{2} h''(0)\mu^2, \tag{36}$$

where $h''(0) > 0$, the optimal learning rate μ^* can be explicitly expressed as

$$\mu^* = -\frac{h'(0)}{h''(0)} \tag{37}$$

and its corresponding $h(\mu^*)$ as

$$h(\mu^*) = h(0) - \frac{1}{2}\frac{[h'(0)]^2}{h''(0)}. \tag{38}$$

So, the slope of the line from $(0, h(0))$ to $(\mu^*, h(\mu^*))$ can be found as

$$\frac{h(\mu^*) - h(0)}{\mu^*} = \frac{1}{2}h'(0).$$

This suggests that when $0 < \tau_1 < \frac{1}{2} < \tau_2 < 1$ holds, (35) is satisfied, so the optimal learning rate thus obtained will be valid.

Based on the preceding analysis, the whole procedure in search of a valid optimal learning rate μ^* can be implemented in four steps:

1. Initialize the line search. Choose τ_1 and τ_2 such that $0 < \tau_1 < \frac{1}{2} < \tau_2 < 1$ is satisfied. Let $\mu_{\max}$ be a large enough upper bound of the learning rate. For $k = 0$, take an arbitrary $\mu_0 \in (0, \mu_{\max})$ as the initial value for the line search. Otherwise, for $k > 0$, set $\mu_0 = \tilde{\mu}^*(k-1)$, the estimated optimal learning rate in the $(k-1)$th iteration. For $j = 0$, set $a_j = 0$ and $b_j = \mu_{\max}$, then go to step 2.
2. Verify the lower bound for the descent δE. Compute $h(\mu_j) = E[W(k) + \mu_j P(k)]$ and compare $h(\mu_j)$ with $h(0) + \tau_1 \mu_j h(\mu_j)$. If the inequality on the left-hand side of (35) holds, perform step 3. Otherwise, let $a_{j+1} \leftarrow a_j$ and $b_{j+1} \leftarrow \mu_j$, go to step 4.
3. Verify the requirement for keeping μ_j away from zero. Make a comparison between $h(\mu_j)$ and $h(0) + \tau_2 \mu_j h'(0)$. If the right-hand-side inequality of (35) is met, then μ_j is a valid learning rate. The line search ends up with $\tilde{\mu}^*(k) = \mu_j$. If not, set $a_{j+1} \leftarrow \mu_j$ and $b_{j+1} \leftarrow b_j$, if $b_{j+1} \leqslant \mu_{\max}$ proceed to step 4, otherwise, let $\mu_{j+1} = 2\mu_j$ and $j \leftarrow j+1$. Return to step 2.
4. Try a new test point. Go back to step 2 with $\mu_{j+1} = (a_{j+1} + b_{j+1})/2$ and $j \leftarrow j+1$.

This procedure will terminate after a few iterations with a valid optimal learning rate obtained.

Another more effective criterion that can be used for making the valid μ less close to zero is to use the derivative information at $\mu = \mu_j$ such that

$$h'(\mu_j) \geqslant \tau_3 h'(0). \tag{39}$$

To guarantee that the optimal learning rate μ^* mentioned in (37) and (38) appears in the valid range, τ_1 and τ_3 should be confined, respectively, in the following intervals:

$$0 < \tau_1 < \tfrac{1}{2} \qquad \text{and} \qquad 0 < \tau_3 < 1. \tag{40}$$

B. Methods 2 and 3: Using a Newton-Like Method to Compute μ

These two methods also use the first two derivatives. As discussed in Section III.A, the optimal learning rate can be explicitly given by (37) if the cost function $h(\mu)$ can be characterized as a convex parabola. And, indeed, in most cases $h(\mu)$ approximately takes a convex quadratic form in the small neighborhood of $\mu \geqslant 0$. Therefore,

$$\tilde{\mu}^*(k) = -\frac{h'(0)}{h''(0)} \tag{41}$$

would be a suitable estimate of the optimal learning rate at iteration k.

As remarked in Section II.A, $h'(0)$ and $h''(0)$ can be computed from (14) and (15) by way of iterating (5) and (6), respectively. They can also be obtained from (18) and (19) if one decides to employ a second-order method [23]. Furthermore, if one wishes to estimate the optimal learning rate with much higher accuracy, (41) should be generalized into the standard Newton's method [20].

In the case that $h''(0) \leqslant 0$ [this happens when $h(\mu)$ decreases sharply at $\mu = 0$] when the Newton-like method fails to provide a valid learning rate, the line search method of Section III.A should be used. We shall distinguish two cases when (41) is used in connection with a specified direction vector $P(k)$ for weights update (to be discussed in Section V): (a) $P(k)$ is simply the negative gradient, referred to as Method 2, and (b) $P(k)$ is the defined conjugate gradient direction, referred to as Method 3.

C. Method 4: Using the Higher-Order Derivatives of μ

Once again, we look at the approximation of the cost function $h(\mu)$ by a Taylor series expansion in the small neighborhood of $\mu = 0$, but assume that the higher-order derivatives are also available at this point, or

$$h(\mu) = h(0)+h'(0)\mu+\frac{1}{2!}h''(0)\mu^2+\frac{1}{3!}h^{(3)}(0)\mu^3+\frac{1}{4!}h^{(4)}(0)\mu^4+\cdots. \tag{42}$$

To estimate the optimal μ that minimizes $h(\mu)$, we calculate the derivative of (42) w.r.t. μ and obtain

$$h'(\mu) = D + C\mu + B\mu^2 + A\mu^3 + \cdots, \tag{43}$$

where the coefficients are, respectively, $A = h^{(4)}(0)/6$, $B = h^{(3)}(0)/2$, $C = h''(0)$, and $D = h'(0)$.

In the following, we recognize four special cases for separate treatment in the light of the numerical values adopted by the coefficients C, B, and A. Note that, because $P(k)$ is a descent direction, there always exists $D = h'(0) < 0$:

Case 1. $B > 0$: In this case, (43) is truncated into a quadratic equation, or a parabola whose arms open upward

$$h'(\mu) \approx B\mu^2 + C\mu + D. \tag{44}$$

Because there exists $D < 0$, hence $C^2 - 4BD > C^2 > 0$ holds. The optimal solution for μ is given by

$$\mu^*(k) = \frac{-C + \sqrt{C^2 - 4BD}}{2B}. \tag{45}$$

Case 2. $B < 0$ and $C > 0$: The condition $C > 0$ implies that $h(\mu)$ itself in (42) can be suitably characterized by a quadratic function for a small μ by leaving out the higher-order terms $\mathcal{O}(\mu^3)$. This makes $h'(\mu)$ a linear function. So the optimal estimation of μ can be achieved from

$$\mu^*(k) = -D/C. \tag{46}$$

Case 3. $B < 0$, $C < 0$, and $A > 0$: In this case, (43) takes an approximate form

$$h'(\mu) \approx A\mu^3 + B\mu^2 + C\mu + D. \tag{47}$$

Noting the assumptions for A, B, and C plus $D < 0$, both conditions $3AC - B^2 < 0$ and $2B^3 - 9ABC + 27A^2D < 0$ are satisfied. According to Press *et al.* [24], the unique positive root of the cubic equation (47) is given by

$$\mu^*(k) = \begin{cases} 2\sqrt[3]{r}\cos\theta - \dfrac{B}{3A}, & \text{for } q^2 + p^3 < 0, \\ \sqrt[3]{u} + \sqrt[3]{v} - \dfrac{B}{3A}, & \text{for } q^2 + p^3 \geqslant 0, \end{cases} \tag{48}$$

where

$$p = \frac{3AC - B^2}{9A^2}, \qquad q = \frac{2B^3 - 9ABC + 27A^2D}{54A^3},$$

$$r = \sqrt{-p^3}, \qquad \theta = \frac{1}{3}\arccos\left(-\frac{q}{r}\right),$$

and

$$u = -q + \sqrt{q^2 + p^3}, \qquad v = -q - \sqrt{q^2 + p^3}.$$

Case 4. $B < 0$, $C < 0$, and $A < 0$: When this case happens, we cannot find a meaningful value from $h'(\mu) = 0$ at all, because all the first four derivatives assume a negative value at $\mu = 0$. Nevertheless, we can try alternative ways to estimate the optimal $\mu^*(k)$ through direct search for the positive root from $h(\mu)$ itself [or the absolute minimum of the cost function $E(W)$]. From this perspective, we rewrite the truncated Taylor series expansion of $h(\mu)$ as

$$h_t(\mu) \doteq h(0) + D\mu + C\mu^2 + B\mu^3 + A\mu^4. \tag{49}$$

Note that, because all four coefficients A, B, C, and D take a negative value, this leads to $h_t'(\mu) < 0$ for $\mu > 0$. Therefore, $h_t(\mu)$ is monotonically decreased for $\mu > 0$. Because $h(0) > 0$, it follows that (49) has a unique positive root, which can be explicitly obtained by solving the following two quadratic equations:

$$\left(y + \frac{\alpha y - \gamma}{\sqrt{8y + \alpha^2 - 4\beta}}\right)x^2 + \left(\alpha + \sqrt{8y + \alpha^2 - 4\beta}\right)\frac{x}{2} + 1 = 0, \tag{50}$$

$$\left(y - \frac{\alpha y - \gamma}{\sqrt{8y + \alpha^2 - 4\beta}}\right)x^2 + \left(\alpha - \sqrt{8y + \alpha^2 - 4\beta}\right)\frac{x}{2} + 1 = 0, \tag{51}$$

where y is a real-valued root of the equation:

$$8y^3 + 4\beta y^2 + (2\alpha\gamma - 8\delta)y + \delta(4\beta - \alpha^2) - \gamma^2 = 0, \tag{52}$$

with the definitions of $\alpha = D/h(0)$, $\beta = C/h(0)$, $\gamma = B/h(0)$, and $\delta = A/h(0)$.

Because $D < 0$, an alternative approach to the solution of (49) is simply to iterate the following expression:

$$\mu_{i+1} = -\frac{1}{D}\left[h(0) + C\mu_i^2 + B\mu_i^3 + A\mu_i^4\right] \tag{53}$$

for $i = 0, 1, \ldots,$ with the initial value $\mu_0 = 0$. It is readily shown that (53) will converge to the unique positive root of (49) because of the monotonicity of $h_t(\mu)$ for $\mu > 0$.

IV. SIMULTANEOUS OPTIMIZATION OF μ AND α

In this section, we consider the modified weights adjustment formula of (20). We propose a method to simultaneously determine, if possible, both the learning rate and the momentum term based on the first two partial derivatives of the cost function $h(\mu, \alpha)$.

A. METHOD 5: USING THE FIRST TWO PARTIAL DERIVATIVES

Consider the bivariate formulation (21) of the cost function $h(\mu, \alpha)$. The Taylor series expansion of $h(\mu, \alpha)$ in the small neighborhood of $\mu = 0,\ \alpha = 0$ can be written as

$$\begin{aligned} h(\mu, \alpha) &= h(0,0) + h'_{\mu}(0,0)\mu + h'_{\alpha}(0,0)\alpha \\ &\quad + \tfrac{1}{2}h''_{\mu^2}(0,0)\mu^2 + \tfrac{1}{2}h''_{\alpha^2}(0,0)\alpha^2 + h''_{\mu\alpha}(0,0)\mu\alpha + \cdots. \end{aligned} \tag{54}$$

Note that, as discussed in Section II.B, the initial conditions (34) hold. So, $h'_{\alpha}(0,0)$, $h'_{\mu}(0,0)$, $h''_{\mu^2}(0,0)$, $h''_{\alpha^2}(0,0)$, $h''_{\mu\alpha}(0,0)$ can be recursively computed via (23)–(33).

As we know, the minimum of the cost function $h(\mu, \alpha)$ can be achieved at (μ^*, α^*) if the following two differential equations are satisfied:

$$h'_{\mu}(\mu^*, \alpha^*) = 0, \tag{55}$$

$$h'_{\alpha}(\mu^*, \alpha^*) = 0. \tag{56}$$

Differentiate (54) with respect to μ and α, respectively, leaving out the higher-order terms, and then substitute the results into (55) and (56). We have

$$\begin{bmatrix} h''_{\mu^2}(0,0) & h''_{\mu\alpha}(0,0) \\ h''_{\mu\alpha}(0,0) & h''_{\alpha^2}(0,0) \end{bmatrix} \begin{bmatrix} \mu \\ \alpha \end{bmatrix} = \begin{bmatrix} -h'_{\mu}(0,0) \\ -h'_{\alpha}(0,0) \end{bmatrix}. \tag{57}$$

Let qd be the determinant of the previous coefficient matrix, or

$$qd = h''_{\mu^2}(0,0)h''_{\alpha^2}(0,0) - \left[h''_{\mu\alpha}(0,0)\right]^2.$$

In the following we recognize three different cases for computing valid learning parameters:

Case 1. $h''_{\mu^2}(0,0) > 0$ and $qd > 0$: In this case, the approximation of (54) is valid, so the best learning parameters (μ^*, α^*) can be calculated from (57) as

$$\begin{aligned} \mu^* &= \left[h'_{\alpha}(0,0)h''_{\mu\alpha}(0,0) - h'_{\mu}(0,0)h''_{\alpha^2}(0,0)\right]/qd, \\ \alpha^* &= \left[h'_{\mu}(0,0)h''_{\mu\alpha}(0,0) - h'_{\alpha}(0,0)h''_{\mu^2}(0,0)\right]/qd. \end{aligned}$$

Case 2. $h''_{\mu^2}(0,0) > 0$ and $qd < 0$: The best learning rate μ^* can be obtained by a Newton-like method (Method 2 or 3), and the momentum term α^* is reset to 0.

Case 3. $h''_{\mu^2}(0, 0) \leqslant 0$: In this case, α^* is clamped to 0, and the best learning rate μ^* is estimated by the learning rate search with an acceptable descent value method (Method 1).

V. SELECTION OF THE DESCENT DIRECTION

So far the specification of the descent direction $P(k)$ has not been mentioned. There can be many choices including the negative gradient [17, 25], conjugate gradient direction [9, 20, 26, 27], Newton direction [11, 12], Gauss–Newton direction [20, 28], among many others. The four choices are listed as follows:

Negative gradient:	$P(k) = -\nabla_w(k)$,
Conjugate gradient:	$P(k) = -\nabla_w(k) + \lambda_k P(k-1)$,
Newton:	$P(k) = -H^{-1}(k)\nabla_w(k)$,
Gauss–Newton:	$P(k) = -T^{-1}(k)\nabla_w(k)$,

where $\nabla_w(k)$ represents the gradient of the cost function $E(W)$ w.r.t. the weight vector W at iteration k; $H(k)$ and $T(k)$ are the Hessian matrix and the Jacobian correlation matrix, respectively; λ_k is an orthogonalizing factor.

Note that although the Newton and Gauss–Newton directions are very effective in that the number of iterations necessary for a learning process can be sharply reduced, they are somehow limited to small-scale applications because of the quadratically increased computational and storage demand with the size of a network with respect to each iteration.

In the standard BPA, the $P(k)$ assumes the steepest descent direction. When the search for optimal weights comes across long and narrow regions of the cost function, the BPA becomes inefficient, exhibiting many zig-zag movements and with only slight progress being made at each iteration [19]. In order to avoid this undesired phenomenon without apparently increasing the computational complexity, we employ the type of conjugate gradient direction stated previously.

Note that the orthogonalizing factor λ_k, usually given as the ratio of the squared gradient vector length between two consecutive iterations, or

$$\lambda_k = \frac{\nabla_w(k)^t \nabla_w(k)}{\nabla_w(k-1)^t \nabla_w(k-1)}, \tag{58}$$

is appropriate only when the cost function can be treated as a quadratic form in the neighborhood of $W = W(k)$. For the purpose of accommodating the nonlinear optimization task of backpropagation learning, the orthogonalizing factor is periodically restarted and/or cleared as

$$\lambda_k = \begin{cases} 0, & k = rN_r \text{ or } \dfrac{|\nabla_w(k)^t \nabla_w(k-1)|}{\|\nabla_w(k)\| \|\nabla_w(k-1)\|} \leqslant 0.2, \\ \dfrac{\nabla_w(k)^t[\nabla_w(k) - \nabla_w(k-1)]}{\|\nabla_w(k-1)\|^2}, & \text{otherwise,} \end{cases} \tag{59}$$

where r is a positive integer; N_r stands for the restarting period which empirically takes the value of $Q/2$ (Q is the dimension of the weight vector W). As can be easily seen, when the optimal dynamic learning rate is obtained, the use of a conjugate gradient direction for weight vector update is straightforward.

It was observed by the authors [9] that the preceding conjugate gradient method employing an optimal learning rate at each iteration is equivalent to a BPA with both an optimal learning rate and an optimal momentum factor. As compared to the conjugate gradient methods reported in [26, 27], the present approach is more effective because a near-optimal learning rate has been utilized, leading to near-orthogonal search directions in the weight space.

VI. SIMULATION RESULTS

In this section, we shall compare our optimized dynamic learning parameters determination methods, Methods 1–5 detailed in previous sections, with several classic backpropagation learning algorithms, including the standard BP algorithm, the delta–bar–delta, and the extended Kalman filtering algorithm. The latter views the training of a neural network as a nonlinear system identification problem.

We start with a brief analysis of the computational complexity associated with each method. As usual, we use P to denote the size of a training set, M the number of layers of a neural network, N_m the number of units including bias in the mth layer, $U = \sum_{m=1}^{M}(N_m - 1)$ the total number of noninput units of the network, and finally $Q = \sum_{m=1}^{M}(N_m - 1)N_{m-1}$ the total number of synaptic weights. The computational and storage demand for each algorithm as regards each training example is listed as follows. For clarity, we use the symbols $+$, $\times$, $\mathcal{F}$, and $\mathcal{S}$ to represent, respectively, the number of additions, multiplications, unipolar sigmoid function evaluations, and storage demand.

The standard BP algorithm (BP):

$+$: $ADD_{BP} = N_M - 1 + \sum_{m=1}^{M-1}(N_m - 1)(N_{m+1} - 1) + Q + Q/P$,
$\times$: $MUL_{BP} = N_M - 1 + \sum_{m=1}^{M-1}(N_m - 1)(N_{m+1} - 1) + Q + Q/P$,
$\mathcal{F}$: $SIG_{BP} = U$,
$\mathcal{S}$: $STR_{BP} = 2Q + 2\sum_{m=0}^{M} N_m$.

Delta–bar–delta:
$+$: $ADD_{BP} + 4Q/P$,
$\times$: $MUL_{BP} + 2Q/P$,
$\mathcal{F}$: SIG_{BP},
$\mathcal{S}$: $STR_{BP} + 2Q$.

EKF:
$+$: $4Q^2N_M + Q(2N_M^2 - 3N_M) + N_M^3 + N_M$,
$\times$: $4Q^2N_M + Q(2N_M^2 + N_M) + N_M^3 + 2N_M^2$,
$\mathcal{F}$: SIG_{BP},
$\mathcal{S}$: $STR_{BP} + Q^2 + 2QN_M$.

Line search (Method 1):
$+$: $\propto ADD_{BP}$,
$\times$: $\propto MUL_{BP}$,
$\mathcal{F}$: $\propto SIG_{BP}$,
$\mathcal{S}$: $STR_{BP} + Q$.

Newton-like method with negative gradient direction (Method 2):
$+$: $ADD_{BP} + 5Q + 5U + 3N_M$,
$\times$: $MUL_{BP} + 5Q + 8U + 3N_M$,
$\mathcal{F}$: SIG_{BP},
$\mathcal{S}$: $STR_{BP} + 2\sum_{m=0}^{M} N_m$.

Newton-like method with conjugate gradient direction (Method 3):
$+$: $ADD_{BP} + 5Q + 5U + 3N_M + 3Q/P$,
$\times$: $MUL_{BP} + 5Q + 8U + 3N_M + 3Q/P$,
$\mathcal{F}$: SIG_{BP},
$\mathcal{S}$: $STR_{BP} + 2\sum_{m=0}^{M} N_m + Q$.

Higher-order derivatives method (Method 4):
$+$: $ADD_{BP} + 9Q + 17U + 8N_M$,
$\times$: $MUL_{BP} + 9Q + 39U + 8N_M$,
$\mathcal{F}$: SIG_{BP},
$\mathcal{S}$: $STR_{BP} + 4\sum_{m=0}^{M} N_m$.

First two partial derivatives of (μ, α) (Method 5):
$+$: $ADD_{BP} + 12Q + 16U + 8N_M$,
$\times$: $MUL_{BP} + 12Q + 24U + 8N_M$,
$\mathcal{F}$: SIG_{BP},
$\mathcal{S}$: $STR_{BP} + 5\sum_{m=0}^{M} N_m$.

There are several observations that can be made from the preceding list. First, the computational demand of the EKF algorithm is quite large, on the order of $\mathcal{O}(Q^2N_M)$, which is against about $\mathcal{O}(Q)$ of the standard BP algorithm, and the storage requirement for the two algorithms is $\mathcal{O}(Q^2)$ versus $\mathcal{O}(Q)$. The main reason is that the EKF algorithm needs substantial $Q \times Q$ and $Q \times N_M$ matrix operations.

Second, the computational complexity of the delta–bar–delta (DBD) method is almost the same as that of the standard BP algorithm, so is our Method 1, the line search method. The DBD, however, does not provide us with a systematic way to adapt the learning parameters. It makes use of local information and heuristics to adjust the learning parameters, though sometimes obtaining a good guess about the possible step length for the next move cannot be justified to achieve the optimal effects. It is sensitive to the setting of initial parameters, too. If the parameters are not set properly, the convergence rate may even be slower than the standard BPA.

Third, unlike other methods, the family of methods we proposed attempts to make full exploration of the error surface to dynamically decide the optimal learning rate and/or momentum term for every iteration step. The increase in computational complexity for the methods is only proportional to that of the standard BP algorithm with respect to each iteration, but a considerably faster convergence rate can be achieved as will be shown in the following examples.

In order to evaluate the performance of these algorithms in backpropagation learning, we have conducted simulations on three typical benchmark problems containing two classification and one approximation tasks. The basis for comparison between different techniques was the learning curves (convergence performance) showing the decrease of the cost function versus CPU running time. For a fair treatment, all the classic methods are carefully tuned with the learning parameters, where appropriate, chosen via a trial-and-error method so as to obtain the fastest possible convergence rate. All the simulations written in C were run on a SPARC-10 Workstation. Unless otherwise stated, the initial values of the synaptic weights were drawn from uniformly distributed random values between -0.1 and 0.1. In all trials, the training process terminated when either of the following three conditions was met: (a) the total number of iterations (epochs) of an algorithm exceeded 10^4; (b) the CPU running time was longer than a preset limit; (c) the cost function reached a prespecified small value $T_d = 1.0 \times 10^{-4}$. The performance of an algorithm is measured by averaging the results of 10 independent trials. The root mean square (rms) error is used in the learning performance plots shown later on. Finally, we adopt the notation $N_i - N_{h_1} - N_{h_2} - \cdots - N_o$ to represent a fully connected feedforward neural network with N_i input units, N_o output units and, in between, (at least one) hidden layers with, respectively, $N_{h_1}, N_{h_2}, \ldots,$ hidden units. All the hidden and output units have a unipolar sigmoid transfer function.

EXAMPLE 1. This is a function approximation problem. A three-layered network having a 1–5–8–1 architecture of 67 synaptic weights was used to perform

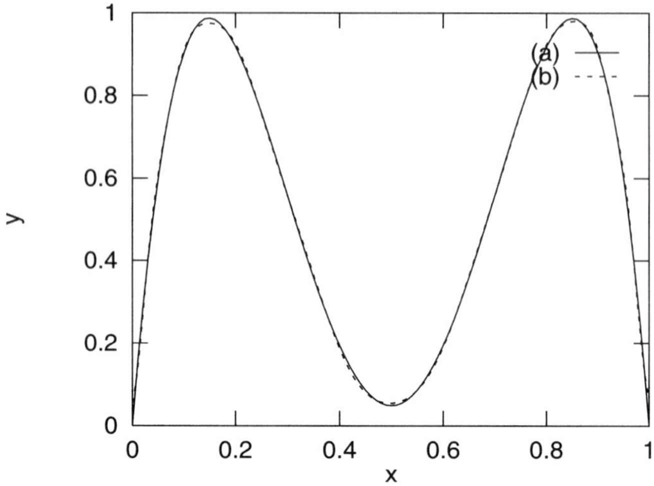

Figure 2 Example 1: (a) original recursive function; (b) function approximation from a 1–5–8–1 neural network trained by Method 4.

the task. The function is given by

$$y = f(f(x)), \quad \text{with } f(x) = 3.95x(1-x), \ 0 \leqslant x \leqslant 1$$

and is plotted in Fig. 2 (solid line). The training set consisted of 100 examples with x being drawn randomly from the interval [0, 1]. Similarly, a second data set of 1000 samples was obtained for testing the convergence performance of each training algorithm. Figures 3 and 4 give the plots of (averaged) test set convergence performance for each method. It can be easily seen that the Newton-like method (with or without conjugate gradient direction), the higher-order derivatives method (Method 4), and the EKF show apparent advantages over the rest in terms of both the fast convergence rate and the finally achieved small root mean square error. The EKF in this case seems to have a very rapid initial decreasing rate. All our five dynamical learning parameters adaptation methods outperform the delta–bar–delta and standard BP algorithm considerably. Figure 2 (dashed line) shows a neural network approximation result of the function adopting the higher-order derivatives method.

EXAMPLE 2. Our second example is to classify two-dimensional meshed and disconnected regions depicted in Fig. 5a. The whole square area is divided into four categories with interlocking subregions. This problem was considered by Singhal and Wu [13], Pushkorius and Feldkamp [14], and Yu *et al.* [9] as

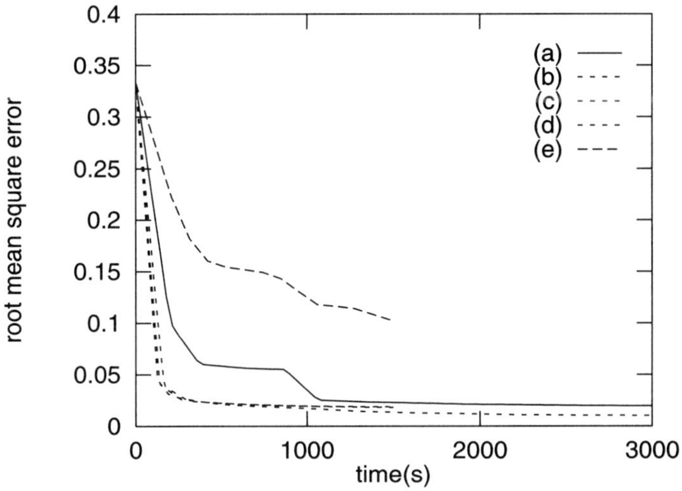

Figure 3 Example 1: averaged test set convergence performance for each of the algorithms proposed. (a) Method 1; (b) Method 2 (Newton-like method with negative gradient direction); (c) Method 3 (Newton-like method with conjugate gradient direction); (d) Method 4; (e) Method 5.

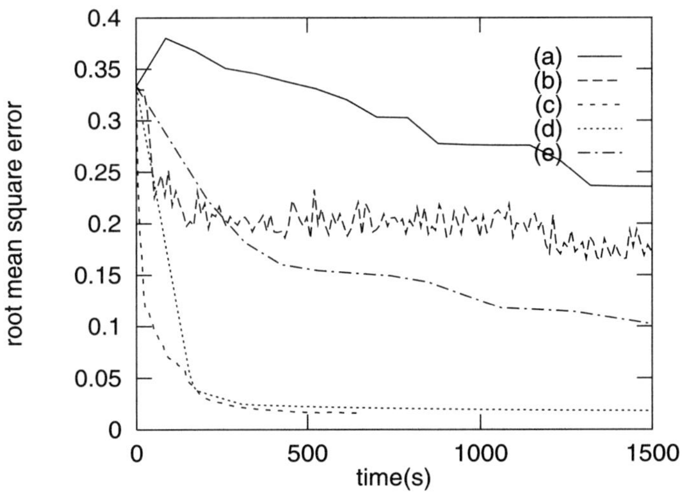

Figure 4 Example 1: comparison of averaged test set convergence performance between the standard BPA, delta–bar–delta, EKF, and the current Methods 4 and 5. (a) Standard BPA with $\mu = 0.45$ and $\alpha = 0$; (b) delta–bar–delta algorithm with $\mu = 0.5, \alpha = 0.4, \kappa = 0.01, \beta = 0.2$, and $\xi = 0.6$; (c) EKF algorithm with $R(k) = I e^{-k/50}$ and $P(0) = 1000I$; (d) Method 4; (e) Method 5.

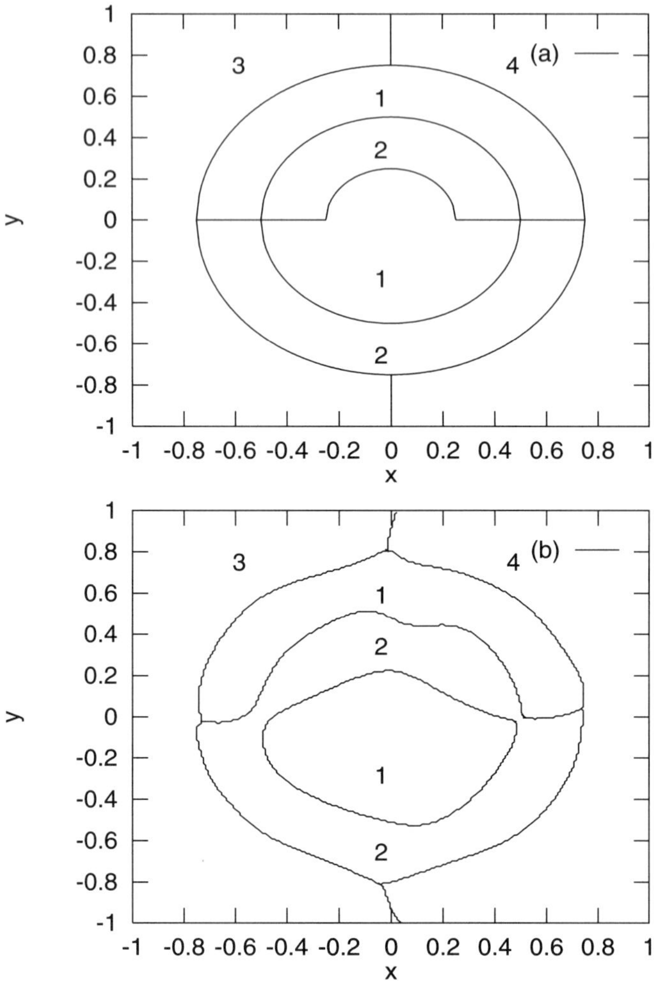

Figure 5 Example 2: (a) ideal decision regions; (b) learned decision regions formed by a 2–10–10–4 neural network trained by Method 5.

a benchmark for testing the convergence rate of backpropagation learning. The network used for this task has a 2–10–10–4 architecture of 184 synaptic weights. A set of 1000 training examples drawn uniformly over the square area was used to train the network. The plots of learning performance for each method are summarized in Figs. 6 and 7. We can observe that Method 5 attempting to simultaneously optimize both the learning rate and the momentum achieves the best convergence

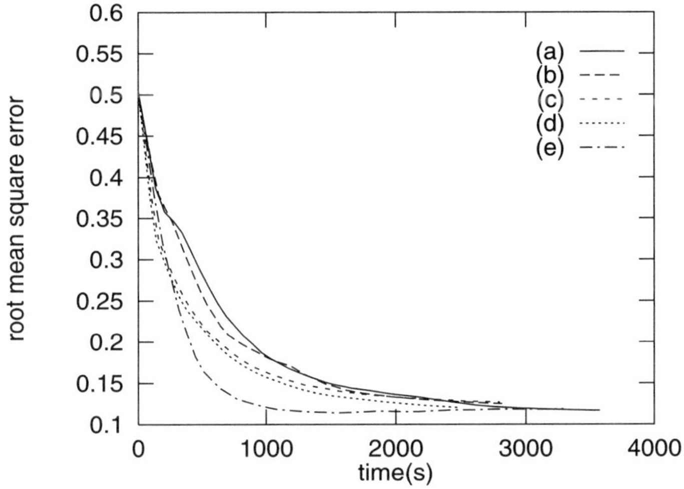

Figure 6 Example 2: averaged learning performance for the algorithms proposed. (a) Method 1; (b) Method 2 (Newton-like method with negative gradient direction); (c) Method 3 (Newton-like method with conjugate gradient direction); (d) Method 4; (e) Method 5.

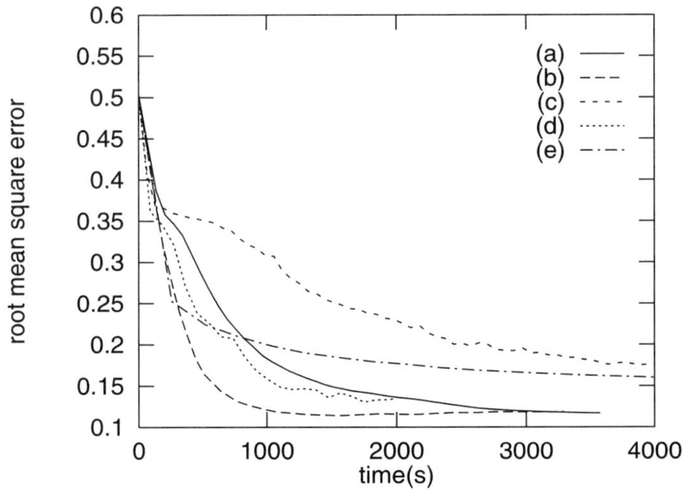

Figure 7 Example 2: comparison of the averaged learning performance between the standard BPA, delta–bar–delta, EKF, and the current Methods 1 and 5. (a) Method 1; (b) Method 5; (c) standard BPA with $\mu = 0.025$ and $\alpha = 0$; (d) delta–bar–delta algorithm with $\mu = 0.01$, $\alpha = 0.7$, $\kappa = 0.001$, $\beta = 0.2$, and $\xi = 0.6$; (e) EKF algorithm with $R(k) = I\mathrm{e}^{-k/50}$ and $P(0) = 1000I$.

performance. The delta–bar–delta method in this case is comparable to other dynamic learning parameters determination methods in terms of convergence rate, though it is quite sensitive to the random weights initialization. The decision region formed by a trained neural network adopting Method 5 is given in Fig. 5b.

EXAMPLE 3. The final task is a six-parity check problem which was investigated in Yu *et al.* [9]. It was found that for this task the performance of a neural network is very sensitive to the initial weights values. For this problem, the input patterns are six-dimensional vectors with each element equal to either 1 or 0. The target output takes the value of 1 if the input pattern is made up of odd 1's; otherwise, the target output takes 0. A neural network having a 6–8–5–1 architecture of 107 synaptic weights was used in the simulations. The training set consisted of all possible 64 training pairs. The learning rate and momentum of the standard BPA were set to 2.0 and 0.8, respectively, for obtaining better learning performance.

It was observed through numerous trials that, when the initial random weights were drawn from the interval $[-d, d]$ with d being less than 1.0, all the methods became more or less unstable. However, our five methods were able to converge in most cases (only one diverged among 30 trials), whereas the delta–bar–delta and EKF methods could hardly succeed in converging. As the interval was enlarged to $[-2, 2]$, or $d = 2$, all methods worked normally except the EKF, with Method 3 achieving the fastest convergence rate. Figures 8 and 9 summarize the averaged performance (among 10 trials) for each method. For the case where the interval was reduced to $[-1.25, 1.25]$, or $d = 1.25$, our five methods converged, whereas the delta–bar–delta and EKF methods diverged for most trials. The simulation results are provided in Figs. 10 and 11. The results clearly indicate that the set of algorithms we proposed is much more robust to the initial weights values.

In addition to the problem of being sensitive to initial weights values, we found that the delta–bar–delta method is also sensitive to the learning parameters μ, α, κ, β, and ξ. This phenomenon can be observed in Fig. 12, where μ, α, κ, β, and ξ take the values of 0.1, 0.6, 0.005, 0.8, and 0.6, respectively, and the delta–bar–delta method converges very rapidly. However, if we vary α from 0.6 to 0.5, β from 0.8 to 0.7, and keep the other three parameters unchanged, the learning process becomes dramatically different from the previous case. This implies that the learning parameters of the delta–bar–delta algorithm are highly problem-dependent and should be very carefully chosen through a trial-and-error method.

VII. CONCLUSION

In this chapter, a family of novel methods is presented for effectively estimating the optimal dynamic learning parameters (learning rate and momentum) to speed up backpropagation learning of neural networks. The exclusive derivatives information necessary for computing the learning parameters is gathered from

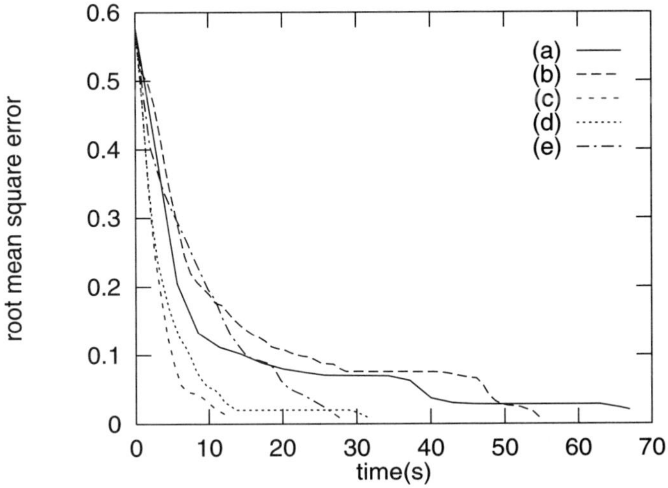

Figure 8 Example 3: averaged learning performance for the algorithms proposed when the initial weights values of the network are drawn randomly from the range [−2, 2]. (a) Method 1; (b) Method 2 (Newton-like method with negative gradient direction); (c) Method 3 (Newton-like method with conjugate gradient direction); (d) Method 4; (e) Method 5.

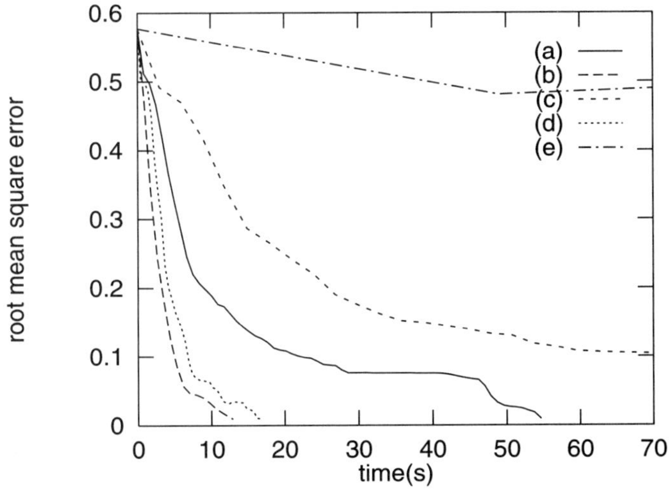

Figure 9 Example 3: comparison of the averaged learning performance between the standard BPA, delta–bar–delta, EKF, and the current Methods 2 and 3 when the initial weights values of the network are drawn randomly from [−2, 2]. (a) Method 2; (b) Method 3; (c) standard BPA with $\mu = 0.12$ and $\alpha = 0.0$; (d) delta–bar–delta algorithm with $\mu = 0.1, \alpha = 0.5, \kappa = 0.005, \beta = 0.7$ and $\xi = 0.6$; (e) EKF algorithm with $R(k) = I\mathrm{e}^{-k/50}$ and $P(0) = 1000I$.

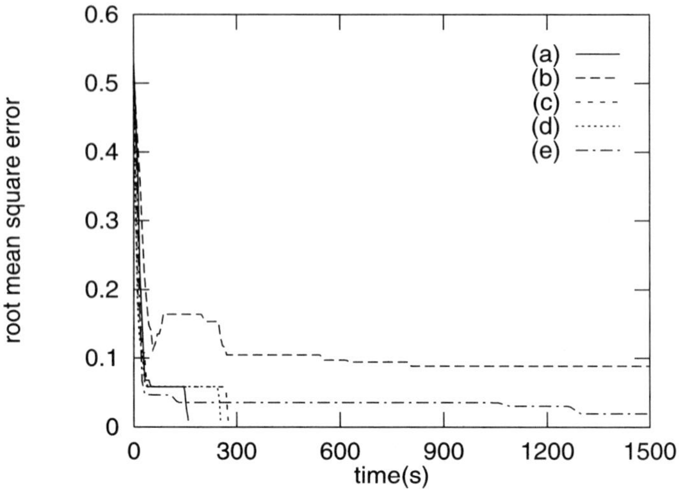

Figure 10 Example 3: averaged learning performance for the algorithms proposed when the initial weights values of the network are drawn randomly from the range [−1.25, 1.25]; (a) Method 1; (b) Method 2 (Newton-like method with negative gradient direction); (c) Method 3 (Newton-like method with conjugate gradient direction); (d) Method 4; (e) Method 5.

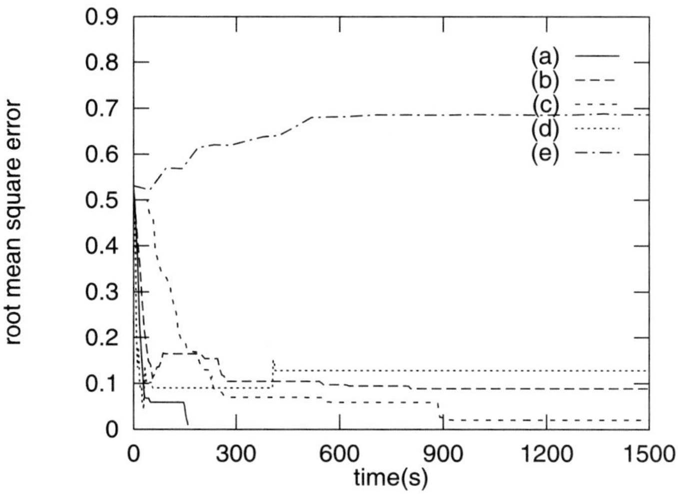

Figure 11 Example 3: comparison of the averaged learning performance between the standard BPA, delta–bar–delta, EKF, and the current Methods 1 and 2 when the initial weights values of the network are drawn randomly from [−1.25, 1.25]; (a) Method 1; (b) Method 2; (c) standard BPA with $\mu = 0.12$ and $\alpha = 0$; (d) delta–bar–delta algorithm with $\mu = 0.1$, $\alpha = 0.5$, $\kappa = 0.005$, $\beta = 0.7$ and $\xi = 0.6$; (e) EKF algorithm with $R(k) = Ie^{-k/50}$ and $P(0) = 1000I$.

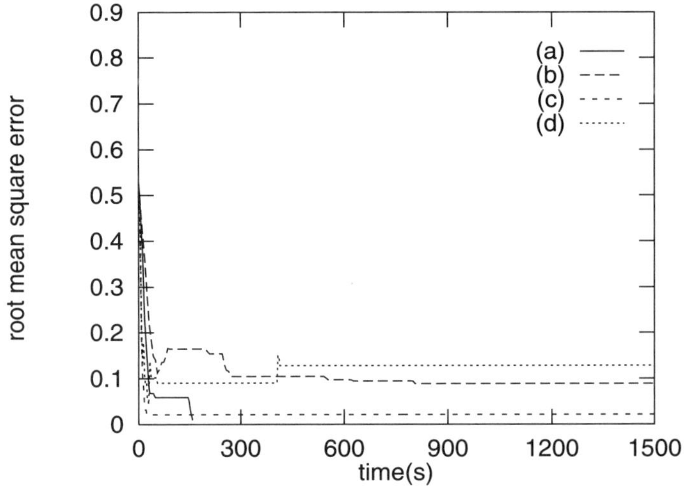

Figure 12 Example 3: illustration of the learning performance for the delta–bar–delta algorithm with different learning parameters. The current Methods 1 and 2 are also included for comparison; (a) Method 1; (b) Method 2; (c) delta–bar–delta algorithm with $\mu = 0.1, \alpha = 0.6, \kappa = 0.005, \beta = 0.8$ and $\xi = 0.6$; (d) delta–bar–delta algorithm with $\mu = 0.1, \alpha = 0.5, \kappa = 0.005, \beta = 0.7$ and $\xi = 0.6$.

some extended feedforward propagation procedures, with the computational and storage overhead for each iteration being limited to the same order as that of the standard BPA. This is in contrast to the previously reported second-order methods exploiting Hessian-like information of the weight vector, which require at least an order of magnitude increase in computational and storage amount. Extensive computer simulations have demonstrated the effectiveness of the set of methods. In general, they can provide rapid convergence and achieve considerable gain in overall running time savings as compared to several classic training methods also used in the experiments including the delta–bar–delta, the extended Kalman filter, and the standard BPA. In addition, the strong dependence of these classic methods on the network weights initialization is largely removed due to the near-optimal learning parameters obtained at each iteration. Finally, the family of methods are well suited to large-scale application problems.

ACKNOWLEDGMENTS

XHY gratefully acknowledges partial support from the Transcentury Talent Foundation of the State Education Commission of China and from the Climbing Programme—National Key Project for Basic Research in China under Grant NSC 92097. LQX would like to thank BT Laboratories for its support and interest in this project.

REFERENCES

[1] D. R. Hush and B. G. Horne. Progress in supervised neural networks. *IEEE Signal Process. Mag.* 10:8–39, 1993.

[2] L. G. Allred and G. E. Kelly. Supervised learning techniques for backpropagation networks. In *Proceedings of the International Joint Conference on Neural Networks,* San Diego, Vol. 1, pp. 702–709, 1990.

[3] Z. Luo. On the convergence of the LMS algorithm with adaptive learning rate for linear feedforward networks. *Neural Comput.* 3:226–245, 1991.

[4] S. E. Fahlman. An empirical study of learning speed in back-propagation networks. Technical Report CMU-CS-88-162, School of Computer Science, Carnegie-Mellon University, 1988.

[5] R. A. Jacobs. Increased rates of convergence through learning rate adaptation. *Neural Networks* 1:295–308, 1988.

[6] T. P. Vogel. Accelerating the convergence of the backpropagation method. *Biol. Cybernet.* 59:257–263, 1988.

[7] F. M. Silva and L. B. Almeida. Speeding up backpropagation. In *Advances of Neural Computers* (R. Eckmiller, Ed.), pp. 151–158. North-Holland, Amsterdam, 1990.

[8] M. Riedmiller. Advanced supervised learning in multi-layer perceptrons—from backpropagation to adaptive learning algorithms. *Internat. J. Comput. Standards Interfaces* 5, 1994.

[9] X.-H. Yu, G.-A. Chen, and S.-X. Cheng. Dynamic learning rate optimization of the backpropagation algorithm. *IEEE Trans. Neural Networks* 6:669–677, 1995.

[10] S. Becker and Y. Le Cun. Improving the convergence of back-propagation learning with second order methods. In *Proceedings of the 1988 Connectionist Models Summer School* (D. Touretzky, G. E. Hinton, and T. J. Sejnowski, Eds.), pp. 29–37. Morgan Kaufmann, San Mateo, CA, 1989.

[11] R. Battiti. First- and second-order methods for learning: between steepest descent and Newton's method. *Neural Comput.* 4:141–166, 1992.

[12] A. R. Webb, D. Lowe, and M. D. Bedworth. A comparison of nonlinear optimization strategies for feed-forward adaptive layered networks. Memorandum 4157, Royal Signals and Radar Establishment, 1988.

[13] S. Singhal and L. Wu. Training feedforward networks with the extended Kalman algorithm. *Proceedings of the International Conference on Acoustics, Speech and Signal Processing,* Scotland, pp. 1187–1190, 1989.

[14] G. V. Pushkorius and L. A. Feldkamp. Decoupled extended Kalman training of feedforward layered networks. *Proceedings of the International Joint Conference on Neural Networks,* Seattle, Vol. 1, pp. 771–777, 1991.

[15] S. Kollias and D. Anastrassiou. An adaptive least square algorithm for the efficient training of artificial neural networks. *IEEE Trans. Circuits Systems* 36:1092–1101, 1989.

[16] M. Monhandes, C. W. Codrington, and S. B. Gelfand. Two adaptive stepsize rules for gradient descent and their application to the training of feedforward artificial neural networks. In *Proceedings of the IEEE International Conference on Neural Networks,* Orlando, Vol. 1, pp. 555–560, 1994.

[17] Y. Le Cun, P. Y. Simard, and B. Pearlmutter. Automatic learning rate maximization by on-line estimation of the Hessian's eigenvectors. In *Advances in Neural Information Processing Systems* (S. J. Hanson, J. D. Cowan, and C. L. Giles, Eds.), Vol. 5, pp. 156–163. Morgan Kaufmann, San Mateo, CA, 1993.

[18] S. Haykin. *Neural Networks—A Comprehensive Foundation*, pp. 138–219. Macmillan, New York, 1994.

[19] C. M. Bishop. *Neural Networks for Pattern Recognition.* Oxford University Press, 1995.

[20] X.-H. Yu and S.-X. Cheng. Training algorithms for backpropagation neural networks with optimal descent factor. *Electron. Lett.* 26:1698–1700, 1990.

[21] X.-H. Yu and L.-Q. Xu. Optimization of dynamic learning rate by its higher-order derivatives in backpropagation learning. Unpublished.

[22] M. A. Wolfe. *Numerical Methods for Unconstrained Optimization*. Van Nostrand Reinhold, New York, 1978.

[23] R. L. Watrous. Learning algorithms for connectionist networks: applied gradient methods of non-linear optimization. In *Proceedings of the First International Conference on Neural Networks*, Vol. 2, pp. 619–628, 1987.

[24] W. H. Press, S. A. Teukolsky, W. T. Vetterling, and B. P. Flannery. *Numerical Recipes in C—The Art of Scientific Computing*, 2nd ed. Cambridge University Press, 1992.

[25] D. E. Rumelhart and J. L. McClelland. *Parallel Distributed Processing: Explorations in the Microstructure of Cognition*, Vol. 1. MIT Press, Cambridge, MA, 1986.

[26] A. H. Kramer and A. Sangiovanni-Vicentelli. Efficient parallel learning algorithms for neural networks. In *Advances in Neural Information Processing Systems* (D. S. Touretzky, Ed.), Vol. 1, pp. 40–48. Morgan Kaufmann, San Mateo, CA, 1989.

[27] M. S. Møller. A scaled conjugate gradient algorithm for fast supervised learning. *Neural Networks* 6:525–534, 1993.

[28] M. T. Hagan and M. B. Menhaj. Training feedforward networks with the Marquardt algorithm. *IEEE Trans. Neural Networks* 5:989–993, 1994.

[29] M. Feigenbaum. Quantitative universality for a class of nonlinear transformation. *J. Statist. Phys.* 19:25–52, 1978.

[30] B. Widrow and M. A. Lehr. *Adaptive Signal Processing*. Prentice-Hall, Englewood Cliffs, NJ, 1985.

Learning of Nonstationary Processes

V. Ruiz de Angulo
Institut de Ròbotica i Informàtica
Industrial (CSIC-UPC)
Edifici NEXUS
08034-Barcelona, Spain

Carme Torras
Institut de Ròbotica i Informàtica
Industrial (CSIC-UPC)
Edifici NEXUS
08034-Barcelona, Spain

I. INTRODUCTION

The degradation in performance of an associative network over a training set when new patterns are trained isolatedly is usually called forgetting or catastrophic interference. Applications entailing the learning of a time-varying function require the ability to quickly modify some input–output patterns while at the same time avoiding catastrophic forgetting. Learning algorithms based on the repeated presentation of the learning set—back propagation being the most popular representative—are only suited to tasks admitting two separate phases, an off-line one for learning, and another one for operation.

If a very different and representative input–output pattern needs to be learned after the training of the main set of patterns has been completed, one gets into trouble. One can train the new pattern isolatedly or retrain a mixture of the new and old patterns. The first option is the best for this kind of application because a quicker adaptation is obtained, but the interference must not be very strong, or at least one must be able to mitigate it. Unfortunately, the forgetting problem turns out to be a very serious drawback of back-propagation networks.

Although some interest has recently emerged on this issue, the connectionist community has not yet paid enough attention to it. Ratcliff [1] and McCloskey and Cohen [2], after many systematic studies, simply arrive at the conclusion that this problem cannot be satisfactorily solved. French [3] claims that the cause of forgetting is the overlap between the representations of the different patterns and, therefore, he modifies back propagation so as to produce semidistributed representations. In these representations, only a few hidden units take the value 1, whereas the majority take the value 0. This type of approach has very serious convergence problems and requires a much larger number of hidden units than straight back propagation [4]. Reducing the distributedness of the representations also has the very undesirable effect of losing some of the most interesting neural network properties, such as generalization (as French himself points out) and damage resistance.

Hetherington and Seidenberg [5] and especially Robins [6] have studied another aspect of the problem, namely the influence of the training regimes. The latter proposes mixing the training of the new pattern with pseudo-patterns, that is, points of the function implemented by the network. The problem is that the number of pseudo-patterns needed to mix with the new one is much higher than the number of old true patterns required to get the same effect.

Brousse and Smolensky [7] hold that there is no forgetting problem in what they call a "combinatorial environment," because in such an environment there are many virtual memories (error-free novel patterns) that do not interfere with old patterns. However, their results can be accounted for by the drastic restriction of the possible input–output patterns that can appear in a combinatorial environment (as Hetherington [8] recognizes) and by the use of autocoder networks. Both facts help generalization, as it actually happens also in some of Hetherington's own experiments regarding the influence of increasing the training set size. Sharkey and Sharkey (1994) also used autocoder networks in their studies of catastrophic interference. We think that results about forgetting obtained with autocoder networks and low-error patterns must not be extrapolated to more general situations.

Kruschke [9] has pointed out that the huge receptive field of the weighted-sum units is responsible for interference in neural networks. Units with a limited receptive field are increasingly being used [10–12]. Locally tuned units that use radial basis functions (RBFs) are the common choice. This can be a valid solution, but an important drawback of RBF units is that they need many more examples than weighted-sum units to generalize well, especially in high-dimensional input spaces. The problem comes from the very local representations formed by this type of unit, which can be avoided (e.g., by allowing large radii). However, it is precisely this locality that allows the prevention of forgetting. The more the receptive fields grow and overlap, the more the interference problem comes back to the scene. There exists a tradeoff between resistance to interference—local representations—and generalization—distributed representations.

Our approach does not require information to be stored in special types of representations. In fact, we even try to take advantage of the distributed ones. We simply investigate what can be reasonably done to introduce a new pattern into a previously trained network, while increasing minimally the error in the recall of the previously trained items. An algorithm, which we call LMD for "learning with minimal degradation," has been developed to accomplish this task efficiently in a general feedforward net. The previous work most closely related to ours is that of Park *et al.* [13]. They state the problem in a very similar way to that in Section III, but their resolution method is considerably different, it being based on the gradient reduced method for nonlinear constrained optimization. Our approach, based on the transformation of the problem into a nonconstrained optimization, has evident advantages. For a more detailed comparison, see [14].

II. *A PRIORI* LIMITATIONS

It is worth noting that success for any procedure with the same objective as LMD must be necessarily limited. For the different settings in which such a procedure can be applied, some existing *a priori* limitations are spelled out in the following discussion.

When a feedforward network with a fixed number of units has enough capacity to encode a fixed set of patterns, there is a bound on how fast learning can take place, because this problem has been proven to be NP-complete [15, 16]. Therefore, we cannot aim at finding a procedure that, in approximately constant time, learns a new pattern while leaving the old ones intact, because by iterating this procedure learning could be carried out in time linear in the number of patterns.

Now suppose that the chosen architecture is unable to encode all the patterns perfectly. Let E be the error function over the patterns $1, \ldots, n-1$, E' be the error function over the patterns $1, \ldots, n$, and E_p be the individual error in pattern p. Applying an ideal procedure to learn pattern n when the network is at the minimum of E, the sole unlikely possibility of arriving at the minimum of E' is that $E_n = 0$ at this minimum. Note that, in general, the value of E in the minimum of E' will be almost surely higher than the minimum of E. Therefore, if the final aim is to arrive at the minimum of E', introducing perfectly the nth pattern can be worse than doing nothing. As shown in Fig. 1, it is possible that E' grows when one forces the surface implemented by the net to pass over a point. The network whose results are displayed in the figure has one input and one bias unit connected to one output unit. The axes in the diagram stand for the input–output coordinates, each point thus representing a pattern. The best approximation the network can do of the old patterns is the continuous line. Learning the new pattern while minimizing the error over the old patterns results in weights giving the dashed-line interpolation. Note that this figure presents the worst possible case: a huge number of points that can be interpolated only with a surface having low frequency–high amplitude oscillations, a network with few parameters that is completely unable to fit the surface, and a new pattern far away from the mean of the old patterns.

The moral of this discussion is that not only is it impossible to devise a "perfect" algorithm for the nondisruptive encoding of a new pattern, but that even

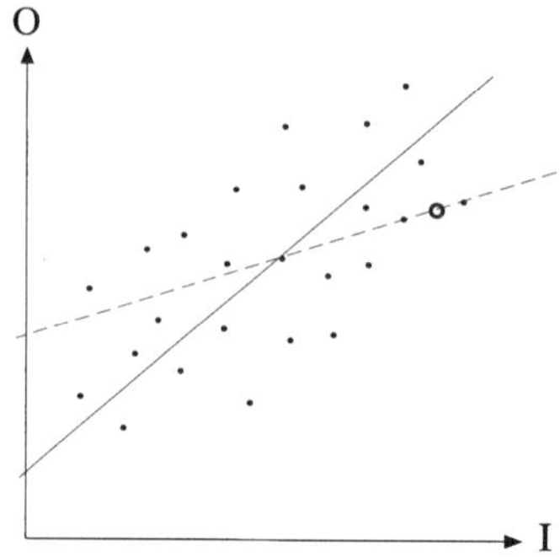

Figure 1 The points represent the learned patterns, which are approximated by the network with the continuous line. Constraining the network to perfectly encode the new association (open circle), while minimizing the error over the old patterns, gives a global approximation (dashed line) worse than the previous one.

if we had such an algorithm, to apply it indiscriminately in an incremental way could be inappropriate in many situations.

The most natural setting for the application of LMD is that in which the set of patterns to be learned is not fixed but time-varying, in the sense that there is a moving window for the error function so that when new patterns arrive, some of the oldest ones are no longer taken into account. In this case, the previous arguments cannot be applied. When introducing a new pattern, the emphasis must be put in quick adaptation and some forgetting of the old patterns is desirable. Typical applications of this kind are time series prediction and some control problems.

III. FORMALIZATION OF THE PROBLEM

The problem addressed in this chapter can be formulated as the minimization of the error over the $n-1$ previously trained patterns, constrained by perfect encoding of the new pattern. It is convenient to consider the current weights (before the application of LMD) as constants and then write each error E_i as a function of weight increments:

$$\begin{aligned} \text{Min} \sum_{i=1}^{n-1} &= E_i(\Delta W), \\ \text{subject to } E_n(\Delta W) &= 0. \end{aligned} \tag{1}$$

Evaluating the E_i's accurately for a given ΔW would entail presenting the whole set of patterns to the network. If the optimum is to be found through some search process, this evaluation has to be performed repeatedly, leading to a high computational cost.

An alternative solution to accurate evaluation is to approximate the error function over the old patterns through a truncated Taylor series expansion, for example, with a second-order polynomial in the weights. A linear model is too rude and not feasible because, as it will be pointed out at the end of Section V, it may turn the problem into an unsolvable one. The coefficients of the polynomial have a direct interpretation in terms of the first and second derivatives of E. Cross-terms are not included because, besides adding great complexity to the algorithm, the calculation of the Hessian requires computations very costly in memory and time.

Thus, the most faithful problem formulation we can reasonably aspire to deal with is

$$\begin{aligned} \text{Min } F(\Delta W) &= \sum_i c_i \Delta W_i^2 + \sum_i b_i \Delta W_i, \\ \text{subject to } E_n(\Delta W) &= 0, \end{aligned} \tag{2}$$

where F is the cost function that estimates the error increment in E (loss function) when the b_i and c_i constants are the first and second weight derivatives of E, respectively.

A usual way to tackle a constrained optimization problem is to linearly combine the cost function with the deviation of the constraint from zero, and then minimize this new error function:

$$\text{Min } \mu F(\Delta W) + \beta E_n(\Delta W).$$

In the minimization of this function, there is a tradeoff between E_n and F that depends on μ and β, and the error in the new pattern will not be 0 unless the β/μ relation tends to infinity with time. In practice, this is impossible and it is approximated through an appropriate schedule for changing μ and β.

The algorithm we have developed avoids this approximation by converting the constrained minimization problem into an unconstrained one, thus tackling the problem in a more direct and efficient manner. A different derivation of the algorithm can be found in [14].

IV. TRANSFORMATION INTO AN UNCONSTRAINED MINIMIZATION PROBLEM

The crucial finding underlying the transformation described in this section, and finally leading to the LMD algorithm, is the existence of a one-to-one mapping (except in very special cases) between a hidden-unit configuration and the best solution in a big subset of the weight increment space, $\Delta \mathbf{W}$. It is therefore possible to disregard all solutions but the best one from each subset, thus drastically reducing the search space. Furthermore, due to the one-to-one relation mentioned previously, the optimum solution can be looked for indirectly by searching through the set of hidden-unit configurations.

To explain this in more detail, we need to introduce some notation. An individual weight will be called w_{ji} (from unit i to unit j). Inc(j) is the index set of the units from which unit j receives direct input. Out(j) is the index set of the units to which unit j sends direct output. The connection graph is supposed to be loop free. Let $x_j = \sum_{i \in \mathrm{Inc}(j)} w_{ji} y_i$ be the total input received by unit j, and $y_j = f_j(x_j)$ be the activation of the same unit, where f_j is the activation function of unit j, which, for the moment, we assume is invertible. **I** and **O** stand for the spaces of input and output vectors, whereas I_d and O_d denote the input vector and the output vector of the new pattern.

Remember that our actual variables are weight increments and thus we consider the current weights as constants. All possible configurations of hidden-unit activations y_j define a vector space **H**, in which each component is limited by the range of its corresponding activation function. Then we define f_1: $\Delta\mathbf{W} \to \mathbf{H}$, the function that produces the vector of hidden-unit activations originated by a certain ΔW, when I_d is presented as input to the network. We also define f_2: $(\Delta\mathbf{W}, \mathbf{H}) \to \mathbf{O}$ as the function that returns the network output vector originated by a set of increments and hidden-unit activations, under I_d as input to the network. Notice that this function is only defined for $(\Delta W, H)$ such that $f_1(\Delta W) = H$.

Finally, we define the one-to-one mapping we have referred to previously:

$$\begin{aligned} D\colon\ & \mathbf{H} \to \Delta\mathbf{W} \\ & H \to \Delta W = D(H) \qquad \text{such that} \quad F\big(D(H)\big) = \min F(\Delta W), \\ & \qquad\qquad\qquad\qquad\qquad\qquad\quad f_1(\Delta W) = H, \\ & \qquad\qquad\qquad\qquad\qquad\qquad\quad f_2(\Delta W, H) = O_d. \end{aligned} \tag{3}$$

$D(H)$ is thus the best solution among those that produce the hidden activation vector H when the new pattern (I_d, O_d) is perfectly encoded. Figure 2 illustrates this mapping graphically, as well as the displacement of search from the original weight-increment space $\Delta\mathbf{W}$ to the space of hidden-unit configurations **H**.

Let $F(H)$ denote the H-dependent function $F(D(H))$. Provided we derive an expression for D and prove that D is a well-defined one-to-one mapping, which we do in the next section, we have transformed the original problem into the following one:

$$\text{Min}\, F(H)$$

because, once the solution to this problem H^* is obtained, the solution to the original problem is just $\Delta W^* = D(H^*)$.

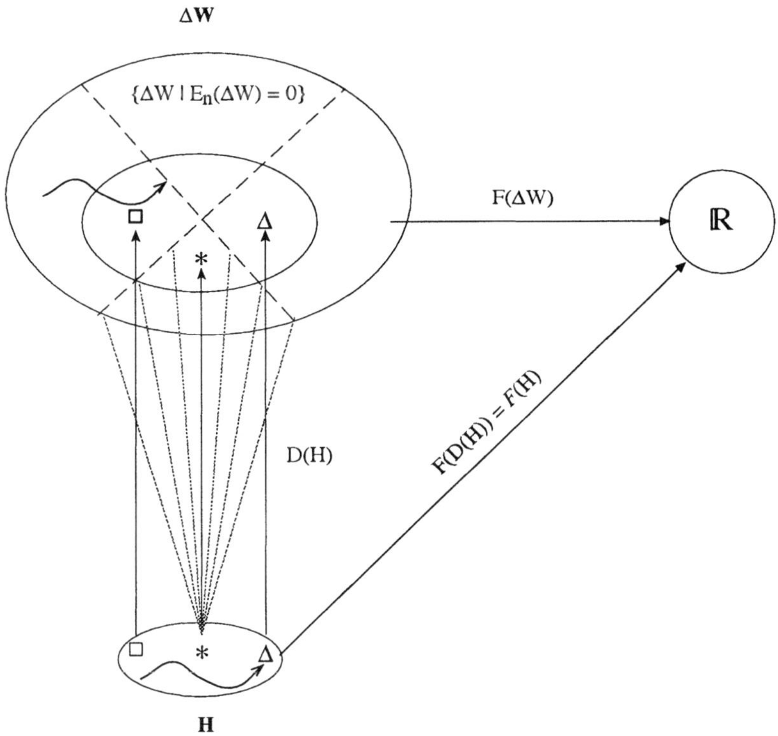

Figure 2 Diagram illustrating the meaning of the mapping D. The dashed lines delimit the zones of $\Delta\mathbf{W}$ leading to the same hidden-unit configuration in response to the new input pattern. In the intersection of one of these zones with the subset of $\Delta\mathbf{W}$ satisfying the new pattern, there is a unique minimum for the function F, which is the one we place in $D(\mathbf{H})$. Thus, only the weight increments in $D(\mathbf{H})$ are worth the search, which can be carried out through the function $F(\mathbf{H}) = F(D(\mathbf{H}))$. The search is thus brought to the space of hidden-unit configurations and, when finished, the solution obtained H^* can be easily translated into its corresponding weight increment $D(H^*)$.

The benefits of this problem transformation are:

- The original constrained minimization problem has been turned into an unconstrained one.
- The new pattern is always perfectly encoded, thus obviating the tradeoff between cost minimization and constraint satisfaction mentioned in the preceding section.
- There are far less variables than in the original formulation.
- The domain of each variable is more restricted, because hidden-unit activation functions are normally of limited range.

V. THE ONE-TO-ONE MAPPING D

The validity of the transformation performed in the preceding section depends on the fact that D is really a function, that is, that for each H there is one and only one ΔW satisfying the conditions in (3). Let us now prove this fact.

We start by noting that the unit activations y_j are components of either H or the new input–output pattern. Due to the assumption that activation functions f_j are invertible, the x_j become fixed as $x_j = f_j^{-1}(y_j)$. Therefore, in what follows, we assume that x_j and y_j are constants and the usual equations $x_j = \sum_i w_{ji} y_i$ do not hold.

Then consider

$$\begin{aligned} F_j &= \sum_i c_{ji} \Delta w_{ji}^2 + \sum_i b_{ji} \Delta w_{ji}, \\ R_j &\equiv \sum_1 (\Delta w_{ji} + w_{ji}) y_i - x_j = 0. \end{aligned} \tag{4}$$

When j runs along all the indices such that $\mathrm{Inc}(j) \neq \emptyset$ (hidden and output units), $F = \sum_j F_j$ and the set of R_j is equivalent to the conditions in the definition of D in (3). Thus, (3) can be rewritten as

$$\begin{gathered} \mathrm{Min} \sum_{j/\mathrm{Inc}(j) \neq \emptyset} F_j(\Delta W), \\ \text{subject to } \{R_j\}_j / \mathrm{Inc}(j) \neq \emptyset. \end{gathered} \tag{5}$$

Because F_j and R_j share all variables and have no one in common with F_k and R_k if $k \neq j$, our problem is reduced to independently minimize each F_j under the R_j constraint.

This can be easily done, for example, using the Lagrange multipliers method. First, we rewrite R_j in a more convenient manner:

$$R_j \equiv \sum_i \Delta w_{ji} y_i = x_j - \sum_i w_{ji} y_i$$

and, calling Δx_j the constant $x_j - \sum_i w_{ji} y_i$, we have

$$R_j \equiv \sum_i \Delta w_{ji} y_i = \Delta x_j. \tag{6}$$

Now we define the functions to be minimized G_j as

$$G_j = F_j + t R_j.$$

All $\partial G_j/\partial \Delta w_{jk}$ must be zero in the solution,

$$\frac{\partial G_j}{\partial \Delta w_{jk}} = 2c_{jk}\Delta w_{jk} + b_{jk} + t y_k = 0.$$

Hence,

$$\Delta w_{jk} = \frac{-b_{jk} - t y_k}{2c_{jk}}. \tag{7}$$

Then, using the constraint R_j in (6),

$$\Delta x_j = \sum_i \Delta w_{ji} y_i = -\sum_i \frac{b_{ji}}{2c_{ji}} y_i - \frac{t}{2} \sum_i \frac{y_i^2}{c_{ji}}.$$

Therefore,

$$t = -2\frac{\Delta x_j + \sum_i (b_{ji} y_i / 2c_{ji})}{\sum_i (y_i^2 / c_{ji})},$$

and, substituting in (7),

$$\Delta w_{jk} = \frac{y_k(\Delta x_j + \sum_i (b_{ji} y_i / 2c_{ji}))}{c_{jk} \sum_i (y_i^2 / c_{ji})} - \frac{b_{jk}}{2c_{jk}}. \tag{8}$$

Observe that for the case in which $c_{ji} = 1,\ b_{ji} = 0$, this is the Widrow–Hoff rule [17]. This formula will be simplified further, but we can conclude now that D is a well-defined function if and only if:

(a) At least one input arriving at each unit is not zero. This is not a real danger unless threshold units that take zero as one of their states are used.
(b) All the c_{ji} are nonzero. We have to take care of this when selecting the parameters of the quadratic function (see Section VIII). This is the reason for the unsuitability of a linear F.
(c) The denominator of the first fraction must be nonzero. The opposite is an unlikely event that is avoided by choosing positive c_{ji} (see again Section VIII).

VI. LEARNING WITH MINIMAL DEGRADATION ALGORITHM

We can now derive a gradient algorithm for carrying out the minimization in (5). Let us begin by giving $F = \sum_j F_j$ a more explicit form. From (8):

$$\Delta w_{jk}^*(Y) = \frac{\mathcal{E}}{c_{jk}\mathcal{M}_j} y_k - \frac{b_{jk}}{2c_{jk}},$$

where

$$\mathcal{E}_j = f_j^{-1}(y_j) - \sum_i w_{ji} y_i + \sum_i \frac{b_{ji} y_i}{2c_{ji}}$$

and

$$\mathcal{M}_j = \sum_i \frac{y_i^2}{c_{ji}} .$$

Substituting this expression for Δw^*_{jk} into (5):

$$\begin{aligned} F_j(Y) &= \sum_i c_{ji} \left(\frac{\mathcal{E}_j}{c_{ji}\mathcal{M}_j} y_i - \frac{b_{ji}}{2c_{ji}} \right)^2 + \sum_i b_{ji} \left(\frac{\mathcal{E}_j}{c_{ji}\mathcal{M}_j} y_i - \frac{b_{ji}}{2c_{ji}} \right) \\ &= \sum_i c_{ji} \frac{\mathcal{E}_j^2 y_i^2}{c_{ji}^2 \mathcal{M}_j^2} + \sum_i c_{ji} \frac{b_{ji}^2}{4c_{ji}^2} - 2 \sum_i c_{ji} \frac{\mathcal{E}_j y_i}{c_{ji}\mathcal{M}_j} \frac{b_{ji}}{2c_{ji}} \\ &\quad + \sum_i b_{ji} \frac{\mathcal{E}_j y_i}{c_{ji}\mathcal{M}_j} - \sum_i \frac{b_{ji}^2}{2c_{ji}} \\ &= \sum_i \frac{\mathcal{E}_j^2 y_i^2}{c_{ji}\mathcal{M}_j^2} - \frac{1}{4} \sum_i \frac{b_{ji}^2}{c_{ji}} \end{aligned}$$

where the definition of $\mathcal{M}_j$ was used in the last step. Observe that $\mathcal{M}_j$ and $\mathcal{E}_j$ can be taken out from the summations:

$$F_j(Y) = \frac{\mathcal{E}_j^2}{\mathcal{M}_j^2} \sum_i \frac{y_i^2}{c_{ji}} - \frac{1}{4} \sum_i \frac{b_{ji}^2}{c_{ji}} = \frac{\mathcal{E}_j^2}{\mathcal{M}_j} - \frac{1}{4} \sum_i \frac{b_{ji}^2}{c_{ji}} ,$$

so finally,

$$F(Y) = \sum_j F_j(Y) = \sum_j \frac{\mathcal{E}_j^2}{\mathcal{M}_j} - \frac{1}{4} \sum_{j,i} \frac{b_{ji}^2}{c_{ji}} . \tag{9}$$

To greatly reduce the complexity, it is convenient to work with the parameters $w'_{jk} = w_{jk} - b_{jk}/2c_{jk}$, because then $\mathcal{E}_j = f_j^{-1}(y_j) - \sum_i w'_{ji} y_i$.

Thus, the first step in the algorithm will be the transformation of all the weights w_{jk} into w'_{jk}, and the last step will be the translation of the optimal hidden-unit configuration into the new increments $\Delta w'^*_{jk}$, which can also be simplified as follows:

$$\Delta w'^*_{jk} = \left(w_{jk} + \Delta w^*_{jk} \right) - w'_{jk} = \Delta w^*_{jk} + \frac{b_{jk}}{2c_{jk}} = \frac{\mathcal{E}_j y_k}{c_{jk}\mathcal{M}_j} . \tag{10}$$

Let us now derive the gradient of F. To prevent that during the search the y's would travel beyond their valid ranges, because the activation functions f_j usually have a limited range, we choose to calculate $\partial F/\partial x_j$. Note that, using $x_j = f_j(x_j)$, x_j appears explicitly in $\mathcal{E}_j$ and therefore in F_j, but also implicitly in all F_s such that $s \in \mathrm{Out}(j)$.

The gradient when j is a hidden-unit index is then

$$\frac{\partial F}{\partial x_j} = -2\frac{\mathcal{E}_j}{\mathcal{M}_j} + 2f_j'(x_j) \sum_{s\in\mathrm{Out}(j)} \frac{\mathcal{E}_s w_{sj}' \mathcal{M}_s + \mathcal{E}_s^2 y_j/c_{sj}}{\mathcal{M}_s^2}. \tag{11}$$

This formula is valid for networks with whatever number of layers, and even with jumps between layers. The gradient formula is easily implemented by an algorithm with a data flow in two phases, which resembles back propagation. In the first forward phase, $\mathcal{E}_j$ and $\mathcal{M}_j$ are computed for all units with incoming connections. Then the first gradient term is easily calculated and each of the second term addends is backpropagated to get the total gradient. Do not be misled by the "forward" and "backward" names, because the similarity with back propagation is limited by the fact that here the information needed by a unit in both phases is already available locally, without any time delay required to wait for information from remote layers. Because of this independence, updating can occur in any order, allowing even total overlapping. This makes complete parallelism in the implementation possible, not only within a layer, but also among layers.

The LMD algorithm is, therefore, as follows:

0. $w_{jk}' \leftarrow w_{jk} - b_{jk}/2c_{jk}$.
1. Fix the input and output pattern in the network input and output and derive x_j for all output units.
   Choose initialization values for all the hidden-unit total inputs x_j.
2. Repeat until a given *stopping criterion*
   a. Calculate $\mathcal{M}_j$ and $\mathcal{E}_j$ for all units with incoming connections (forward pass).
   b. Back-propagate the second term in (11) and update x_j for all hidden units using $x_j \leftarrow x_j + \mu(\partial F/\partial x_j)$ (backward pass).
3. Change weights using (10).

Remember that every time x_j is changed, y_j must also be updated because they are binded variables.

It may seem excessive to dedicate a sequence of cycles for only one pattern. Perhaps a solution could be approximated, for example linearizing the network or with another heuristic. It all depends on the difference between a solution of this kind and the true minimum, how this difference is reflected in E, and how the increase in damage has repercussion on the recovery learning time over the previous patterns and the new pattern. It can be supposed that it is worthwhile to spend some more time with only one pattern, trying to trim a bit E, if in exchange one avoids some cycles over the whole learning set.

VII. ADAPTATION OF LEARNING WITH MINIMAL DEGRADATION FOR RADIAL BASIS FUNCTION UNITS

Section II demonstrated that the forgetting problem can be insurmountable for fixed architectures in some applications. In this case, it would be convenient to use incremental architectures, and RBF units seem the best candidates to be the new units, because they are suitable to modify locally the learned function. In this section, the derivations needed to apply the LMD algorithm to RBF units are spelled out.

We shall limit the study to the case of $b_{ji} = 0$ which, as we will see in the next section, is the one used in practice, because the inclusion of the b_{ji} reduces to a preprocessing step with a standard second-order algorithm. The solution for arbitrary c_{ji} is difficult to find analytically. For this reason, the c_{ji} of the weights impinging on a Gaussian unit j are all taken to be equal and are denoted by c_j.

The propagation function of the weights outcoming from an RBF unit is a scalar product and the activation function of the output units is linear or sigmoidal. Thus, the cost function associated with these weights is the same as in back-propagation networks.

RBF units, however, introduce an element that does not fit in the solution framework stated in Sections IV and V. The activation function of these units is Gaussian, and thus not invertible, which was one of the assumptions made to show that D is a one-to-one function. Nonetheless, this difficulty can be easily overcome. Although we base all the framework in the hidden-unit activations for the sake of clarity in the exposition, it is also possible to base it on the total inputs to the hidden units, D being a function of these, in such a way that the invertibility assumption can be dropped.

Consider C_j and R_j when j is an RBF unit:

$$C_j = \sum_i c_j \Delta w_{ji}^2,$$

$$R_j \equiv \sum_i \left((\Delta w_{ji} + w_{ji}) - y_i\right)^2 - a_j^2 = 0.$$

The independence of the problems associated with each pair C_j and R_j still holds, due to the absence of common variables. Thus, it suffices to solve each subproblem using the Lagrange multipliers. The function G_j that must be optimized is again $G_j = C_j + tR_j$, and thus its derivatives must be null:

$$\frac{\partial G_j}{\partial \Delta w_{jk}} = 2c_j \Delta w_{jk} + 2t(w_{jk} + \Delta w_{jk} - y_k) = 0.$$

Hence,

$$\Delta w^*_{jk} = \frac{-t(w_{jk} - y_k)}{c_j + t}.$$

Now we substitute the increments in R_j to obtain t:

$$\sum_i \left((w_{ji} - y_i) + \frac{-t(w_{ji} - y_i)}{c_j + t} \right)^2 - a_j^2 = 0,$$

$$\left(1 - \frac{t}{c_j + t}\right)^2 \sum_i (w_{ji} - y_i)^2 = a_j^2,$$

$$\left(1 - \frac{t}{c_j + t}\right)^2 = \frac{a_j^2}{\sum_i (w_{ji} - y_i)^2}.$$

From this we cannot obtain t easily. We shall do the variable change $t' = -t/(c_j + t)$. In this way, $\Delta w^*_{jk} = t'(w_{jk} - y_k)$ and

$$1 - \frac{t}{c_j + t} = \pm \frac{a_j}{\sqrt{\sum_i (w_{ji} - y_i)^2}} \Rightarrow t' = \pm \frac{a_j}{\sqrt{\sum_i (w_{ji} - y_i)^2}} - 1,$$

so

$$\Delta w^*_{jk} = (w_{jk} - y_k)\left(\pm \frac{a_j}{\sqrt{\sum_i (w_{ji} - y_i)^2}} - 1 \right)$$
$$= \pm \frac{(w_{jk} - y_k) a_j}{\sqrt{\sum_i (w_{ji} - y_i)^2}} - (w_{jk} - y_k).$$

The increments Δw^*_{jk} can take two values. We must select those corresponding to the minimum of C_j. Because a_j is always positive, it is easy to check that

$$\Delta w^*_{jk} = \frac{(w_{jk} - y_k) a_j}{\sqrt{\sum_i (w_{ji} - y_i)^2}} - (w_{jk} - y_k)$$

is always the option with lower absolute value, and thus, this is the desired solution.

Let us now calculate C_j:

$$\begin{aligned} C_j &= c_j \sum_k \left(\frac{(w_{jk} - y_k)a_j}{\sqrt{\sum_i (w_{ji} - y_i)^2}} - (w_{jk} - y_k) \right)^2 \\ &= \frac{c_j a_j^2 \sum_k (w_{jk} - y_k)^2}{\sum_i (w_{ji} - y_i)^2} + c_j \sum_k (w_{jk} - y_k)^2 - \frac{2 c_j a_j \sum_k (w_{jk} - y_k)^2}{\sqrt{\sum_i (w_{ji} - y_i)^2}} \\ &= c_j a_j^2 + c_j \sum_k (w_{jk} - y_k)^2 - 2 c_j a_j \sqrt{\sum_k (w_{jk} - y_k)^2}. \end{aligned}$$

The gradient of C_j with respect to a_j is

$$\frac{\Delta C_j}{\Delta a_j} = -2 c_j a_j + 2 c_j \sqrt{\sum_k (w_{jk} - y_k)^2}.$$

We can finally write the total gradient of an RBF unit with respect to C. Remember that the second term of (11) corresponds to the gradients of the outcoming weights, which are the same in this case:

$$\begin{aligned} \frac{\Delta C}{\Delta a_j} &= -2 c_j a_j + 2 c_j \sqrt{\sum_k (w_{jk} - y_k)^2} \\ &\quad + 2 f_j'(a_j) \sum_{s \in \mathrm{sal}(j)} \frac{\mathcal{E}_s w_{sj}' \mathcal{M}_s + \mathcal{E}_s^2 y_j / c_{sj}}{\mathcal{M}_s^2}. \end{aligned}$$

Because each RBF unit is directly connected to the input of the network, $y_k = x_k$, and thus $\sqrt{\sum_k (w_{jk} - y_k)^2}$ (which would be a_j in the forward operating mode of the network) is constant during the search and must be computed only once.

VIII. CHOOSING THE COEFFICIENTS OF THE COST FUNCTION

The most obvious election for the coefficients b_{jk} and c_{jk} is that yielded by an instantaneous second-order approximation of the error function $E(\Delta W)$ ignoring the off-diagonal terms. Then $c_{jk} = \partial^2 E / \partial w_{jk}^2$ and $b_{jk} = \partial E / \partial w_{jk}$.

With this choice, F is exactly the local estimation of E assumed by several authors [18–20] to justify this simple pseudo-Newton rule for optimizing E:

$$\Delta w_{jk} = -\mu \frac{\partial E}{\partial w_{jk}} \Big/ \frac{\partial^2 E}{\partial w_{jk}^2}.$$

Note that step 0 in the LMD algorithm, when $c_{jk} \neq 0$, is exactly one of these pseudo-Newton steps. The implicit aim of this step is to bring the network to the minimum of F, cancelling out first derivatives. Unfortunately, the minimum of F normally is not the minimum of E, and first derivatives could still remain significant. The conclusion is that this step is scarcely useful and, in fact, what we really need is to minimize the first derivatives of E by whatever means before the application of LMD. Later, in Section XI.B, we will give some advice to reduce first derivatives during training.

Then, if step 0 is taken out of the algorithm, we are minimizing the cost function:

$$F(\Delta W) = \sum_{j,k} \frac{\partial^2 E}{\partial w_{jk}^2} \Delta w_{jk}^2.$$

In the minimum, this is still a diagonal second-order approximation. In an arbitrary point, it can be shown [21] that, for a function of the type

$$\sum_{jk} c_{jk} \Delta w_{jk}^2,$$

this choice of coefficients is, on average, the best for estimating E.

In order to prevent the nullity of the c_{jk} to fulfill condition (c) in the last section, we can add a very small constant (range of the hidden activation function/1000, for instance) to every one of the second derivatives. Even when some c_{jk} are not null, but very close to zero, this helps to ameliorate the behavior of the algorithm. Because, as we said, it is necessary to have positive c_{jk}, we should take absolute values of the second derivatives. Note that when the network is in the minimum, these second derivatives are already positive.

There is another way of using the coefficients to realize a coarser estimation of the damage to the net. By making all the c_{jk} parameters equal to 1 and all the b_{jk} equal to 0, the algorithm is somewhat simpler and permits either saving the cost of calculating c_{jk} (though it is relatively cheap) or working when they are not available. What LMD is calculating in this case is the nearest solution for the new pattern in the weight space. This is a good heuristic to look for the intersection of the solution space of patterns $1, \ldots, n-1$ and the solution space of pattern n. Under total uncertainty about the shape of the solution space for the new pattern, using this version amounts to introducing white noise into the network with a uniform probability distribution. This type of noise seemed to do little injury in a study performed by Hinton and Sejnowski [22].

In sum, two versions of LMD have been mainly explored in the experiments: the *standard* one that uses $c_{jk} = |\partial^2 E / \partial w_{jk}^2|$ and the *coarse* one that uses $c_{jk} = 1$.

IX. IMPLEMENTATION DETAILS

One of the features that must characterize the application of the algorithm is total automation. We can neither expect to test and correct parameters each time we introduce a new pattern, nor to watch over to decide when convergence has been completed. Besides, we need a reliable algorithm in all situations to bring the network to the minimum without risk of catastrophe. Therefore, in the following subsections we describe the rationale underlying the determination of parameters and initialization values.

A. ADVANCE RATE

The current implementation of LMD follows pure gradient descent but with adaptive step size. The strategy is very simple. When the last step is beneficial (i.e., it leads to a decrement in the cost function), the advance rate μ is multiplied by a number μ^+ slightly larger than 1. In the opposite case (increment in F), the step is partially undone, the advance rate is reduced by a factor of μ^-, and then the step is redone. This can be implemented with little more computation than a forward pass. The advance rate control routine, which after step 2b, is as follows:

if $\Delta F < 0$ then $\mu \leftarrow \mu\mu^+$
while $\Delta F \geqslant 0$
  $x_j \leftarrow x_j - (1 - \mu^-)\mu(\partial F/\partial x_j)$
  $\mu \leftarrow \mu\mu^-$
  Forward pass 2a
end while

We have always taken $\mu^- = 0.5$ and $\mu^+ = 1.3$. Observe that, according to (9), ΔF can be easily calculated from the old value of F and the new values of $\mathcal{E}_j$ and $\mathcal{M}_j$ obtained in the forward pass. The $\partial F/\partial x_j$ are always those computed in step 2b and, thus, they do not need to be recomputed. The initial value of μ is only important for a quicker convergence, because convergence is guaranteed. The most convenient value depends on the size of the network and, therefore, for the scaling experiment we also scale the initial μ; in the other experiments the networks are of comparable size and a value of 2 is always used. As a refinement to the basic algorithm, it is possible to use a much larger μ^+ and a very small μ^- in the first iteration of LMD until a mistaken step is done (if the initial step is not overshot) or a valid step is done (if the initial step is overshot), and afterward use the normal values.

B. Stopping Criterion

The search is finished when

$$\sqrt{\frac{\sum_{j/\mathrm{Inc}(j)\neq\emptyset}(\partial F/\partial x_j)^2}{n_H}} < G_{\min},$$

where n_H is the total number of hidden units and $G_{\min}$ is a constant that regulates the accuracy in finding the minimum. Dividing by $\sqrt{n_H}$ seems convenient to obtain values independent of the network dimensionality. In normal practice, $G_{\min} = 0.005$ seems a good choice, and this is the value used in all the experiments reported.

C. Initial Hidden-Unit Configuration

We have used several architectures with different weights to test the existence of local minima. This was done by using a great number of random initial hidden-unit configurations and measuring the cost function in the final points reached. The result has been that networks that are able to learn; that is, those with moderately saturated units rarely lead to landscapes with local minima. Only using random weights of increasing magnitude, local minima begin to appear more numerous and higher. Because of the scarcity of local minima, the initial hidden-unit configuration is not a crucial question, but the number of cycles required for convergence can increase with a bad selection of the initial point.

There seem to be two privileged initial points: one is the hidden-unit configuration that results from propagating activity through the network and the other is the one with all the unit activations in the middle of the activation range. The first is good in the case where the error in the new pattern is low, because then the weight modifications needed to get the new pattern are small, and, as a consequence, the ideal hidden-unit configuration is near this point. The second point has the advantage that each hidden unit is completely free to go in one direction or another (in fact, this is also useful to avoid local minima) and the gradient of the activation function is maximum (at least for sigmoids). Here all the experiments use the second option. A more elaborate decision could be to switch to the first option when the error in the new pattern is small.

X. PERFORMANCE MEASURES

In this section, we develop tools for evaluating the computational savings provided by LMD, which turn out to have wider application.

One method of measuring the benefits of using LMD would be comparing E', the global error over the patterns $1, \ldots, n$, before and after applying LMD to the nth pattern. This should be indicative of how much the search of the E' minimum is facilitated. Surprisingly, it is not like this. For instance, if the error after the application of LMD is slightly higher, we have systematically observed that LMD helps nevertheless to shorten learning times. Probably this phenomenon is similar to the accelerated relearning times registered in [22] when some disturbance is introduced in trained networks.

It is a knotty problem to measure the computation costs of arriving at a minimum from two different points. Here we suggest two measures that are independent of any algorithm parameters, and rely only on the landscape of the error function. Therefore, they reflect objectively some aspect of the difficulty in finding a minimum from different initial points. We have checked that both give results qualitatively similar if it is not required to arrive very accurately at the minimum. For instance, the phenomenon mentioned previously appears independently of the measure used.

The first measure is the time a dynamical system driven by the system of equations

$$\frac{\partial E}{\partial W} = \frac{\partial W}{\partial t}$$

would take to go from one point to another of its trajectory. We call this measure the *back-propagation time*, because these are the learning equations of back propagation as a continuous dynamical system. To estimate the back-propagation time, we constrain pure gradient back propagation to take only steps that produce error decrements that can be predicted by a linear approximation of the error function; that is, we take the criterion:

$$\frac{|\mathrm{Est}\,\Delta E^{k)} - (E^{k+1)} - E^{k)})|}{\sqrt{(E^{k+1)} - E^{k)})\mathrm{Est}\,\Delta E^{k)}}} < \text{exigence} \tag{12}$$

to accept the learning rate $\mu^{k)}$ that produced $\Delta w_{jk}^{k)}$ as correct, where $\mathrm{Est}\,\Delta E^{k)}$ is the linear estimation of the error increment based on the first derivatives:

$$\mathrm{Est}\,\Delta E^{k)} = \sum_{j,i} \Delta w_{ji}^{k)} \frac{\partial E^{k)}}{\partial w_{ji}}.$$

If all the steps along the learning process satisfy this criterion, we can say that the trajectory followed by the algorithm approximates the trajectory that would follow the preceding dynamical system. The fidelity to the continuous path will be controlled by the exigence parameter.

If step k satisfies the criterion, $W(t)$ is approximately linear in the section between $W^{k)}$ and $W^{k+1)}$, and, thus, we can estimate the time to perform $\Delta W^{k)}$

by dividing directly the distance by the instantaneous velocity of the system, the gradient norm:

$$t\left(W^{k)}, W^{k+1)}\right) = \frac{\|\Delta W^{k)}\|}{\|\partial E/\partial W\|} = \mu^{k)}.$$

Therefore, the time to complete a trajectory is $\sum_k \mu^{k)}$ when all the steps satisfy the linear constraint (12).

It is possible to adapt μ near optimally during the training with an algorithm similar to the one presented in the following discussion.

Using this measure, we have observed that the back-propagation time required to eliminate the last residuals of the error is much larger than that needed to eliminate the main part of the error. This is due to the fact that the velocity slows down in flat regions and especially in the neighborhood of a minimum. Algorithms more sophisticated than raw back propagation are expected to behave in a rather different manner.

We propose another measure that overcomes the previous shortcoming, while at the same time allowing quicker computation. The measure is the standard curve length defined for rectifiable functions:

$$\Lambda(t_1, t_2) = \int_{t_2}^{t_1} \left\| \frac{\partial E}{\partial W} \right\| (t)\, dt,$$

where t_1 is the initial time point and t_2 is the final time point. We could compute it in a similar way to the one used for the back-propagation time, taking only linearly predictable steps. Instead, we have developed a more efficient, although more complicated, algorithm, which will be used later for other purposes.

The idea is that to calculate the curve length approximating the system trajectory, we can relax the linearity constraint of magnitude predictability to only angle continuity between two steps; that is, we only accept one step if the angle with the next step is close enough to zero. In this way, we profit from the fact that, unlike the back-propagation time, the curve length is independent of the velocity with which E comes down, and can always be computed in one step in the zones where the gradient direction does not change. The only inconvenience, which adds some complexity to the algorithm, is that to know whether one step is too large to be accepted, we must know the direction of the next step. This is the algorithm used to estimate the curve length:

Repeat until stopping criterion
$\Delta W \leftarrow \mu \partial E/\partial W$
$W \leftarrow W + \Delta W$
While $\text{ang}(\Delta W, \partial E/\partial W(W)) >$ exigence
    $W \leftarrow W - \alpha \Delta W$
    $\Delta W \leftarrow (1 - \alpha)\Delta W$

$$\mu \leftarrow (1-\alpha)\mu$$
$$\Lambda \leftarrow \Lambda + \|\Delta W\|$$
$$\mu \leftarrow \beta\mu$$

Exigence regulates the fidelity to the continuous system trajectory. μ is adapted to grow slightly at a geometrical rate of $\beta > 1$ when one step is accepted, and to decrease more quickly at a rate of $1-\alpha$, $1 < \alpha < 0$, when it is not. This is to keep μ near the highest values allowed by the angle continuity constraint. In our simulations, $\alpha = 0.5$ and $\beta = 1.2$. To compute $\text{ang}(\Delta W, \partial E/\partial W(W))$, a complete back-propagation cycle is required to get the E gradients (which, when the while condition fails, can be reused without further computations for the next step), but weights and weight increments must not be updated.

XI. EXPERIMENTAL RESULTS

A. SCALING PROPERTIES

Figure 3 gives an idea of the number of steps needed to reach the minimum and the scaling properties of LMD. Networks of several sizes were used in this experiment, all with the same proportion of units in each layer. The smallest was an 8–3–1 network and the others were gotten by multiplying by $10, 20, \ldots$ the number of units in each layer of this network. The initial μ for the 8–3–1 network was 1, and this was multiplied in the same manner as before for the rest of the networks. The abscissas indicate the number of connections. The ordinates show the average of forward–backward cycles and forward steps alone (due to mistaken steps) for 20 random new patterns. The cost function was $F = \sum \Delta w_{jk}^2$ and each weight w_{jk} in the network was obtained randomly with the uniform distribution $[-1.5/|\text{Inc}(j)|, 1.5/|\text{Inc}(j)|]$.

It can be seen that the number of steps is not excessive, indicating the great simplicity of the unconstrained search space. Scaling properties are good, making less indispensable the development of an acceleration algorithm. Other experiments also indicate that scaling with the number of layers is better than for back propagation.

B. SOLUTION QUALITY FOR DIFFERENT COEFFICIENT SETTINGS

Table I presents results for three different networks trained with two different numbers of patterns. It shows the average error increment in the output units for the previously trained pattern when LMD is applied, with $c_{jk} = 1$ and $c_{jk} =$

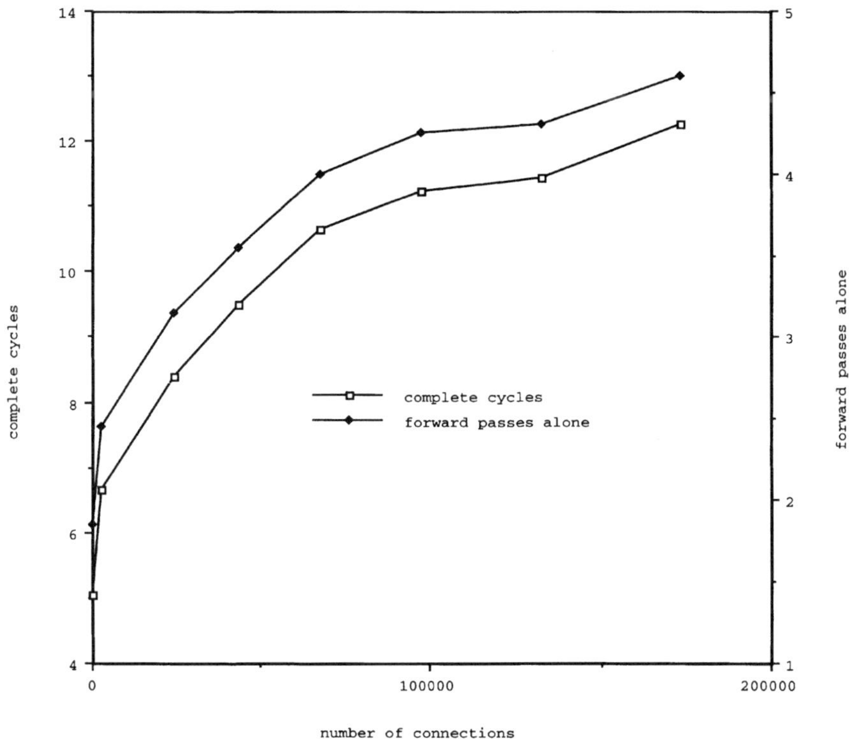

Figure 3 Scaling experiment: number of LMD complete cycles and feedforward steps alone (due to overshot steps) as a function of network size (number of connections).

$|\partial^2 E/\partial w_{jk}^2|$. The error measure used is

$$E = \frac{1}{2}\frac{1}{n\,\text{No}} \sum_{\substack{j=1,\ldots,n \\ i=1,\ldots,\text{No}}} \varepsilon_{ji}^2,$$

where No is the number of output units and ε_{ji} is the difference between the desired value and the network response value in ouput unit i for pattern j. The net structure is expressed in the obvious way (net 8–4–1 has eight input units, four hidden units, and one output unit). Both the previously learned patterns and the new ones were generated randomly from a uniform distribution in $[-1.2, 1.2]$. A symmetric sigmoid ranging from -1.712 to 1.712 was used for the hidden-unit activations. The activation functions of the output units were linear. Take into account that these conditions are bad to reduce the error increment: linear output

Table I

Error Increments after Applying LMD with $c_{jk} = 1$ and $c_{jk} = |\partial^2 E/\partial w_{jk}^2|$

Architecture	8–3–1		32–12–4		300–30–10	
Number of patterns	5	10	10	25	50	150
Coarse LMD	0.064	0.111	0.018	0.32	0.002	0.002
Standard LMD	0.038	0.087	0.012	0.028	0.001	0.002

units allow the error to grow unlimitedly, and the large ranges of the input–output patterns and hidden-unit activation functions lead to a great variance in the output of the network. The table shows averages over 20 new patterns. When the networks have been trained with a lot of patterns, the error increments are larger for the two versions. This is due to the fact that a network with more random patterns has more output variance, and thus the error average of the new learned random patterns is higher (more than double in the network 32–12–4). Also the advantage of the standard version over the coarsest one is lower with more patterns, because second derivatives are more uniform, indicating that the network is more saturated with information, parameters are less free to vary, and, as a consequence, there are no privileged directions. When the patterns are not random, the networks become saturated more gradually. We can also guess that, in the case of many patterns, the quadratic estimation of the error function is poorer, because the very large errors force the network to look for solutions far from the present position in weight space.

Finally, another experiment was made to test the solution quality provided by different settings of the coefficients in networks outside the vicinity of a minimum of the learning set. Table II presents results for the 8–3–1 architecture trained with five patterns until some fixed error is reached. Five networks with the error levels shown in the table were used, and each of them underwent the introduction of a new pattern with the coarse and standard versions of LMD. As before, results were averaged over 20 new patterns. It can be seen that the standard version of LMD still gets significantly better results than the coarse one.

Table II

Error Increments out of the Minimum

Old patterns error	0.1	0.05	0.01	0.001	0.0005	0.0001
Coarse LMD	0.049	0.54	0.057	0.061	0.62	0.063
Standard LMD	0.023	0.025	0.029	0.033	0.034	0.035

C. Computational Savings Derived from the Application of Learning with Minimal Degradation

We present now some experimental results regarding the improvement in training times provided by the use of LMD. The networks are the same as those in the preceding experiment, with the same already learned patterns and the same 20 new patterns. Let $W^*_{1,\ldots,n-1}$ be the weight configuration at which the minimum error $E^*_{1,\ldots,n-1}$ for the set of patterns $1, \ldots, n-1$ is attained, and E be the error function over patterns $1, \ldots, n$. For each network and new pattern, we first calculate very accurately the minimum error E^* for the set of patterns $1, \ldots, n$. Then we measure the curve length from $W^*_{1,\ldots,n-1}$ to the first point in the trajectory with an error less than $E^* + 0.2(E(W^*_{1,\ldots,n-1}) - E^*)$, and after from the same original point transformed by applying LMD with pattern n, to the first point with the same level of error as before. Table III shows the results. The differences in curve length should be indicative of the advantage of using LMD to tune the network before applying any algorithm of the back-propagation type, because length only depends on the error function shape. The computational effort to introduce the new pattern with LMD is negligible, because the number of LMD cycles is usually less than the number of patterns in the training set. The table reveals that it is more advantageous to use LMD when there are few patterns than when there are many. The main reason is the same one pointed out in the preceding subsection: growth of the new pattern errors leads to growth of the damage to the network. However, this is a very abnormal situation because in typical applications the error in the new patterns tends to decrease when the number of learned associations grows. Moreover, to be always advantageous under this incremental scheme, LMD (and any other algorithm) needs to be presented with new patterns yielding progressively smaller errors, because, in the long run, a highly erroneous new pattern can lead to a situation of the type discussed in Section II and exemplified in Fig. 1. This shortcoming could be overcome, for example, by reducing wisely only a fraction instead of the complete error of the new pattern. We have left these refinements for future work.

Table III
Length Traveled with and without LMD to Reach the Minimum of the Patterns 1, . . . , *n*

Architecture	8–3–1		32–12–4		300–30–10	
Number of patterns	5	10	10	25	50	150
Length (without LMD)	0.335	0.546	0.334	0.456	0.182	0.169
Length (after LMD)	0.088	0.397	0.059	0.262	0.002	0.061

D. Learning with Minimal Degradation versus Back Propagation

We present now some experiments comparing LMD and standard back propagation with different advance rates μ. This parameter turns out to be a very important factor in the comparison.

In Fig. 4a, the net 32–12–4, trained with 10 random patterns, undergoes the introduction of a new pattern with different back-propagation learning rates. The

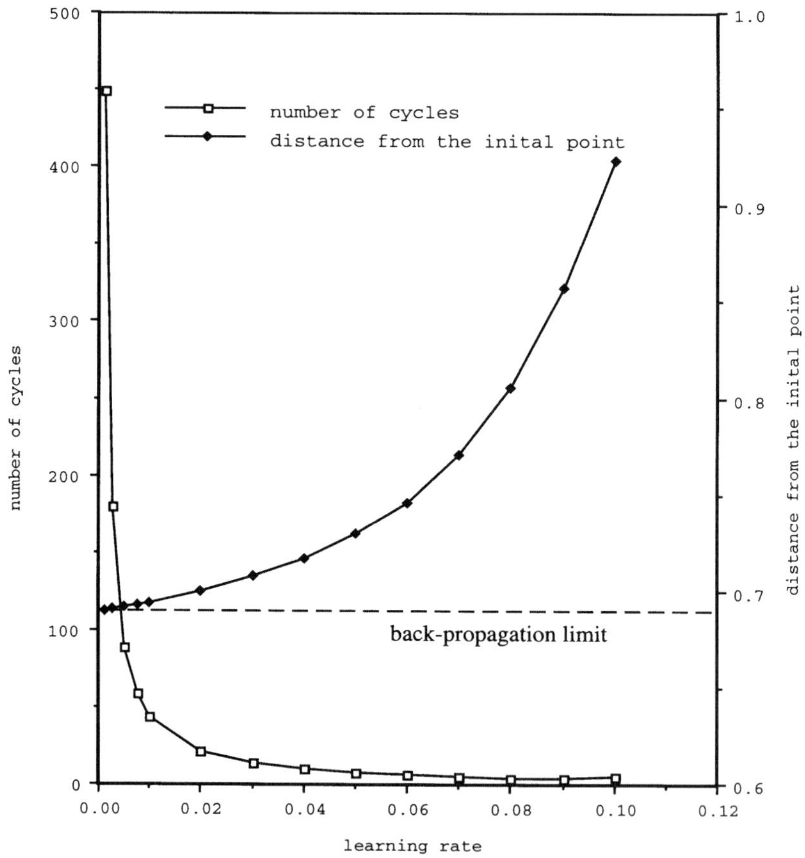

Figure 4a Starting with a network in the minimum of the error function for the old patterns, a new pattern is introduced with different back-propagation learning rates. The graphic shows the number of cycles needed to learn the pattern and the total distance covered from the initial point in each case. It is clear that, when the learning rate tends to zero, the distance tends to a limit.

pattern is considered to have been learned when the average absolute error in the output units is 0.01. The distance from the starting point to the final one in the weight space and the number of cycles needed are shown. The limit to which the distance tends when $\mu \rightarrow 0$ can only be approximated with large compu-

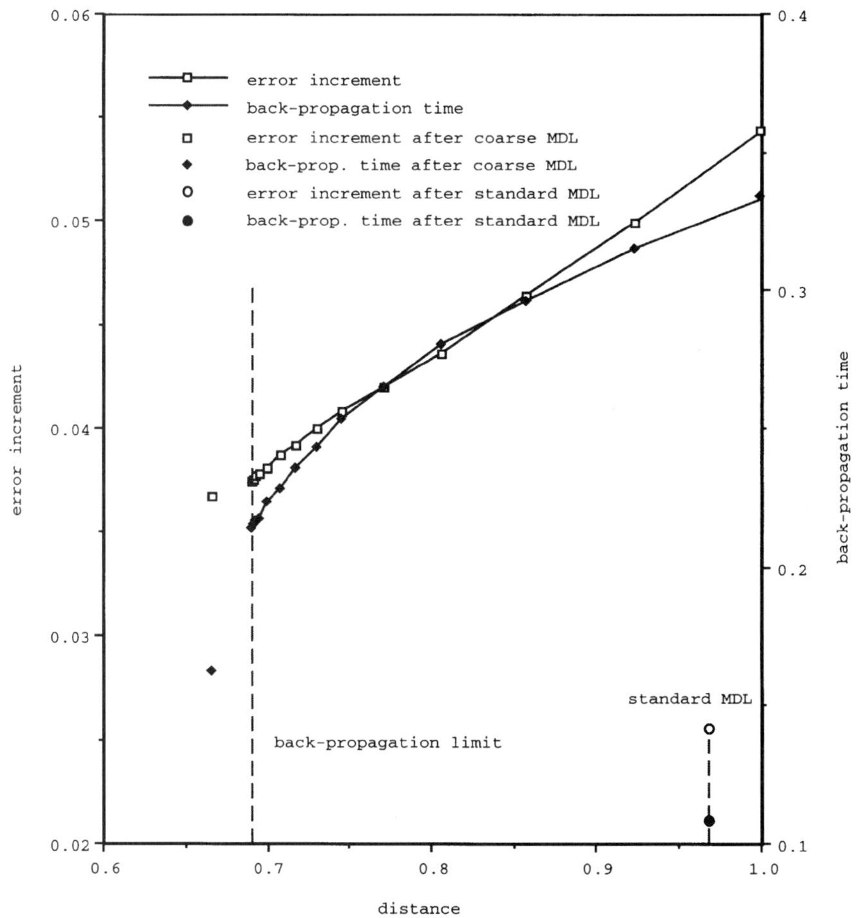

Figure 4b Another aspect of the experiment described in Fig. 3a. The distances represented in Fig. 3a are now on the axis of the abscissa. The two curves represent the error increments and the back-propagation times beginning after the introduction of the new pattern and ending in the global minimum of the error function including the new pattern. Coarsest LMD ($c_{jk} = 1$) minimizes explicitly the distance and so it must obtain results better than the back-propagation limit. However, standard LMD ($c_{jk} = |\partial^2 E/\partial w_{jk}^2|$), which does not take into account the distance, gets the best results.

tational costs. Figure 4b displays how the distances to the starting point shown in Fig. 4a affect the error in the old patterns, and how this, in turn, affects the recovery time (measured in terms of back-propagation time) needed to relearn both the new and the old patterns. The distance lower bound for back propagation when $\mu \rightarrow 0$ (0.69), as well as results with the coarsest ($c_{jk} = 1$) and standard ($c_{jk} = \partial^2 E/\partial w_{jk}^2$) versions of LMD, is also shown here.

The different weight solutions provided by back propagation are in random directions with respect to the starting point, because they were obtained independently of E; however, a neat linear relation appears between the distance and the error increment. An important finding for our investigation is that the recovery time follows an exponential-like curve, implying that it is very important, with respect to recovery time, to trim as much as possible the damage caused to the old patterns, even if the possible reductions are small.

This example also shows how the distance for the coarsest version of LMD must be, by definition of the cost function, always the shortest one. The difference between the true minimum distance found by the coarsest version of LMD and the approximation made in the limit by back propagation is variable, especially for highly erroneous new patterns, but usually the two points are close.

Finally, note that the standard version of LMD provided by far the best results (very small error increment and recovery time) and it did so by moving a long distance away from the initial point, thus proving the existence of a privileged direction.

XII. DISCUSSION

A. Influence of the Back Propagation Advance Rate on Forgetting

One of the main results we have obtained is the dependence of back-propagation forgetting on the learning rate. Cohen [2] already observed this in a systematic variation of all back-propagation parameters. We can now provide an explanation of this dependence. The gradient of the error function for the new pattern can lead the network anywhere in principle, but it is a good heuristic for finding one of the nearest solutions. The problem is that back propagation with usual learning rates does not follow the true gradient line, because of its discrete nature. Because the solutions for the new pattern pervade the weight space, a too large step may lead to a point crossed by a gradient line driving the network to a different and farther solution. If back propagation is forced to closely follow the true gradient descent line (as is the case when the advance rate tends to zero), it becomes a reasonable application of the coarsest version of LMD, but with high computational cost. Usual accelerating algorithms, taking bold steps, can only worsen forgetting.

Contrarily, the algorithm for measuring the curve length follows the gradient line with the desired accuracy, but with the highest learning rates possible in each step, alleviating the inefficiency–catastrophic forgetting tradeoff in back propagation, thus finding another use complementary to the one for which it was designed in Section IX. For instance, applying the curve length algorithm with exigence $= 0.99$ in the last experiment of the preceding section, the distance obtained was 0.693 — which is only slightly higher than the back-propagation limit — and only 27 cycles were used — almost half of those needed by back propagation with the appropriate μ to get the same distance. The algorithm to compute the back-propagation time could also do the job in a simpler but more inefficient way.

B. How to Prepare a Network for Damage or the Relation of Learning with Minimal Degradation with Fault Tolerance

A conclusion from the exponential-like curve for the recovery time reported in Section XI.D, as well as the reasoning about the convenience of dispensing with the b_{jk} parameters presented in Section VIII, is that one must wait as long as possible to the completion of the learning of the previous patterns (by second-order, standard back propagation or whatever means) before introducing the new patterns.

The overtraining effect had been observed, but not explained, in studies of forgetting [2] and fault tolerance [23]. Avoidance of forgetting and increasing fault tolerance can be seen as intimately related goals. Both try to minimize the effect of weight perturbations on the information stored in the network. The only difference is that, in order to avoid forgetting, one can control somewhat the form of the perturbation. This similarity can be profited from directly. Here, for example, the explanation for damage reduction after overtraining can be easily transferred from one domain to the other.

Through the Taylor series expansion of the error-increment function produced by a perturbation, it is evident that minimizing the first derivatives of E (the b_{jk} parameters) is the first priority to reduce the error increment. A minimum of E is also a minimum of the absolute value of its first derivatives, but moving across the weight space while decreasing the error function does not imply decreasing derivatives, unless the network is really near the minimum of E. Examining the curves produced by the learning rate $= 0.002$ in Fig. 5a and b, it can be seen that, when the training is finished, the level of error is very good, practically 0, but the gradient norm is still relevant, of the same order of that found in the middle of learning. This is the effect of the velocity difference in minimizing E and its

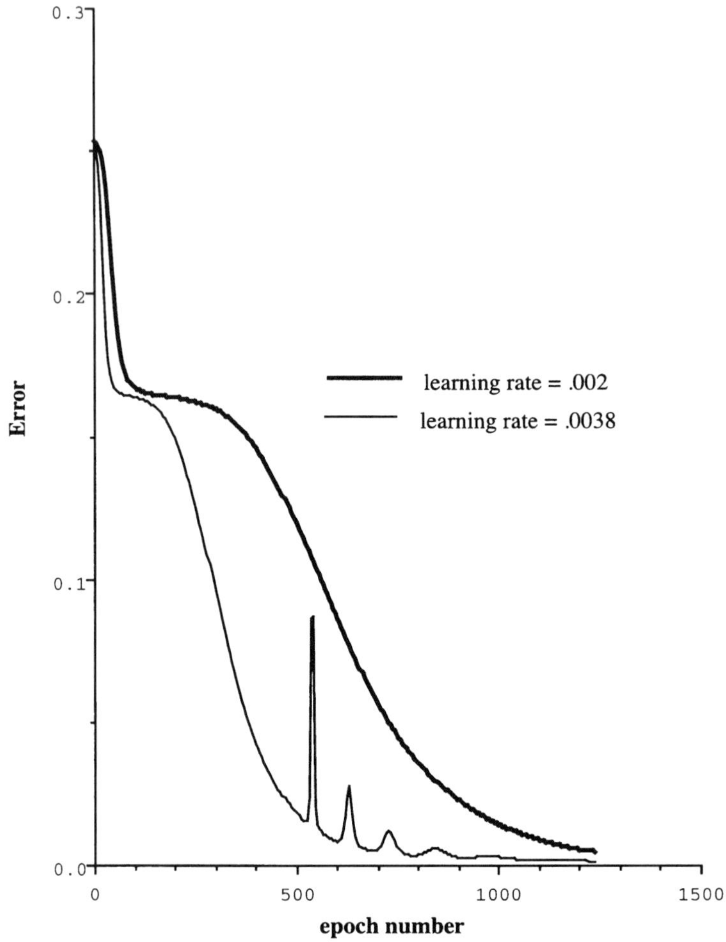

Figure 5a Error evolution in a network with two different learning rates.

derivatives. With enough training (overtraining) both values can be brought to zero.

On the other hand, with a higher learning rate of 0.0038, the learning curve fluctuates but arrives faster at the minimum. Notice that, in the last part of the training, the curve stabilizes and descends uniformly, arriving at a level three times lower than before, but nevertheless the final gradient norm is huge. Even a minimal modification of the weights will produce catastrophic forgetting. The different versions of back propagation are normally used with the highest learning rates that allow faster training in the long term. As a consequence, on many occa-

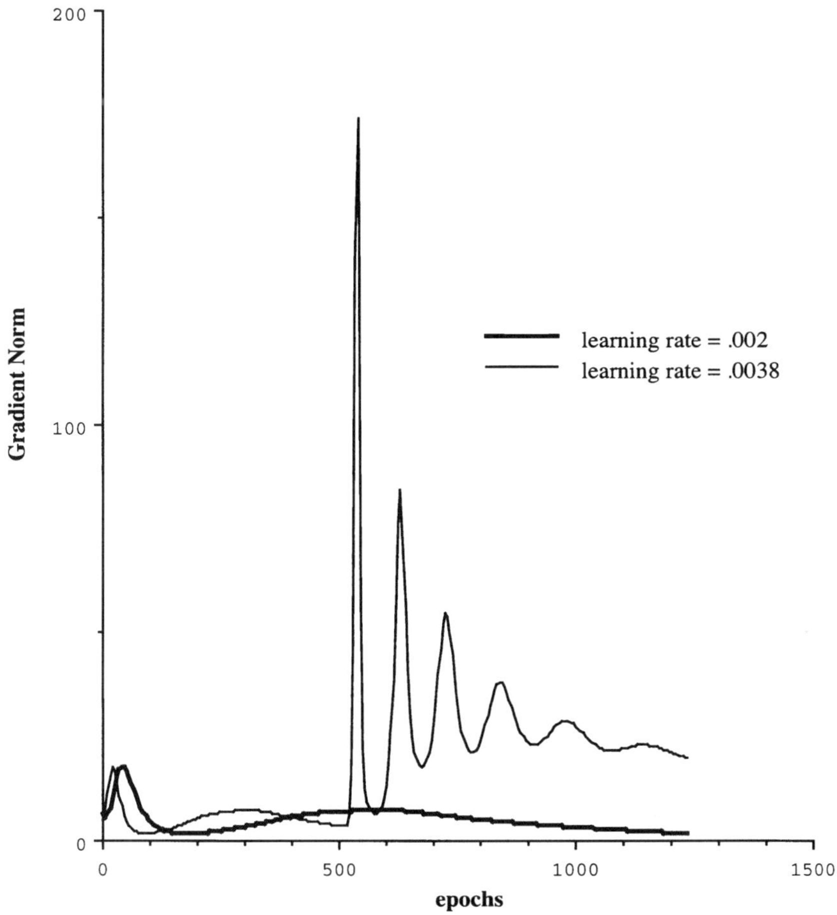

Figure 5b Gradient norm evolutions along the learning corresponding to the training curves in Fig. 5a.

sions (even if the error function is always decreasing) the network is often out of the bottom of the error function valleys, in points with high derivatives. Then if one wants to alleviate the perturbation effects in a given stadium of the learning, the best one can do is to minimize locally the derivatives of E following for a certain time the true gradient of E. This will bring the network to the bottom of the current valley. To follow the gradient line, one can make some steps of back propagation with a very cautious learning rate or, more efficiently and safely, one can use one of the algorithms presented in Section IX, which find thus here still another use.

C. Relation of Learning with Minimal Degradation with Pruning

The standard version of LMD, without b_{jk} and with c_{jk} as second derivatives in the minimum of E, minimizes the same function (constrained by the new pattern) as Le Cun *et al.* [24] in their pruning procedure, our c_{jk} being their sensitivities. In a certain sense, our technique can be seen as opposed to pruning. Pruning detects the less profited weights to eliminate them. Instead, LMD uses them to introduce new information. The relation with pruning suggests that advances in pruning techniques can be incorporated into LMD. For example, some authors have recently used weight sensitivities that go beyond the strict locality of second derivatives [25].

XIII. CONCLUSION

There are easy ways to overcome the catastrophic interference problem by using incremental architectures. In fact, a local hidden unit can be added each time an erroneous new pattern arrives, centering its receptive field in the input pattern and assigning an appropriate value to the hidden-output weight in order to correct the error. It is enough to make the receptive field as small as necessary to mitigate interference up to a desired degree.

However, there are reasons to develop methods compatible with fixed architectures:

1. In an always changing environment, the network would tend to grow indefinitely.
2. More importantly, even when adding a new unit, the new information should somehow be shared with the rest of the network. Otherwise, if each unit only assumes responsibility for a point, one gets a kind of look-up table network. When adding a new unit is considered necessary, it is advisable to correct the new pattern as much as possible with the rest of the network by applying the methods developed to minimize damage over the previous patterns in the case of fixed architectures. For this reason, it is important that these methods are compatible with the use of local units.

The problem of catastrophic interference can be divided into two subproblems. The first (retroactive interference) arises when one tries to introduce new information in a previously trained network, without disrupting the stored information. The second is how to encode a learning set of patterns in such a way that the retention of this set is maximized when some new unknown information has to be stored. The second problem is tackled in [21, 26], where its relation with other connectionist problems is discussed.

This chapter has been concerned with the retroactive interference subproblem, which is solved with the LMD algorithm, based on the relation found between the constrained minimization that represents this problem and the unconstrained minimization of an implicit function over the hidden-unit activations. Full parallelism, both within and between layers, is one of the features of LMD. We implemented two versions of LMD, the coarse one, which searches for the nearest solution in the weight space, and the standard one, based on the information provided by the second derivatives of the weights. We have demonstrated the good scaling properties of the algorithm, as well as the solution quality and computational savings derived from its application. The experiments also revealed that when the learning rate tends to zero, back propagation approximates the coarse version of LMD. The latter algorithm can control the comparative importance of the previous data by weighing each pattern through the coefficients c_{jk} in the error function, that is, using $c_{jk} = \sum_p \alpha(t_p)(\partial^2 E_p/\partial w_{jk}^2)$, where $\alpha(t_p)$ is the weight of each pattern p, depending on the time elapsed since the pattern was presented to the network for the first time. This feature can be useful in applications of the moving-window type, strengthening, for example, the remembrance of the last presented patterns. Furthermore, if the coefficients are not rigidly used to estimate the error function, LMD allows a great flexibility in controlling how much of a pattern is learned by each individual connection, each unit, or each layer.

Two measures and algorithms for measuring the costs of going from one point to another of the weight space have been developed. These algorithms have characteristics that make them interesting for some other purposes, some of which have been pointed out. They could be interesting also for evaluating the goodness of some initial weights for faster learning [27].

The variety of strategies possible for the use of LMD is worth further study. A less raw application of LMD could be beneficial or even necessary, as Section II suggests. The question is what fraction of the new pattern error should be reduced with LMD? The answer when tackling the minimization of $E(W)$ will depend on the number of patterns already stored and on the average error increment expected to be required to modify the network outputs for the pattern, but in any case the new pattern should not be forced to be less erroneous than the previous patterns. On the other hand, we normally want to optimize generalization, which provides another constraint: the error should not be reduced beyond the variance of the noise of the output data. LMD could also be combined with back propagation, by applying it in each back propagation cycle to the most erroneous pattern, which would help to avoid local minima and, as we are beginning to experience, seems to favor learning enormously.

Finally, a general-purpose algorithm based on LMD can be devised. This algorithm would benefit from direct manipulation of the hidden-unit representations, just as the algorithms presented in [28–30] do. The rationale is that, because LMD can exactly control the network output for each pattern, it is possible to implement a gradient descent algorithm at this level. If these variables were independent, due to the quadratic shape of $E(W)$ with respect to the output of the network for the training patterns, the algorithm would bring the network to the minimum in a single step. However, the network outputs for the different patterns are dependent and, although LMD allows us to change some of them while minimizing the effect on the others, shorter steps should be taken. Let Y_p be the network output for pattern p. Then the fraction of $(O_p - Y_p)$ reduced in each cycle would be regulated by a general parameter for all patterns, which should be enlarged or reduced depending on how much the result of applying LMD to all patterns in the previous step deviated from the straight line (in network output space) leading to the minimum.

XIV. ACKNOWLEDGMENT

Carme Torras acknowledges partial support from the Commission of the European Union under contract 8556 (NeuroColt).

REFERENCES

[1] R. Ratcliff. Connectionist models of recognition memory: constraints imposed by learning and forgetting functions. Psychological Rev. 97:235–308, 1990.
[2] M. McCloskey and N. J. Cohen. Catastrophic interference in connectionist networks: the sequential learning problem. In *The Psychology of Learning and Motivation* (G. H. Bower, Ed.). Academic Press, New York, 1989.
[3] R. M. French. Using semi-distributed representations to overcome catastrophic forgetting in connectionist networks. Technical Report 51-1991, Center for Research on Concepts and Cognition, Indiana University, 1991.
[4] J. M. J. Murre. Categorization and learning in neural networks. Ph.D. Thesis, University of Leiden, 1991.
[5] P. A. Hetherington and M. S. Seidenberg. Is there catastrophic interference in connectionist networks? In *Proceedings of the 11th Annual Conference of the Cognitive Science Society*, pp. 26–33. Erlbaum, Hillsdale, NJ, 1989.
[6] A. Robins. Catastrophic forgetting, rehearsal and pseudorehearsal. *Connection Sci.* 7:123–146, 1995.
[7] O. P. Brouse and P. Smolensky. Virtual memories and massive generalization in connectionist combinatorial learning. In *Proceedings of the 11th Annual Conference of the Cognitive Science Society*, pp. 380–387. Erlbaum, Hillsdale, NJ, 1989.
[8] P. A. Hetherington. The sequential learning problem in connectionist networks. Master's Thesis, Department of Psychology, McGill University, Montreal, 1990.
[9] J. K. Kruschke. Human category learning: implications for backpropagation models. *Connection Sci.* 5:3–36, 1993.
[10] J. Moody and C. Darken. Fast learning in networks of locally tuned processing units. *Neural Comput.* 1:281–294, 1989.
[11] T. Poggio and F. Girosi. Networks for approximation and learning. *Proc. IEEE* 78:1481–1497, 1990.

[12] J. Platt. A resource-allocating network for function interpolation. *Neural Comput.* 3:213–225, 1989.
[13] D. C. Park, M. A. El-Sharkawi, and R. J. Marks II. An adaptively trained neural network. *IEEE Trans. Neural Networks* 2:334–345, 1991.
[14] V. Ruiz de Angulo and C. Torras. On-line learning with minimal degradation in feedforward networks. *IEEE Trans. Neural Networks* 5:657–668, 1995.
[15] A. Blum and R. L. Rivest. Training a 3-node neural network is NP-complete. In *Proceedings of the 1988 Workshop on Computational Learning Theory*, pp. 9–18. Morgan Kaufmann, San Mateo, CA, 1988.
[16] S. Judd. *Neural Network Design and the Complexity of Learning*. MIT Press, Cambridge, MA, 1990.
[17] B. Widrow and R. Winter. Neural nets for adaptative filtering and adaptative pattern recognition. *Computer*, 1988.
[18] S. Becker and Y. Le Cun. Improving the convergence of back-propagation learning with second order methods. In *Proceedings of the 1988 Connectionist Models Summer School* (D. Touretzky, F. Hinton, and T. Sejnowski, Eds.), pp. 29–37. Morgan Kaufmann, San Mateo, CA, 1988.
[19] R. Scalatter and A. Zee. Emergence of grandmother memory in feedforward networks. Learning with noise and forgetfulness. In *Connectionist Models and Their Implications, Readings from Cognitive Science* (D. Waltz and J. H. Feldman, Eds.), pp. 309–323. Ablex, Norwood, NJ, 1988.
[20] L. P. Ricotti, S. Ragazzini, and G. Martinelli. Learning word stress in a suboptimal second order back-propagation neural network. In *Proceedings of the IEEE International Conference on Neural Networks*, San Diego, Vol. 1, pp. 355–364. IEEE, New York, 1990.
[21] V. Ruiz de Angulo. Interferferencia catastrófica en redes neuronales: soluciones y relación con otros problemas del conexionismo. Ph.D. Dissertation, Basque Country University, 1996.
[22] G. E. Hinton and T. J. Sejnowski. Learning and relearning in Boltzmann machines. In *Parallel Distributed Processing: Explorations in the Microstructure of Cognition. Foundations* (D. E. Rumelhart and J. L. McClelland, Eds.), Vol. 1. MIT Press, Cambridge, MA, 1986.
[23] J. Mijhuis, K. Hofflinger, V. S. Haik, and L. Spaanenburg. Limits to the fault-tolerance of a feedforward neural network with learning. In *Proceedings of the 20th International Symposium on Fault Tolerance Computing*, Newcastle upon Tyne, pp. 228–235, 1990.
[24] Y. Le Cun, J. S. Denker, and S. A. Solla. Optimal brain damage. In *Advances in Neural Information Processing Systems*. (D. S. Touretzky, Ed.). Morgan Kaufmann, San Mateo, CA, 1990.
[25] E. D. Karnin. A simple procedure for pruning back-propagation trained neural networks. *IEEE Trans. Neural Networks* 1:239–242, 1990.
[26] V. Ruiz de Angulo and C. Torras. Random weights and regularization. In *Proceedings of the International Conference on Artificial Neural Networks* (G. Marinaro and P. Morasso, Eds.), pp. 1456–1459. Springer-Verlag, Berlin/New York, 1994.
[27] S. Gavin. Designing multilayer perceptrons from nearest-neighbour systems. *IEEE Trans. Neural Networks* 3, 1992.
[28] T. Grossman, R. Meir, and E. Domany. Learning by internal representations. *Complex Systems* 2:555–575, 1988.
[29] A. Krogh, C. J. Thorbergsson, and J. A. Hertz. A cost function for internal representation. In *Advances in Neural Information Processing Systems* (S. Touretzky, Ed.). Morgan Kaufmann, San Mateo, CA, 1990.
[30] R. Rohwer. The moving target training algorithm. In *Advances in Neural Information Processing Systems* (D. S. Touretzky, Ed.). Morgan Kaufmann, San Mateo, CA, 1990.

Constraint Satisfaction Problems

Hans Nikolaus Schaller
DSITRI
D-80798 Munich, Germany

I. CONSTRAINT SATISFACTION PROBLEMS

How do we usually solve problems like the seating arrangement for a banquet (Fig. 1), scheduling a meeting, finding time tables, or placing modules onto a printed circuit board (Fig. 2)? Other than in certain special cases, we have to start to examine tryout solutions and modify them until everything fits. If we begin to do this by hand, we rather quickly make compromises or give up because the problem is too difficult. Then we begin to ask for machine support.

But what makes this type of problem much more difficult to solve than, for example, the calculation of a square root or solving a linear equation system? It is more difficult because it is only known which properties solutions must have and which properties they do not have. And a general algorithm to find solutions other than by more or less systematically trying to verify all possible solutions is not known.

Problems of this type are called *constraint satisfaction problems* (CSPs). If the variables are restricted to a discrete domain, they are called more specifically *discrete constraint satisfaction problems* (DCSPs). CSPs and DCSPs have been studied intensively in the artificial intelligence (AI) literature and many solution procedures have been described. Several neural networks have been proposed to overcome drawbacks of the AI methods. This was initiated by a publication by

Optimization Techniques

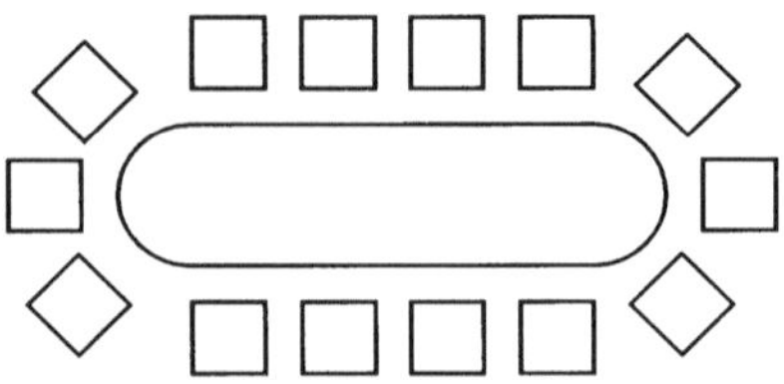

A not neighbour of B
D not opposite to E
H more than two places away from B
if A neighbour of D then E not neighbour of K
two of A, F or G at both ends of the table
H on the right hand of C
...

Figure 1 Seating arrangement problem.

Hopfield in 1982 [1]. The driving factor of this research is the paradigm of inherently parallel computation, the resemblance of the human brain, and the ability to handle soft decisions which could enable them to find CSP solutions that the discrete AI algorithms do not find. In this chapter, we will discuss and compare all of them to understand the relationship of the paradigms and to identify common ideas.

CSPs are closely related to *optimization problems* (OPTs) where we want to know the best solution among many, according to a given optimization criterion. Including CSPs, we can classify by the variable domain (discrete or continuous) the existence and type of contraints and the type of optimization target function (Table I).

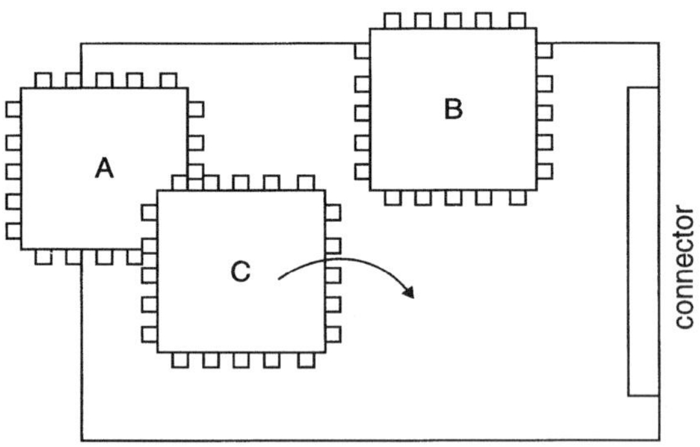

Figure 2 Placement of modules.

Table I

Some Common Optimization Problem Groups

Problem	Abbreviation	Variable domain	Constraints	Target function
Constraint satisfaction	CSP	Continuous	Yes	—
Discrete constraint satisfaction	DCSP	Discrete	Boolean relation	—
(General) optimization	OPT	Continuous, discrete	—	Any function
Linear optimization	LINOPT	Continuous	Linear inequalities	Linear function
Quadratic optimization	QOPT	Continuous	Linear inequalities	Quadratic function
Nonlinear optimization	NL-OPT	Continuous	Linear inequalities	Nonlinear function
Constrained optimization	COPT	Discrete	Boolean relation	Any function
0/1-integer optimization	0/1-OPT	Continuous, discrete	—	0/1

The classification is not unique, as, for example, constraints can be added to the target function in the form of penalty functions, and CSPs are, in fact, optimization problems where solutions are optimal and nonsolutions are worse. Additionally, optimality can be viewed as an additional constraint. Because of this, it is difficult to tell which of both classes of problems is more difficult.

Now, let us assume that we are working on a technical product development project and have identified that a certain task of the job can be easily described as solving a CSP. This may be, for example, the problem of finding a placement for N optical processing elements in a two-dimensional-plane so that all elements can be reached by light beams from eight directions (Fig. 3) [2].

Because the constraints (the size N, the number of directions, the dimensionality, etc.) of this task may have to be dynamically modified in an unpredictable manner, we have to abandon the development of a specialized algorithm.

So we are now trying to find a general solution strategy for this CSP that can be run effectively on a microcontroller or some special eletronic hardware. We will have to search through the literature and then do an assessment of the methods we find. To get information on the methods, we will take a look at traditional state space search methods (depth first search), more advanced stochastic methods like simulated annealing and genetic algorithms, and last, but not least, neural network techniques.

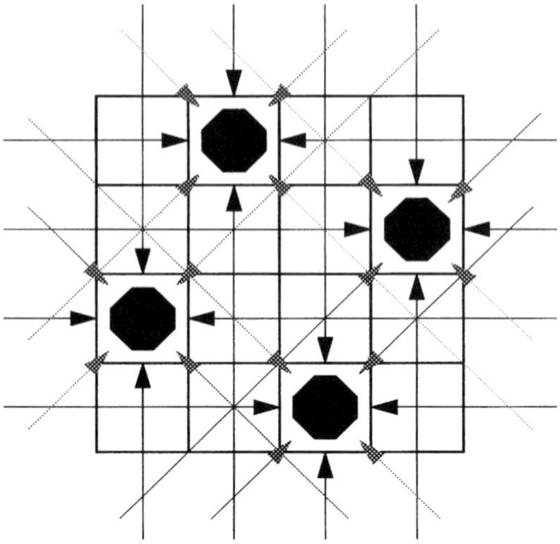

Figure 3 Optical processor placement.

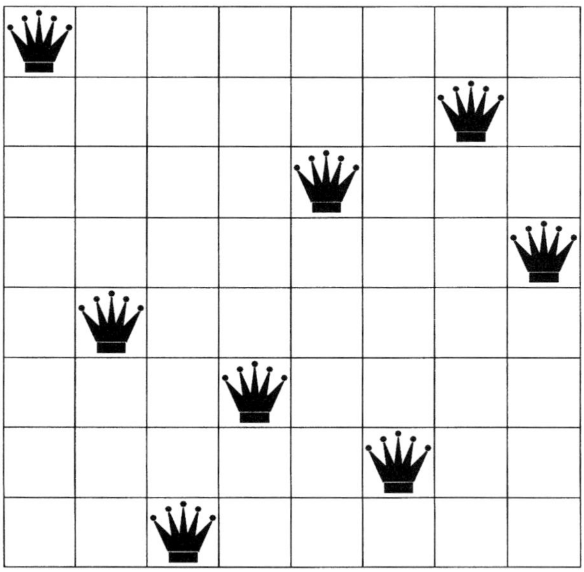

Figure 4 One of the 92 solutions of the 8-queens problem. Reprinted from H. N. Schaller, Design of neurocomputer architectures for large-scale constraint satisfaction problems, *Neurocomputing* 8:315–339, 1995 with kind permission from Elsevier Science–NL, Sara Burgerhartstraat 25, 1055 KV Amsterdam, The Netherlands.

To make this assessment fair, we take the optical processor placement as a benchmark problem to compare the quality of the methods. It is structurally identical to the well-known N-queens problem (NQP) which was posed in 1850 by Carl Friedrich Gauss [3] (according to [4] by Nauck). In this problem, we take a chess board and want to know how to place N queen figures on the board so that they do not attack each other according to the rules of the game. These nonattacking requirements are the constraints for the possible placements.

It is known that the problem has solutions for $N = 1$ and all values of $N \geq 4$. An example of such a placement for $N = 8$ is shown in Fig. 4. Seven more solutions result by mirroring and rotation and all eight build a family. The other solutions belong to 11 families with either four or eight solutions. Together, there are 92 solutions for $N = 8$ [5].

II. ASSESSMENT CRITERIA FOR CONSTRAINT SATISFACTION TECHNIQUES

Before we start to look at any of the proposed methods, we must clarify the properties we want to have for a practically useful constraint satisfaction method. We can get ideas for the development of criteria by studying the theory of computational complexity, the resource requirements (memory capacity and circuit complexity) and speed, techniques of parallel computation, computer architecture, and examples of applications.

A. P and NP Problems, Complexity Theory

We must always be aware that there are principal limitations described by the theory of computational complexity. Besides a better understanding of the limitations, a look into this theory will give us a metric for differentiating the resource requirements and the speed [6].

The computation time and the requirements for space resources, which means the memory capacity and the circuit complexity, usually grow with the size of the problem. This dependency can be given exactly, but it is commonly summarized by the O notation which describes only the most significant component. For example, $O(n^k)$ means that the time or resource depends on a problem parameter n (e.g., the number of variables) by a polynomial of the degree k. $O(2^n)$ stands for exponential dependency on n, and $O(1)$ for independence.

Computational complexity theory classifies problems to be computed as polynomial (P) and nondeterministic polynomial (NP) problems. This means a separation of problems that can be solved on a deterministic computer and problems that require a nondeterministic computer (i.e., one with an oracle). A nondeterministic computer does not exist physically, of course, so it is interesting how fast NP problems can be solved on a deterministic computer.

Computational complexity theory now tells us in a well-known, but yet unproven, conjecture, which is a in a popular version:

> A deterministic algorithm that finds solutions of a problem of class NP with polynomial complexity, that is $O(n^k)$, does not exist or P = NP.

This conjecture was published for the first time about 25 years ago, but until now it has neither been proven nor been rejected by presenting a polynomial algorithm for an NP problem. In this direction, much has been anticipated from neural network techniques, as it is assumed that they have a nondeterministic component. Some publications suggest algorithms with polynomial complexity for NP problems [7–9], but the results are not general and have not been verified independently. Therefore, we have to expect exponential growth of the computation time or processor complexity for some of the CSPs, because we cannot classify arbitrary CSPs in advance as P or NP. Remember that it is sometimes simple and sometimes difficult to solve them by hand. Therefore, general large-scale (approximately, $N > 10$) constraint satisfaction problems may require too much processing time or too large a processor to be solved by a machine.

However, we just notice that this conjecture does not say that all problems are difficult nor does it say that a specific CSP must be difficult. Therefore, the picture is not as dark as it may look. We have learned that the general CSP belongs to NP, and there is no general efficient, that is, polynomial algorithm, for NP problems. However, there are still many practical problems belonging to P that can be described as a CSP, and we will look for efficient algorithms to solve them.

B. Scaleability and Large-Scale Problems, Empirical Complexity

To assess the speed of a given algorithm or CSP solving method, the distinction of the underlying problem between P and NP is not of primary importance, but rather the behavior of the algorithm for different problems and problem sizes. This is described more exactly by the *scaleability*.

The scaleability describes how the computation time increases for large-scale variants of a given problem. This requires a problem class, whose members are differentiated by a size parameter, like the N of the N-queens problem. If we now take a problem of size N, run the algorithm to solve the problem, and measure the number of computation steps, we will get the empirical complexity of the problem for a given algorithm.

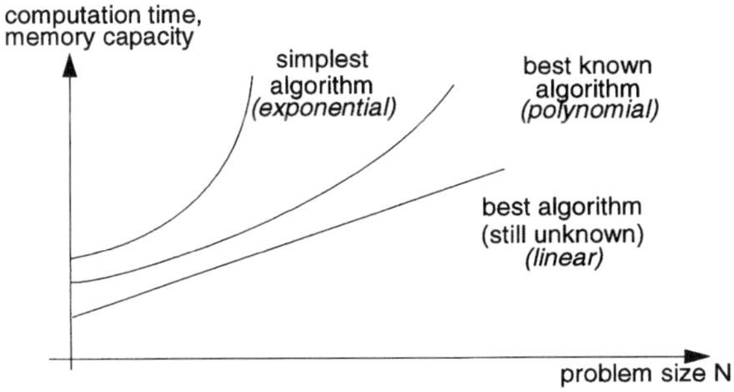

Figure 5 Complexity of algorithms for a given problem.

We can also use the same problem class and compare the speed of several algorithms to determine their relative speed. Or we can vary both the size and the algorithm. All these approaches help us to determine the best and most general algorithm among several.

Possible outcomes for the solution-finding time plotted over the problem size for several algorithms are shown in Fig. 5. Using the appropriate coordinate system, we can easily distinguish between linear, polynomial, and exponential behavior (Fig. 6). We also have to take into account here that some of the algorithms we will compare have a stochastic component, that is, a random noise generator, that throws a die at certain decision points. This random noise introduces unpredictability similar to the oracle of a nondeterministic computer. Note, however, that nondeterminism is not the same as stochastic.

To make the results comparable, we have to repeat several runs for the same problem size and then take some mean value of the individual solution-finding time measurements. Neither the arithmetic nor the geometric mean value is appropriate in this case. Rather, we should use the median (i.e., the value separating the sorted list of values in two equally sized halves) and the span (not the standard deviation) to describe the variance.

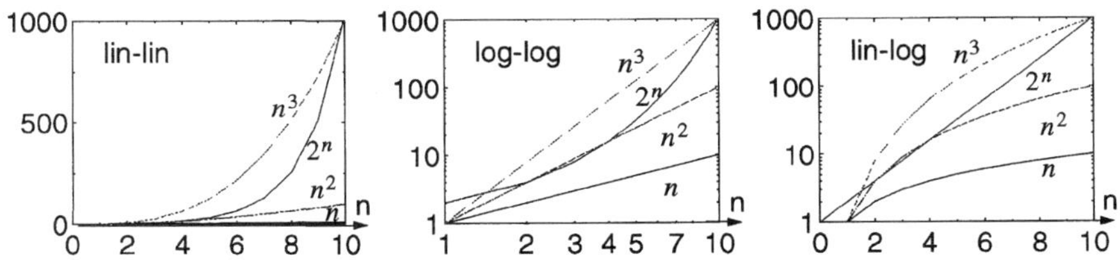

Figure 6 Exponential, polynomial, and linear dependency in different coordinate systems.

From this discussion, we have the first criterion:

(C1) polynomial empirical complexity

Some algorithms, especially those based on stochastic components like neural networks, simulated annealing, and genetic algorithms, introduce control parameters (e.g., the random initialization, the cooling curve, the mutation rate). These parameters have to be selected carefully to make the algorithm converge at all. Unfortunately, the optimal parameter may depend on the given problem and there does not exist a method to determine the parameter automatically. Otherwise, it could become part of the algorithm and will no longer be a parameter. Therefore, applying these algorithms contains a "good luck" component, which makes them very unreliable for practical purposes. Therefore, we require a

(C2) low number of control parameters

The stochastic component itself makes the result not repeatable. Therefore, we require a

(C3) low influence of stochastic components

C. Parallelization

A well-known paradigm of electronic computation is to use several processors if a single one is too slow. This requires parallelization of the algorithm to solve the given problem. In the last few years, there has been much success in the parallelization of computational algorithms, as supercomputers have shown with their TeraFLOPs computing power. And there are also clusters of workstations or multiprocessor personal computers (PCs) on the market, especially database servers.

Classical parallelization is based on the paradigm of communicating sequential processes. The limitation lies in the growing demand for communication with an increasing number of processors. In many cases, the increase in speed is limited to a factor in the range of 4 to 10 even with a large number of processors. Therefore, much research has been done to improve the communication network, of which the Internet has become the most prominent.

Today, the parallelization of algorithms is supported on the operating system level by tasks and threads, semaphores, mailboxes, communication channels, pipes, shared variables, clients and servers, etc. A different approach tries to hide the communication from the programmer and adds parallel operators to programming languages to give the compiler hints to generate parallel code. Finally, operations can be parallelized on the machine instruction level by pipelining. This technique is used in most modern microprocessors. Modern compilers also try to identify code that has inherent parallelism (e.g., vector operations) and generate appropriate parallel code.

The neural network paradigm, also called *parallel distributed processing* (PDP), has opened a very different approach to parallel operation. It assumes a large number of rather simple processors, called neurons, and an interconnection network that also participates in the operation by weighting the values sent over the network. Another major difference to sequential computers is the lack of a program that defines the steps of the calculation. The main hope is that the communication bottleneck can be overcome.

This leads to a new criterion

(C4) simple algorithm structure that can be parallelized without scaling loss

D. Design Principles for Computer Architectures

For practical applications, a well-defined architecture makes the understanding, integration, and modification of a computing device much more simple. Therefore, we must also take a look at the quality of the architecture of the CSP solving technique. Blaauw [10] defines three computer *design levels*:

1. programming model (architecture)
2. implementation
3. realization

The *programming model* describes the logical view of the machine for a programmer. The programmer need not be concerned about how the commands are implemented. The *implementation* is the view of the functional organization. Most notably, there may be different implementations of the same programming model (compare, for example, the 80386 and 80486 microprocessors) which makes software portable. And the same implementation may have a different *realization* by electronic devices, for example, different silicon technology. Although these levels were originally introduced for mainframe computers and then used for microprocessors, they are very general for all computation devices. Because of this, we will compare whether the CSP solving technique follows this architectural view. Remember, we are interested in solving CSPs on a machine, and perhaps we want to replace the machine with a newer or different one but we do not want to reformulate the CSP again. From this, we formulate the criterion

(C5) well-defined architecture

Blaauw has also defined *design principles* for the design of a high-quality programming model. From the principles of orthogonality, symmetry, generality, transparency, open-endedness, and completeness, the generality (appropriateness for many different problems) and open-endedness (extensibility) can be easily transferred to a CSP language. Therefore, we want to have a

(C6) general and open-ended CSP language

E. Examples of Constraint Satisfaction Problems

We have now collected many criteria for the assessment of constraint satisfaction methods, but have not yet taken a look at examples and applications. Let us check whether we can learn additional criteria.

Table II lists applications that can be mapped to constraint satisfaction problems taken from the neural network literature. An example for the polyominoes placement puzzle is shown in Fig. 7. The N-coloring problem and the NQP are described in the following sections.

Table II

Constraint Satisfaction Problems from the Neural Network Literature

Problem	Application area	Reference
Analog–digital converter	Circuit design	[11, 12]
Frequency planning for cellular mobile radio	Telecommunications	[13, 14]
Switching	Telecommunications	[15–19]
Error correction	Circuit design	[20]
Circuit layout	CAD	[21–23]
Steiner tree for circuit layout	CAD	[24]
Digital circuit test pattern generation	CAD	[25, 26]
Composition of harmonic music	Music	[27]
N-coloring of a graph	Scheduling	[28–30]
Assignment problem	Scheduling	[31, 32]
Scheduling	Scheduling	[10, 33–35]
Polyomino placement puzzle	Puzzle	[36–38]
3-satisfaction problem	Mathematical logic	[39]
Knight's tour problem	Puzzle	[7, 40]
N-queens problem (NQP)	Puzzle	References to Table V
Jigsaw puzzle	Puzzle	[41, 42]
Factorization of large numbers	Cryptography	[41, 42]
RNA analysis	Biochemistry	[8]

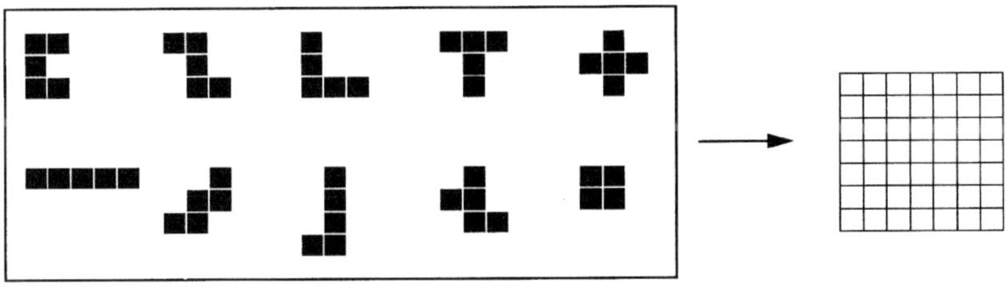

Figure 7 The 7 × 7 polyominoe puzzle.

1. *N*-Coloring Problem

Many technical resource allocation problems, like the register assignment in compilers, scheduling in automation, switching etc., are similar to the coloring problem in map drawing. The question is how to assign colors to a set of areas so that adjacent areas do not get the same color assigned. Otherwise, they cannot be distinguished visually any more. The structure of the areas can be represented by an adjacency graph and colors are assigned to the nodes (Fig. 8).

This problem brings up the topic of the solvability of problems. For a long time, it has been an open question of mathematics whether the four-coloring problem is really solvable for arbitrary graphs. Recently, there has been a proof, but we still can formulate a three-coloring problem as a CSP that has no solutions. Therefore, we also want to have our solver indicate if the given problem is unsolvable, that is, overconstrained. Thus, we require the

(C7) detection of unsolvable problems

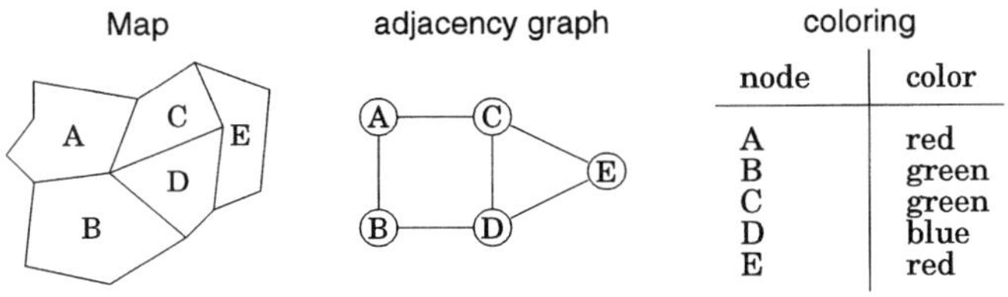

Figure 8 A map-coloring problem.

2. N-Queens Problem

What if there is more than a single unique solution to the problem? We already know this for the NQP. We could expect that the solver identifies all solutions, one after the other. However, we have to note that the number of solutions may grow exponentially with the size of the problem, and, for most practical applications, a single schedule, a single circuit placement etc. sufficies. Therefore, we expect that the solver generates one of the solutions, but, to be fair, with the same probability as any other solution. Thus, we require

(C8) fair selection between several solutions

The NQP is good to describe the aspect of variable encoding [43]. For the NQP, which asks for the placement of N queen figures on an $N \times N$ chess board, we have several options. Three of them are:

(Q1) N variables for all N rows that indicate the column position $(1, \ldots, N)$ of the queen figure in that row.
(Q5) N variables for the figures; the domain, that is, the possible values, is the (x, y)-positions of the board.
(Q6) $N \times N$ binary variables for each position that indicate whether a queen is placed on the position (value 1) or not (value 0).

For all these different encodings, a different set of contraints for the values of the variables has to be formulated. For Q6, for example, the constraints are

(NQP1) exactly one queen per row
(NQP2) exactly one queen per column
(NQP3) at most one per falling diagonal
(NQP4) at most one per rising diagonal

All the encodings describe the same problem and will result in the same set of solutions. However, we can expect that the speed of the algorithm differs largely between these different encodings. And an algorithm may even not be appropriate for a given encoding or fail to find a solution at all. So we have an additional criterion, the

(C9) unaffectedness by different problem formulations

F. SUMMARY

Finally, we are interested in hardware implementations, and therefore we want to check if

(C10) a hardware implementation is reported

Table III

Assessment Criteria for CSP Solving Techniques

C1	exp.	Polynomial (empirical) complexity
C2	dir.	Low number of control parameters
C3	dir.	Low influence of stochastic component
C4	dir.	Simple algorithm structure that can be parallelized without scaling loss
C5	dir.	Well-defined architecture
C6	dir.	General and open-ended CSP language
C7	dir.	Detection of unsolvable problems
C8	exp.	Fair selection between several solutions
C9	exp.	Unaffectedness by different problem formulations
C10	dir.	A hardware implementation is reported

Now, let us summarize all the assessment criteria we have defined (Table III). The second row indicates how we can check these criteria. We either have to do experiments (exp.) or can check it directly (dir.).

III. CONSTRAINT SATISFACTION TECHNIQUES

The literature is rife with methods for solving problems. They can be separated into three major groups (Fig. 9): algebraic methods and two numeric methods. The numeric methods can be subdivided into global search and local search methods.

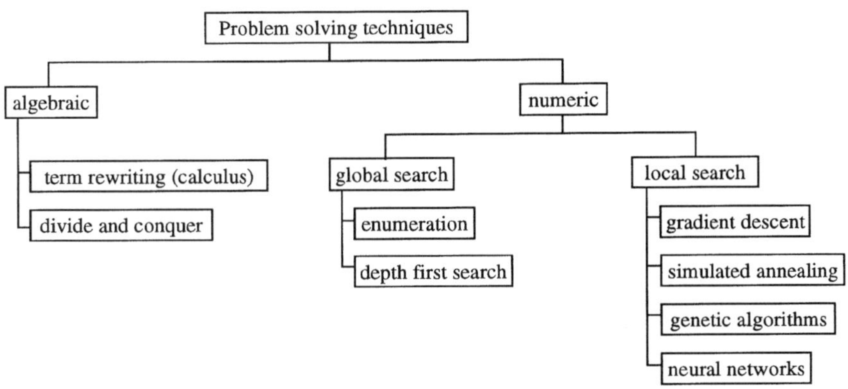

Figure 9 Problem-solving techniques.

We will skip the algebraic methods (term rewriting, divide and conquer) here and concentrate on the search methods.

Before we describe the constraint satisfaction techniques in detail, we define a CSP formally as:

1. a set of variables z_i,
2. a domain of values D_i for each variable z_i,
3. a set of relations over these variables (constraints).

For a discrete CSP, the domains have a discrete set of values. Note, however, that the discreteness itself may be seen as a constraint to a continuous problem. The set of relations can be described by formulas. An example of a constraint set over $\{z_1, z_2, z_3, z_4\}$ is

$$1 \leq z_1 + z_3 \leq 3, \qquad 2 \leq z_1 + z_2 + z_4 \leq 3. \tag{1}$$

A. Global Search

1. Enumeration

Enumeration is a very simple, safe, but very inefficient method to solve discrete CSPs. It simply tries all combinations of values out of the variable domain—one after the other—until a solution has been found. This process requires an exponentially growing number of steps for larger problems, as the full state space will be enumerated. Therefore, it is practically useless.

2. Depth First Search

To speed up the enumeration, the depth first search (DFS) method has been developed. It assigns the first value to the first variable and then checks if any constraint has been violated even by this partial assignment. If not, it assigns the first value to the second variable and so on, until all variables are assigned. Then a solution has been found. However, if any constraint is violated during a step, the search method goes back (backtracking) to the previous variable and tries the next value. This is also called pruning of the search tree, as all these partial assignments can be ordered as the nodes of a tree structure.

The order in which variables are assigned and the order in which values are taken from the domain dramatically influence the number of steps until the first solution is found. Unfortunately, there is no method to define this order optimally. Therefore, heuristics have to be applied, for example, the "most constrained variable first" rule. With these rules, the DFS method is successful for many small-scale problems and has found an application in expert system shells.

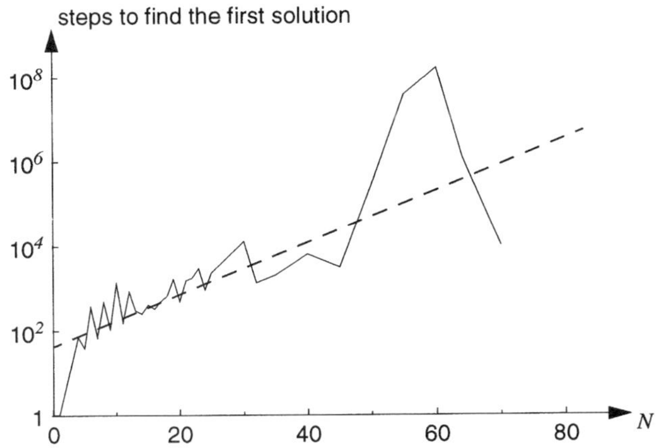

Figure 10 Number of steps required to find the first solution of the N-queens problem by sorted, pruned depth first search in lin-log coordinates. Reprinted from H. N. Schaller, Design of neurocomputer architectures for large-scale constraint satisfaction problems, *Neurocomputing* 8:315–339, 1995 with kind permission from Elsevier Science–NL, Sara Burgerhartstraat 25, 1055 KV Amsterdam, The Netherlands.

The depth first search method (backtracking) can be parallelized [44]. In some cases, the speed-up may be superlinear, which can be explained by the nonuniform distribution of solutions within the search space. Generally, however, it is limited because the communication overhead required for effective load balancing methods increases with the number of processors [45].

Experimental results with the NQP have shown that the empirical complexity of an optimized depth first search grows exponentially with N (Fig. 10). We could not find, for example, solutions for $N = 80$ within 10^9 trials (several days of computation time on a RISC workstation) [46]. The same result has been reported by other authors [47].

B. Local Search

A different approach is based on the paradigm of minimization of the number of errors E (number of conflicts, number of violated constraints, "energy") in a certain test assignment $\mathbf{z}$. This is described by the generate and test principle shown in Fig. 11. When started, the generator produces vectors $\mathbf{z}$, which are evaluated by the tester. The tester returns the value E which also indicates that a solution has been found and the search process can stop (finished). When we are interested in the next solution, we can continue the search process. Ideally, the solver can indicate unsolvable problems. The problem to be solved is directly put

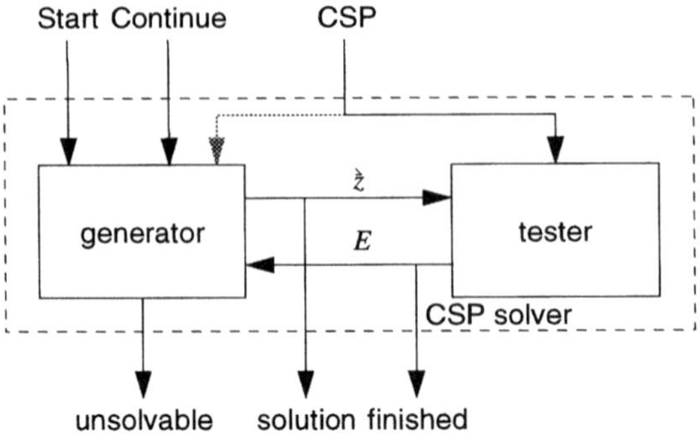

Figure 11 CSP solver based on the generate and test paradigm.

into the tester. To do this, we can implement an algorithm or define a nonlinear function consisting of penalty terms for violated constraints.

Unfortunately, there are methods for the generator that need some information in the form of control parameters about the problem to be solved. This connection is indicated in Fig. 11 by the dotted arrow.

1. Gradient Descent

To reduce the number of errors E, the *gradient descent* starts at a certain initial state. Then the state is modified to reduce the error. As differential calculus shows, the error is reduced by the largest amount if the method makes a step in the direction of the negative gradient. After several such state change steps, the dynamic system comes to rest in a state with minimal error. This state is also called the *equilibrium*.

There are some difficulties to be overcome. First of all, applying the gradient descent to a discrete error function is not possible. Therefore, the gradient is usually replaced by an error difference when changing a single variable by the value 1. Then the minimum of the error may be local; that is, the system settles in a state with some constraints still violated. To avoid this, a stochastic component is added.

2. Simulated Annealing

Simulated annealing (SA) [48, 49] resembles the annealing of a metal in which a new atomic structure with minimal energy is rebuilt, after heating and then slowly cooling down.

This method starts with an initial state $\mathbf{z}$, generates a new, neighboring state $\underline{\mathbf{z}}$, and compares the value of the target function E for both by building the difference ΔE. If the new energy is lower, the new state is accepted, moving the state toward a minimum of E. If the energy is higher, the new state is accepted with an acceptance probability of

$$P_{\underline{\mathbf{z}}} = \exp\left(-\frac{\Delta E}{kT}\right). \tag{2}$$

Several of these steps are repeated in a cycle at a certain temperature level T. Then the temperature is reduced according to a cooling schedule and a new cycle is started. It can be shown that the probability P to find the system in the state $\mathbf{z}_0$ is described by the Boltzmann distribution

$$P(\mathbf{z} = \mathbf{z}_0) \sim \exp\left(-\frac{E(\mathbf{z})}{kT}\right). \tag{3}$$

As we can see, the probability becomes maximal for those states that have minimal E if the temperature is lowered $T \to 0$. Therefore, the system relaxes to global minima.

The simulated annealing method has some drawbacks. First of all, it lacks a programming model and does not show how to define the error function for a given CSP. Then the selection of a cooling schedule is not straightforward. However, because the schedule determines a compromise between the probability of finding a solution and the computation time, it is a system parameter with unpredictable influence. Finally, the algorithm is generally slow, because the number of cycles at a certain temperature level is high. The algorithm can be easily parallelized, however, which will be described in Section IV.E as the Boltzmann machine.

3. Genetic Algorithms

Genetic algorithms (GAs) simulate the population of individuals. Their properties are described by the set of genes. Three operators are applied to this population: mutation, recombination, and selection. Mutation randomly modifies parts of the genes of an individual. Recombination of the genes of two individuals makes a mixture, while selection removes individuals that have a worse performance than others. By repeatedly applying these operators to a set of random initial individuals, the population becomes more and more homogeneous at best fitness.

Like simulated annealing, this method can be formulated as an optimization algorithm. The genes represent trial assignments to variables and the fitness is the target function. This eliminates bad assignments and favors good solutions. The NQP has been solved by this method for $N = 200$ [50]. Like simulated annealing, genetic algorithms require the selection of system parameters that control the convergence: the mutation rate, the recombination rule, and the number of individuals.

4. Probabilistic Local Search

Rok Sosic and Jun Gu have developed a specialized algorithm called *probabilistic local search* to solve the NQP [2, 4, 51, 52]. They use the variable encoding Q1, which describes the column number of each queen, which is tied up to a certain row. The values of these variables are initially a permutation of the numbers $1, \ldots, N$. This ensures that there is only one queen in each row and column (NQP1, NQP2).

The algorithm calculates the number of conflicts, that is, the number of queens that attack each other on any of the diagonals. Then the algorithm exchanges for all pairs of conflicting queens the column numbers to test, if this exchange reduces the number of conficts. If the conflicts are reduced, the new state is accepted and the next pair is taken.

This algorithm rapidly converges to a solution for larger problems, while it may get stuck in a local minimum for smaller ones. In this case, the algorithm is restarted with a different initial permutation. The authors report that they have solved up to $N = 3000000$ [52].

Although this method is very successful, it is not clear how to transfer it to other CSPs. It benefits from a specialized variable encoding that automatically fulfills two of the constraint sets (rows and columns).

C. Neural Networks

A promising approach pioneered by John Hopfield [1, 53–57] uses recurrent neural networks. He discovered in 1982 that it is possible to decribe the dynamic behavior of a single-layer, symmetrically fed-back network of neurons as the minimization of a quadratic energy function.

This energy function depends on the state of the neurons and is determined by the weights of the feedback synapses. By defining the synapse weights appropriately, it is possible to encode a quadratic optimization problem or a discrete constraint satisfaction problem into the network. After relaxing the network state into an equilibrium, the individual neuron states describe the solution that has been found.

The work of John Hopfield and David Tank became popular, because they could approximately solve a difficult optimization problem that belongs to NP: the *traveling salesperson problem* (TSP) [54, 55].

IV. NEURAL NETWORKS FOR CONSTRAINT SATISFACTION

Neural networks for constraint satisfaction problems are recurrent networks based on a technical neuron model. These technical neurons (units, processing units, processing elements) try to imitate the average spike rate (activity) of biological neurons.

A neuron is generally defined as follows (Fig. 12):

- The *input singnals* (e_s) of a neuron are weighted by *synapses* (w_s) and are superimposed linearly to a *threshold value* w_0.
- The resulting *activation* (x) modifies the *soma potential (state u)* of the neuron with a certain time dependency (T).
- The *output signal* (z) is derived from the soma potential by a *nonlinear function* $g(u)$.

This is described by a set of equations:

$$x = \sum_s w_s e_s + w_0, \tag{4a}$$

$$u(t + \Delta t) = f_T\big(u(t), x(t), \Delta t\big), \tag{4b}$$

$$z = g(u). \tag{4c}$$

The weighting factors w_s determine the input–output transfer function of the neuron, and the function f describes the dynamic behavior (delay line, integrator, low-pass) of the neuron. The nonlinear function $g(u)$ (Fig. 13) is usually a threshold function (Heaviside function) or a differentiable sigmoid function, for example, $g(u) = \frac{1}{2}(1 + \tanh u)$.

An algorithm that can be represented by such neurons is called a *neural algorithm*.

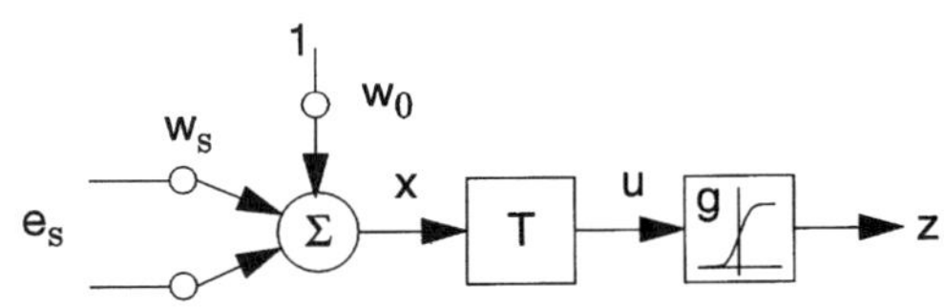

Figure 12 General definition of a technical neuron.

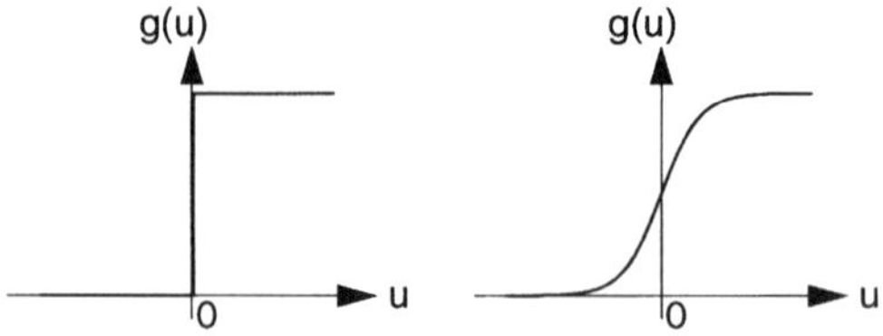

Figure 13 Threshold and sigmoid function.

A. Hopfield Networks

1. Continuous Hopfield Network

A *Hopfield network* [53] (Fig. 14) consists of a single layer of technical neurons that are fully and mutually interconnected and have a bias input I_j (w_0 in the neuron model). The weights are described by the symmetrical weight matrix T_{ij}. Additionally, the diagonal vanishes ($T_{ii} = 0$) to make the dynamical system stable. The neurons multiply the output voltages V_j by the weights T_{ij} and change their soma potential u_i accordingly. The output voltage $V_i^0 \leq V_i \leq V_i^1$ (z in the neuron model) depends on the soma potential and is described by the function $g_i(u_i)$. Neurons with a higher voltage level are called active, with a lower they are called inactive.

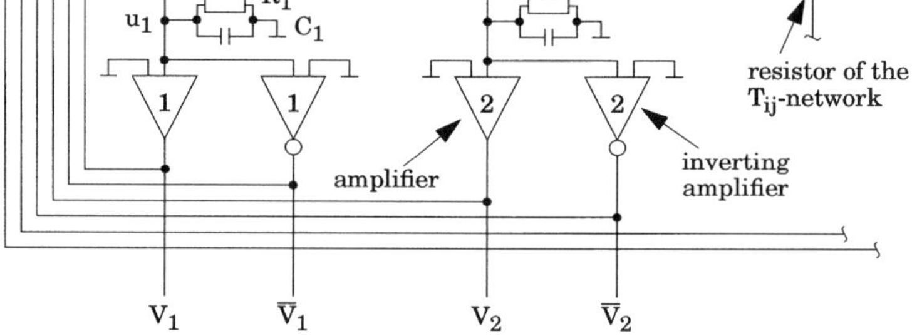

Figure 14 Hopfield network.

In most discussions, the physical units are ignored and the output swing is defined as $V_i^0 = 0$ and $V_i^1 = 1$. The definition $V_i^0 = -1$ is also in use and is called the Ising spin model [58].

The dynamic behavior is described by the differential equation system

$$C_i \frac{du_i}{dt} = \sum_j T_{ij} V_j - \frac{u_i}{R_i} + I_i, \qquad u_i = g_i^{-1}(V_i), \tag{5}$$

where C_i is the capacity of the cell membranes, and R_i is the self-discharge. In many cases, they are assumed to be the same for all neurons and therefore $C_i = C$, $R_i = R$, and $g_i(u_i) = g(u_i)$. For the continuous model, $g(u)$ is the sigmoid function. All states of the neurons can be associated with a computation energy E [57]:

$$E(V_1, \ldots, V_N) = -\frac{1}{2} \sum_{i,j} T_{ij} V_i V_j + \sum_i \frac{1}{R_i} \int_0^{V_i} g_i^{-1}(V)\, dV - \sum_i I_i V_i. \tag{6}$$

The special property of this definition of E is that it is a *Lyapunov function* of Eq. (6). This means [53]

$$\frac{dE}{dt} \leq 0 \qquad \text{and} \qquad E \geq E_0. \tag{7}$$

Therefore, the Hopfield network is stable, cannot oscillate, and must converge into a state with low energy. Note that the equation of motion defined by Eq. (5) describes a modified gradient descent in which the gradient of E without the integral term ($\sum_j T_{ij} V_j + I_i$) is low-pass filtered. Finally, let us comment on the name *continuous Hopfield network* (CHN). It is because the time is continuous as well as are the neuron states.

2. Discrete Hopfield Network

Although the *discrete Hopfield network* (DHN) was published earlier [1] than the CHN, it was extended in [53] and the description follows the latter reference. The discrete model assumes that the slope of the sigmoid function $g'(0)$ is made infinitely large and $g(u)$ becomes a threshold function

$$V = \begin{cases} V_i^0 & \text{for} \sum_{j \neq i} T_{ij} V_j + I_i < U_i, \\ V_i^1 & \text{for} \sum_{j \neq i} T_{ij} V_j + I_i > U_i. \end{cases} \tag{8}$$

U_i is an additional threshold parameter that is usually set to zero.

The time is also made discrete by stochastic sampling. This means that individual neuron states are updated within the time interval Δt with a probability W [1]. If not updated, they retain their value unmodified. Simulations of the discrete Hopfield network usually do a cyclic update, but in a randomly permutated order [56].

The discrete Hopfield network also has an associated energy function

$$E(V_1, \ldots, V_N) = -\tfrac{1}{2}\sum_{i \neq j} T_{ij} V_i V_j - \sum_i I_i V_i + \sum_i U_i V_i. \tag{9}$$

If just a single neuron changes its state within the interval $t, \ldots, t + \Delta t$, then it can be shown that

$$\Delta E = E(t + \Delta t) - E(t) \leq 0, \tag{10}$$

which means that the network does not oscillate and relaxes into a (local) minimum of E.

3. Defining Constraint Satisfaction and Optimization Problems

Hopfield networks can be used to solve constraint satisfaction problems and quadratic optimization problems as Hopfield and Tank have shown. For this purpose, the problem must be formulated as a quadratic function over Boolean variables V_i of the form Eq. (6) or Eq. (9), paying attention to the conditions $T_{ij} = T_{ji}$, $T_{ii} = 0$ and minimality for the solutions of the problem. For example, an n-flop (extended flip-flop) is defined by mutual inhibitory feedback [57].

Table IV

Examples of the Weight Matrix or Energy Function for the NQP

Author	Variables	Weight matrix T resp. energy function E
Tagliarini and Page [59]	v_{ij}	$T_{ij,kl} = A(1 - \delta_{j,1})\delta_{i,k}$ $+B\delta_{j,1}(1 - \delta_{i,k}) + C$ $+D(\delta_{i+j,k+l} + \delta_{i-j,k-l})(1 - \delta_{i,k})$
Shagrir [60]	$a_{x,y}$	$E = \frac{1}{2}A\sum_x\sum_k\sum_{k\neq j} a_{xk}a_{xj}$ $+\frac{1}{2}B\sum_k\sum_x\sum_{y\neq x} a_{xk}a_{yk}$ $+\frac{1}{2}C\left(\sum_x\sum_k a_{xk} - n\right)^2$ $+\frac{1}{2}D\sum_x\sum_k\sum_{m\neq 0} a_{xk}a_{x+m,k+m}$ $+\frac{1}{2}F\sum_x\sum_k\sum_{m\neq 0} a_{xk}a_{x+m,k-m}$

Short Time Memory Problems*

M. Daniel Tom
GE Corporate Research and Development
General Electric Company
Niskayuna, New York 12309

Manoel Fernando Tenorio
Purdue University
Austin, Texas 78746

I. INTRODUCTION

Ever wondered why we remember? Or rather why we forget so quickly? We remember because we have long term memory. We forget quickly because recent events are stored in short term memory. Long term memory has yet to be constructed. Most computational neuron models do not address the issue of short term memory. Each artificial neuron is a memoryless device that translates input to output in a nonlinear fashion. A network of such neurons is therefore memoryless, unless memory devices external to the neurons are used in the network.

For example, the time-delayed neural network [2] uses shift registers to hold a time series in the input field. Elman's recurrent neural network [3, 4] uses a register to hold the hidden layer node values to be presented at the input in the next time step, akin to state automata. The registers in these devices constitute the "short term memory" of the network. The neurons are still memoryless devices. Long term memory is stored in the weights as the network is trained. If these models can achieve amazing results with a memory device external to the neural

*Based on [1]. © 1995 IEEE.

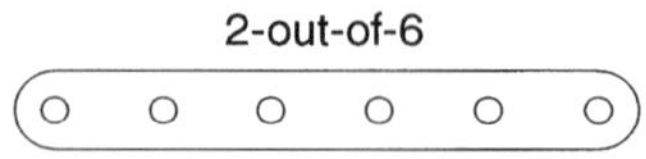

Figure 15 Example of a k-out-of-n constraint for neurons (represented by small circles).

effectively eliminates the arbitrary parameters A, B, ... of other approaches. Additionally, it makes the detection of solutions (global minima) very simple: E is zero only for solutions and positive otherwise. The introduction of slack neurons (Fig. 16) makes the design even more flexible by extending to between-k-and-l-out-of-n constraints. However, the authors still use the standard Hopfield network which gets stuck in undesirable equilibria [68].

Tohru Nakagawa *et al.* developed the *strictly digital neural network* (SDNN) [71–75], which is based on the discrete Hopfield network. The SDNN introduces a time-dependent self-feedback of the neurons (T_{ii}), which results in a hysteresis. This dynamically controlled hysteresis avoids oscillations when several neurons change their state simultaneously to speed up convergence [72].

A *traveling on a hypercube* (TOH) rule controls the dynamic behavior of each individual neuron into a feasible and stable state. For this purpose, the SDNN includes a global detector of stable and feasible states. This allows us to distinguish uniquely between nonminima, local minima, and global minima (solutions). Timers control the hysteresis so that local minima are avoided. Extensive simulations have been conducted and hardware implementations are reported [73].

Nakagawa and Kitagawa have also extended the design rules by more complex building blocks, the neural logic gates (Fig. 17). They represent logical conjuction, disjunction, etc. between Boolean variables and permit us to formulate larger-scale problems.

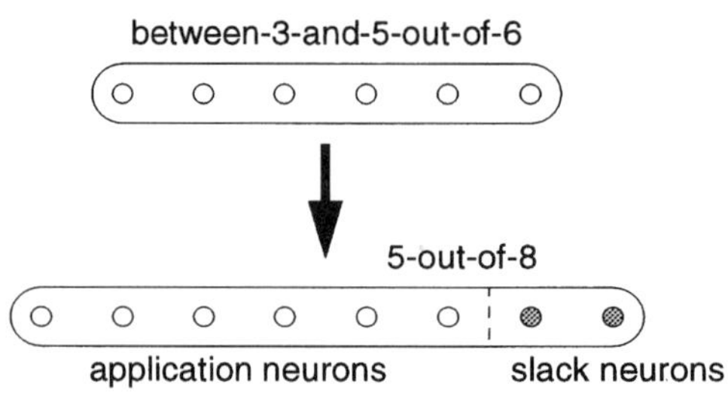

Figure 16 Introducing slack neurons to represent range constraints. Reprinted from H. N. Schaller, Design of neurocomputer architectures for large-scale constraint satisfaction problems, *Neurocomputing* 8:315–339, 1995 with kind permission from Elsevier Science–NL, Sara Burgerhartstraat 25, 1055 KV Amsterdam, The Netherlands.

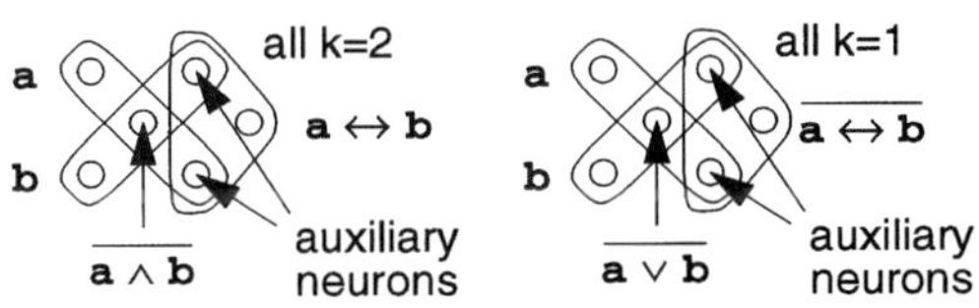

Figure 17 Neural logic gates. Reprinted from H. N. Schaller, Design of neurocomputer architectures for large-scale constraint satisfaction problems, *Neurocomputing* 8:315–339, 1995 with kind permission from Elsevier Science–NL, Sara Burgerhartstraat 25, 1055 KV Amsterdam, The Netherlands.

The SDNN can solve the NQP for $N = 3000$ within constant $[O(1)]$ time and was also applied successfully to test pattern generation, N-coloring, and other problems [25, 28, 72–76].

C. Neural Computing Networks

Yoshiyasu Takefuji *et al.* have applied recurrent neural networks to a very large number of problems [7, 8, 30, 38, 40, 65, 77–81]. For convenience, their networks will be called *neural computing networks* (NCNs), based on the titles of two books [7, 8]. Although the networks are tailored to the given problem, all are based on Hopfield networks and the following modifications can be found in many of them:

- discrete output function $V = f(u) = 1$ for $u \geq 0$, otherwise 0 (i.e., a threshold function like in the discrete Hopfield network), or
- maximum neurons with $V = 1$ only if $u > 0$ and $u = \max_i u_i$ within a group of neurons [80, 81],
- a hill-climbing term $h(x) = 1$ for $x = 0$, otherwise 0 (to detect and escape from local minima) [30, 38, 78],
- asynchronously sampled equation of motion based on the Euler integration $du/dt = -\partial E/dV$ (i.e., no low-pass filter as in the continuous Hopfield network),
- different control parameters A, B, C, ω, etc.,
- periodical modulation of control parameters or contributions to E (to stimulate the escape from local minima) [30, 38].

Neural computing networks are built from these parts for the N-coloring problem, various puzzles like the NQP, the knight's problem, the polyominoe puzzle [38], and many others from very large scale integration (VLSI) design to RNA structure identification [8].

Although the reported results are very encouraging, even indicating the solvability of an NP problem (maximum clique problem) [7, 8], a major drawback of this approach is the lack of a general design principle. The definition of the energy functions or equations of motion, the application of the hill-climbing term, and the selection of the parameters $A, B, C, \ldots$ are only justified by the positive simulation results. This makes a comment on the results difficult.

D. Guarded Discrete Stochastic Net

The *guarded discrete stochastic net* (GDS net) was developed out of the discrete Hopfield network by Hans-Martin Adorf and Mark Johnston [64, 82–84]. The basic idea is to decode certain local minima by guard neurons. The outputs of the guard neurons are fed back to the guarded neurons so that the network cannot stabilize in such a minimum. This does not necessarily guarantee that the network finally reaches a global optimum. Therefore, a timer supervises the progress and it is assumed after time-out that a cycle between local minima has been reached. In this case, the neuron states are randomly reinitialized and the process starts over. It must be noted that the duration of the time-out is a parameter that influences the speed and solution quality.

To speed up convergence into a minimum, a most active first rule is applied to groups of neurons. Only the most active neuron (largest absolute value of the activation x) may change its state. In a tie situation, a random-number generator is consulted to break the tie.

Application of the GDS are the NQP (N up to 1024, resp. 6001 [85]), N-coloring problems, and the experiment planning module of the Hubble Space Telescope [84]. For all these applications, construction rules for the weights have been given. The dynamic behavior of the GDS network was analyzed in detail by Minton *et al.* [86] and redefined as a (nonneural) *heuristic repair algorithm* that was able to solve the NQP for $N = 1000000$.

E. Boltzmann Machine

The *Boltzmann machine* (BM) is the neural implementation of simulated annealing. It can be seen as a discrete Hopfield network with a stochastic neuron dynamic defined as follows:

$$x_\alpha = \sum_\beta T_{\alpha\beta} z_\alpha + I\alpha, \tag{11a}$$

$$P\bigl(z_\alpha(t + \Delta t) = 1\bigr) = \frac{1}{1 + \exp(-x_\alpha / T)}. \tag{11b}$$

These equations describe that the probability of a neurons becoming active [$z_\alpha(t + \Delta t) = 1$] is determined by a nonlinear function of the weighted sum of all neuron outputs. The temperature T plays the same role as in simulated annealing and is gradually lowered. The schedule may be, for example, $T(t) = T(0)\exp(-t/\tau)$. For the NQP the weights T_{xyXY} and biases I_{xy}

$$\begin{aligned} T_{xyXY} &= -A\delta_{x,X}(1-\delta_{y,Y}) - B\delta_{y,Y}(1-\delta_{x,X}) \\ &\quad -C\delta_{x+y,X+Y}(1-\delta_{x,X}) - D\delta_{x-y,X-Y}(1-\delta_{x,X}), \qquad (12) \\ I_{xy} &= \frac{A}{2} + \frac{B}{2} \end{aligned}$$

have been successful for N up to 1000 [37].

An open issue is the definition of the parameters A, B, C, D and the cooling schedule. Both are more or less arbitrarily selected to determine the success of the method. Similar to the Boltzmann machine are the Cauchy machine [87] and the Gauss machine [36] which use a different probability distribution Eq. (11b) as the Boltzmann machine. The benefit of this change seems to be marginal.

A different method has been developed to speed up the Boltzmann machine. Because the Boltzmann machine requires many steps to stabilize at a certain temperature level, *mean field annealing* (MFA) [24, 49, 88, 89] simply calculates the probability that a neuron of the Boltzmann machine would be active. This is very similar to a continuous Hopfield network, in which the temperature controls the slope of the sigmoid function. Consequently, the slope is increased as time passes.

F. *K*-WINNER-TAKE-ALL

Winner-take-all (WTA) networks are neural networks in which the neurons compete with each other for the largest activity. The initially most active neuron becomes more active and suppresses the activity of all the others. Finally, only a single neuron (the winner) remains active. A recurrent network shows this behavior if the feedback weights are negative (inhibition).

This model was extended by Majani *et al.* [90] to the KWTA in which the K neurons with the largest initial activity become more active. Several of these KWTA neuron groups may overlap and form more complex networks [91, 92]. Brown has proposed applying such networks to switching problems in digital telecommunications systems [15, 16]. An additional extension was done by Eberhardt *et al.* in [31] to form a (k, m)-WTA and was proposed for general assignment problems.

It turns out that the WTA describes the same feedback structure as Hopfield's "n-flop" and the KWTA has the same structure as the k-out-of-n design rule. Therefore, the encoding of CSPs into the network structure has the same potential as the SDNN and DBNN (see following section) design.

On the other hand, the WTA feedback dynamic is just the basic continuous Hopfield network with all drawbacks. Therefore, it will get stuck in local minima and it has no way to escape.

G. Dynamic Barrier Neural Network and Rolling Stone Neural Network

1. Programming Model

The *dynamic barrier neural network* (DBNN) [41, 42] and the *rolling stone neural network* (RSNN) [41, 93] are both based on a programming model with between-k-and-l-out-of-n constraints [29], shown in Fig. 18. The constraints take clusters of n neurons (described by $C_{i\alpha}$) and try to enforce the number of active neurons to be between k and l. The values k and l may be externally controlled by input data. The variables z_α may be part of the output. Variables with $o_\alpha = 0$ are auxiliary variables (slack). This is very similar to the constraints used in the SDNN, but has additionally been described as a universal programming model, independent of any neural network method and hardware.

As in all programming environments, higher-level language elements have been described [29, 94], following a declarative programming style. They include input–output interfaces, rules like "at-most-one-out-of-n", "not-all-of-n", etc., complex IF THEN rules (Fig. 19), arbitrary Boolean expressions [95], con-

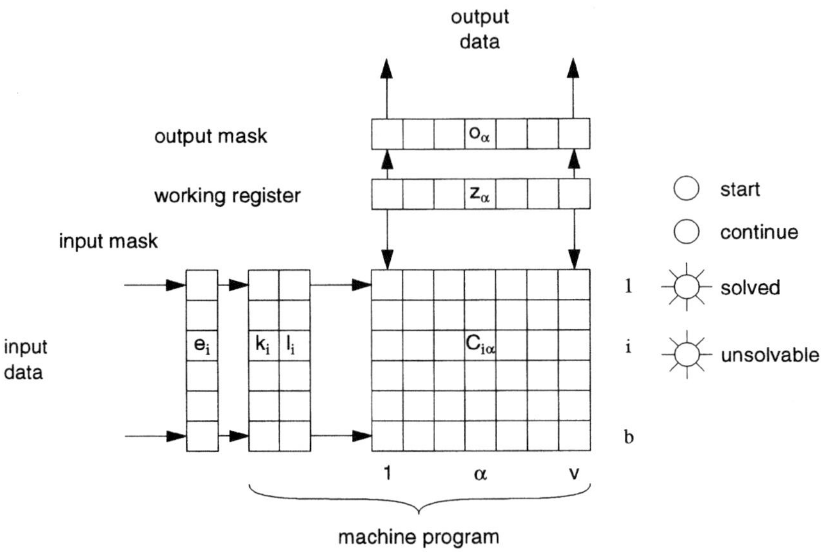

Figure 18 Programming model with between-k-and-l-out-of-n constraints. Reprinted from H. N. Schaller, Design of neurocomputer architectures for large-scale constraint satisfaction problems, *Neurocomputing* 8:315–339, 1995 with kind permission from Elsevier Science–NL, Sara Burgerhartstraat 25, 1055 KV Amsterdam, The Netherlands.

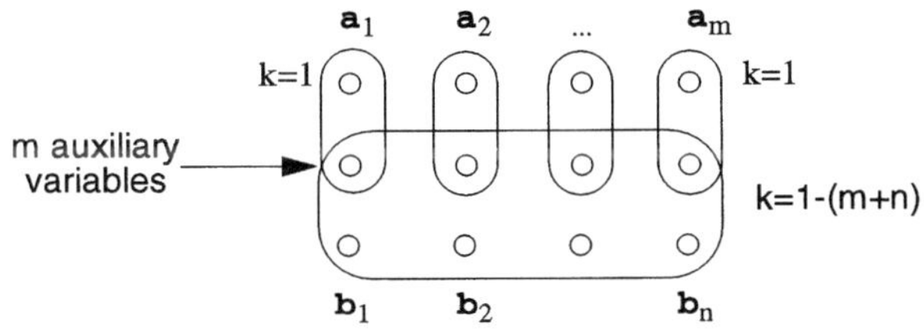

Figure 19 Constraint representation of an IF ALL **a** THEN ANY **b** rule. Reprinted from H. N. Schaller, Design of neurocomputer architectures for large-scale constraint satisfaction problems, *Neurocomputing* 8:315–339, 1995 with kind permission from Elsevier Science–NL, Sara Burgerhartstraat 25, 1055 KV Amsterdam, The Netherlands.

version between different representations, and arithmetic constraints for addition and multiplication. Additionally, a compiler technique has been described [96] to convert high-level declarations into a set of the basic constraints.

Several large-scale problems have been formulated using this technique: a switching problem (Fig. 20), the factorization of large numbers (Fig. 21), and—partially—a complex jigsaw puzzle (Fig. 22) [41].

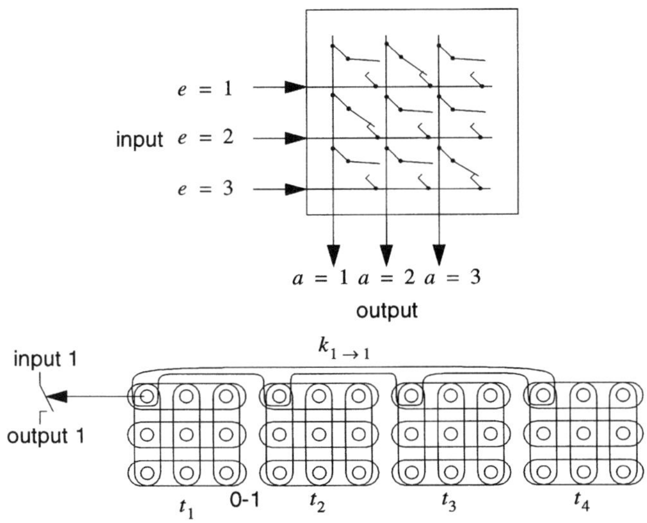

Figure 20 Constraints for a 3 × 3 crossbar switch (only a single connection demand constraint for $k_{1 \to 1}$ is shown). Reprinted from H. N. Schaller, Design of neurocomputer architectures for large-scale constraint satisfaction problems, *Neurocomputing* 8:315–339, 1995 with kind permission from Elsevier Science–NL, Sara Burgerhartstraat 25, 1055 KV Amsterdam, The Netherlands.

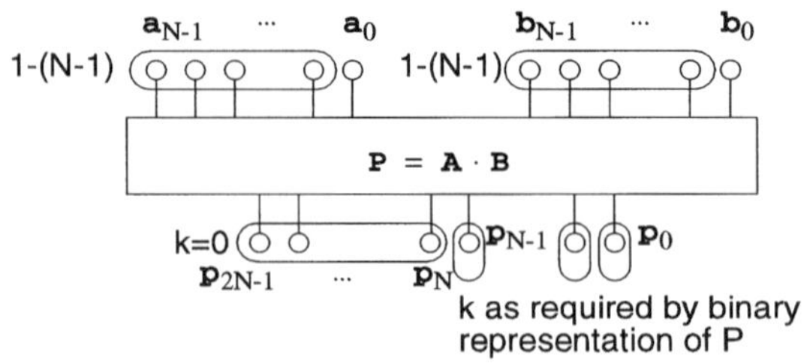

Figure 21 Factorization of P in the binary system using a multiplier constraint macro. Reprinted from H. N. Schaller, Design of neurocomputer architectures for large-scale constraint satisfaction problems, *Neurocomputing* 8:315–339, 1995 with kind permission from Elsevier Science–NL, Sara Burgerhartstraat 25, 1055 KV Amsterdam, The Netherlands.

Although summarized in just some sentences, the method is, in fact, a toolbox to systematically develop quadratic error functions over Boolean variables with the special property that the error is minimal and zero only for solutions of the problem. And, besides the applicability of any optimization method, large-scale problems can be uniquely encoded into the weight matrix and bias vector of a recurrent neural network [41].

2. Dynamic Barrier Neural Network

The *dynamic barrier neural network* (DBNN) is an implementation of a processor to efficiently solve problems formulated in the between-k-and-l-out-of-n language. It is based on a conversion of these constraints into a weight matrix

	7		3		7	
9	=	9	*	1	*	1
	5	=	5	*	1	
7	+	9	=	2	*	8
	6	/	6	=	1	
5	/	1	+	2	=	7
	3		9		7	

Figure 22 Solution of the 7 × 7 jigsaw puzzle. Reprinted from H. N. Schaller, Design of neurocomputer architectures for large-scale constraint satisfaction problems, *Neurocomputing* 8:315–339, 1995 with kind permission from Elsevier Science–NL, Sara Burgerhartstraat 25, 1055 KV Amsterdam, The Netherlands.

encoding of a discrete Hopfield network. Like other approaches, it contains a convergence speed-up, a local minima detection, and a local minima treatment component.

The convergence is accelerated by simultaneously updating the state of several neurons in a discrete Hopfield network. This is done by a rule to select the most active variable within each constraint. In the case of a tie, a random-number generator selects between alternatives. The activity defined by the deviation of the activation x of a neuron from a reference value determined by the connection weights and the neuron state describes how the neuron participates in constraint violations. It is defined by the deviation of the activation x from a reference value determined by the connection weights and the neuron state. By this most active first rule, oscillations are avoided while having rapid convergence into minima of E.

Minima are detected when all of the neurons are stable; that is, they do not change their state any more. In this case, all activity values of the neurons are checked. If all of the neurons are inactive, a global minimum has been reached and a solution is found. Otherwise, the DBNN has fallen into a local minimum.

In this case, a dynamic barrier is built up. The dynamic barrier disables for a certain number of iteration steps all of the neurons from state changes that have participated in the local minimum.

This destabilizes the local minima and forces the DBNN to increase the error E. It must then find a different path to lower E again. Therefore, the DBNN will search through several minima until a global minimum is found.

Experiments with the NQP have been successful for N up to 200. Figure 23 shows the minimum, median, and maximum number of steps required to find solutions in eight repetitions for each different value of N. The results indicate $O(1)$ empirical complexity, that is, constant speed. This result is very important, because the DBNN has not at all been optimized for the NQP but derived from general design principles to improve the discrete Hopfield network and to solve arbitrary between-k-and-l-out-of-n-constraint sets.

The factorization of large numbers and the jigsaw puzzle have been reported to be too difficult for the DBNN (Fig. 22 was found by a specialized DFS algorithm [97]). The DBNN could not find global minima within reasonable simulation time. This may indicate that they are, in fact, NP problems.

3. Rolling Stone Neural Network

The *rolling stone neural network* (RSNN) is a modification of the continuous Hopfield network model. Beginning with the observation that the gradient descent method moves in the direction of the gradient with a velocity proportional to the gradient and comes to rest in a (local) minimum, the RSNN simply makes the speed constant. This has the effect that the RSNN cannot get stuck at minima

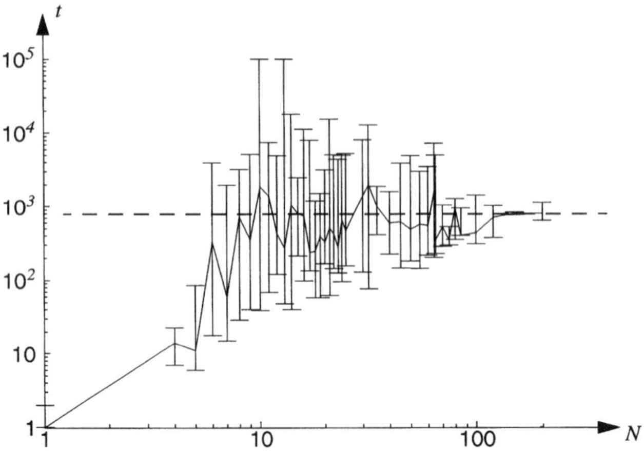

Figure 23 Number of steps to find the first solution of the N-queens problem by the DBNN in log–log coordinates. Reprinted from H. N. Schaller, Design of neurocomputer architectures for large-scale constraint satisfaction problems, *Neurocomputing* 8:315–339, 1995 with kind permission from Elsevier Science–NL, Sara Burgerhartstraat 25, 1055 KV Amsterdam, The Netherlands.

but must alternatingly search for maxima and minima. Because the influence of small disturbances of the movement direction in minima and maxima becomes infinitely large (chaotic behavior), the RSNN follows a trajectory through the error surface E that passes through all extreme points, including global minima. By using the same weight matrix definition as for the DBNN, the RSNN can principally find solutions of CSPs. However, the model has not been studied in more detail. Therefore, solutions of the NQP have not been reported.

V. ASSESSMENT

A. N-QUEENS BENCHMARK

The results of a comparison between many different constraint satisfaction methods for the N-queens problem are shown in Table V, which gives the maximum problem size that was reported by the authors to be solved by their method.

About one-half are based on a neural network method. The most successful are, of course, algebraic methods. Among the numerical methods, specialized algorithms are most successful, but neural network techniques are also very good. One of the worst is the depth first search method.

Table V

Success in Finding Solutions to the *N*-Queens Problem

Author	Method	Neural network?	N_{max}	Year	References
Gauss and Nauck	Trial and error	—	8	1850	[3, 4]
Several	Construction	—	$N \geq 4$	> 1850	[2, 60, 85, 98, 99]
Stone and Stone	Depth first search	—	96	1987	[47]
Page and Tagliarini	Hopfield network	×	10	1987	[59]
Kajura *et al.*	Boltzmann machine	×	1000	1989	[37]
Adorf and Johnston	GDS	×	1024	1989	[64, 82, 83]
Abramson and Yung	Divide and conquer	—	$N > 39$	1989	[98]
Sosic and Gu	Probabilistic local search with conflict minimization	—	3000000	1990	[2, 51, 52]
Chen and Wu	Parallel PROLOG	—	?	1991	[100]
Nakagawa and Kitagawa	SDNN	×	3000	1991	[71–74]
Miller	Depth first search	—	64	1992	[46]
Shagrir	Hopfield network	×	100	1992	[60]
Mandziuk and Macukow	Hopfield network	×	8	1992	[101]
Ali *et al.*	Genetic algorithm	—	200	1992	[50]
Minton *et al.*	Heuristic repair	—	1000000	1992	[86]
Takefuji	Neural computing network	×	100	1992	[7]
Lorenz	GDS	×	6001	1993	[85]
Schaller	DBNN	×	200	1994	[41, 42]

B. Comparison of Neural Techniques

All neural network techniques are based on the original Hopfield networks. A characterization of the network models is shown in Table VI, whereas Table VII summarizes the modifications that have been done to make the network more successful than the Hopfield network on which they are based.

C. Comparison of All Techniques

Now, finally, we compare the methods according to the criteria developed in Section II. The criterion (C1) got nine credits if the NQP was solved for $N \geq 100$; that is, polynomial complexity can be assumed. The DFS still got five credits, as it

Table VI

Characterization of Neural Networks for Constraint Satisfaction

Network	Based on	Random component	System parameters
CHN	—	Initialization	Weight matrix
DHN	—	Initialization	Weight matrix, update rule
SDNN	DHN	— (?)	Weight matrix
NCN	DHN	Initialization	Weight matrix, several
GDS	DHN	Initialization, break ties in most active first rule	Weight matrix, time-out
BM	DHN	New state	Weight matrix cooling schedule
MFA	DHN, BM	—	Weight matrix cooling schedule
KWTA	CHN	Initialization	Weight matrix
DBNN	DHN	Break tie in most active first rule	Weight matrix
RSNN	CHN	Small disturbances	Weight matrix

could solve for $N = 96$ but seems to exhibit exponential behavior. (C2)–(C7) and (C10) give the personal view of the author. (C8) and (C9) have not been included, because there is not enough information in the literature. Finally, the credit points are multiplied by a weighting factor and the scores are summarized in Table VIII.

Table VII

Modifications to the Hopfield Network

Network	(I1) E design	(I2) Convergence acceleration	(I3) Diagnosis of local minima	(I4) Treatment of local minima
SDNN	k-out-of-n	Most active first	Yes	TOH rule
NCN	Good guess	?	Yes	Hill-climbing term
GDS	Systematical	Most active first	Timeout	Random reset
BM	Out of the scope	—	Implicit	Implicit
KWTA	KWTA rule	—	No	No
DBNN	Systematical	Most active first	Yes	Dynamic barrier
RSNN	Systematical	—	Implicit	Constant speed

Table VIII

Assessment According to the Criteria C1–C10

	Criterion								
	C1	C2	C3	C4	C5	C6	C7	C10	Weighted
Factor	10	6	5	9	8	7	3	4	score
Term rewriting	0	9	9	0	0	0	5	0	114
Divide and conquer	9	9	9	0	0	0	5	0	204
Enumeration	0	9	9	9	0	0	9	0	207
Depth first search	5	9	9	9	0	0	9	0	257
Gradient descent	0	9	9	9	1	0	0	0	188
Simulated annealing	9	5	0	7	2	0	0	0	199
Genetic algorithms	9	3	0	7	2	0	0	0	187
Probabilistic local search	9	9	3	5	3	1	0	0	235
CHN, DHN	0	8	8	9	3	1	0	9	236
SDNN	9	9	9	9	7	8	0	9	418
NCN	9	7	9	9	6	5	0	0	341
GDS	9	8	8	9	7	6	0	0	357
BM	9	5	0	9	3	1	0	0	232
KWTA	0	8	8	9	5	6	0	0	251
DBNN	9	9	9	9	9	9	0	0	405
RSNN	0	9	9	9	9	9	0	0	315

D. Summary

Apparently, the between-k-and-l-out-of-n design rule seems to be a very successful method to represent constraint satisfaction problems. The language based on them allows us to program even large-scale problems like test pattern generation, IF THEN rule sets, switching problems, N-coloring, factorization of numbers, and large, complex puzzles. Although independent of any solution-finding method, they are simple enough to be easily converted into the weight matrix of recurrent neural networks to implement a parallel processor in hardware.

Furthermore, neural network techniques are much more successful in finding solutions to the N-queens problem than the traditional depth first search algorithm. On appropriately sized neural hardware, they are even capable of solving the problem in $O(1)$ (nearly constant) time, independent of the parameter N.

Although it has not yet been proven that neural network techniques can really solve NP problems within polynomial time, the SDNN and the DBNN are especially good candidates for solving a large range of practical large-scale constraint satisfaction problems including all those suggested by the NCNs.

To finally find an answer to (C8) and (C9), all methods should pass a more complex standard benchmark—a set of solvable problems such as the N-queens problem. This should give an empirical proof of the independence from parameters by repeating with several problem sizes, different problem and variable encodings, different random seeds, and different random generators. And, if the median number of steps to find a solution is approximately the same (which may be more exactly specified by a statistical test), independent of the problem size, the maximum number of steps is finite, and all solutions are found with equivalent probability, then we have finally found a robust constraint satisfaction method for all our problems.

However, returning to the initially posed technical problem of the optical processor placement we want to solve by a machine, the best solution would appear to be to encode the problems using the design rules of the DBNN paradigm and buy a SDNN chip—if it were available.

REFERENCES

[1] J. J. Hopfield. Neural networks and physical systems with emergent collective computation abilities. *Proc. Nat. Acad. Sci.* 79:2554–2558, 1982.

[2] R. Sosic and J. Gu. Fast search algorithms for the N-queens problem. *IEEE Trans. Systems Man Cybernet.* 21:1572–1576, 1991.

[3] Duden "Informatik." BI Wissenschaftsverlag, Mannheim, 1988; korrigierter reprint 1989.

[4] J. Gu. On a general framework for large-scale constraint-based optimization. *SIGART Bull.* 2:8, 1991.

[5] R. Shinghal. *Formal Concepts in Artificial Intelligence*. Chapman & Hall, London, 1992.

[6] R. Sedgewick. *Algorithms*, 2nd ed. Addison-Wesley, Reading, MA, 1988.

[7] Y. Takefuji. *Neural Network Parallel Computing*. Kluwer Academic, Boston, 1992.

[8] Y. Takefuji and J. Wang. *Neural Computing for Optimization and Combinatorics*. World Scientific, Singapore, 1996.

[9] C. Petersen and J. R. Anderson. Neural networks and NP-complete optimization problems; a performance study on the graph bisection problem. *Complex Systems* 59–89, 1988.

[10] G. A. Blaauw. Computer architecture. *Elektronische Rechenanlagen* 4:154–159, 1972.

[11] B. W. Lee and B. J. Sheu. Modified Hopfield neural networks for retrieving the optimal solution. *IEEE Trans. Neural Networks* 2:137–142, 1991.

[12] D. W. Tank and J. J. Hopfield. Simple "neural" optimization networks: an A/D converter, signal decision circuit, and a linear programming circuit. *IEEE Trans. Circuits Systems* 33:533–541, 1986.

[13] M. Duque Antón and D. Kunz. Parallel algorithms for channel assignment in cellular mobile radio systems: the neural network approach. In *Parallel Processing in Neural Systems and Computers* (R. Eckmiller, G. Hartmann, and G. Hauske, Eds.), pp. 265–268. Elsevier, Amsterdam, 1990.

[14] D. Kunz. Suboptimum solutions obtained by the Hopfield–Tank neural network algorithm. *Biol. Cybernet.* 65:129–133, 1991.

[15] T. X. Brown. Neural networks for switching. *IEEE Comput. Mag.* 72–81, 1989.

[16] T. X. Brown and K.-H. Liu. Neural network design of a Banyan network controller. *IEEE J. Select. Areas Comm.* 8:1428–1438, 1990.

[17] J. Ghosh, A. Hukkoo, and A. Varma. Neural networks for fast arbitration and switching noise reduction in large crossbars. *IEEE Trans. Circuits Systems* 38:895–904, 1991.

[18] A. Marrakchi and T. Troudet. A neural net arbitrator for large crossbar packet-switches. *IEEE Trans. Circuits Systems* 36:1039–1041, 1989.
[19] T. P. Troudet and S. M. Walters. Neural network architecture for crossbar switch control. *IEEE Trans. Circuits Systems* 38:42–56, 1991.
[20] J. Bruck and M. Blaum. Neural networks, error-correcting codes, and polynomials over the binary n-cube. *IEEE Trans. Inform. Theory* 35:976–987, 1989.
[21] D. D. Caviglia *et al.* Neural algorithms for cell placement in VLSI design. In *Proceedings of the IEEE First IJCNN*, Washington, DC, pp. 573–580, 1989.
[22] K. H. Kim *et al.* Neural optimization network for minimum—via layer assignment. *Neurocomputing* 3:15–27, 1991.
[23] J. Naft. A modified Hopfield net approach to multiobjective design optimization for printed circuit board component placement. *Internat. J. Neural Networks* 1:78–84, 1989.
[24] U. Peine and H. P. Siemon. Optimization of the rectilinear Steiner tree using a mean field theory model. In *Artificial Neural Networks: Proceedings ICANN'92* (I. Aleksander and J. Taylor, Eds.), Vol. 2, pp. 1043–1046. Elsevier, Amsterdam, 1992.
[25] M. Arai *et al.* An approach to automatic test pattern generation using strictly digital neural networks. In *IJCNN'92*, Baltimore, Vol. 4, pp. 474–479, 1992.
[26] M. Arai *et al.* A neural inverse function for automatic test pattern generation using strictly digital neural networks. In *11th IEEE VLSI Test Symposium*, Atlantic City, pp. 238–243, 1993.
[27] M. Bellgard and C. P. Tsang. Harmonizing music using a network of Boltzmann machines. In *Proceedings Neuro-Nîmes'92*, Nanterre Cedex, Vol. 2, pp. 321–322, 1992.
[28] K. Murakami *et al.* Solving four-coloring map problems using strictly digital neural networks. In *IJCNN'91*, Singapore, pp. 2440–2443, 1991.
[29] H. N. Schaller. A collection of contraint design rules for neural optimization networks. In *Artificial Neural Networks II: Proceedings ICANN'92* (I. Aleksander and J. Taylor, Eds.), pp. 1039–1042. Elsevier, Amsterdam, 1992.
[30] Y. Takefuji and K. C. Lee. Artificial neural networks for four-coloring map problems and K-colorability problems. *IEEE Trans. Circuits Systems* 38:326–333, 1991.
[31] S. B. Eberhardt *et al.* Competitive neural architecture for hardware solution to the assignment problem. *Neural Networks* 4:431–442, 1991.
[32] S. Silven. A neural approach to the assignment algorithm for multiple-target tracking. *IEEE J. Oceanic Engrg.* 17:326–332, 1992.
[33] P. Bourret *et al.* Optimal scheduling by competitive activation: application to the satellite antennae scheduling problem. In *Proceedings of the IEEE First IJCNN*, Washington, DC, Vol. 1, pp. 529–532, 1989.
[34] L. Fang *et al.* A neural network for job sequencing. In *Parallel Processing in Neural Systems and Computers* (R. Eckmiller, G. Hartmann, and G. Hauske, Eds.), pp. 253–256. Elsevier, Amsterdam, 1990.
[35] P. W. Protzel. Artificial neural network for real-time task allocation in fault-tolerant, distributed processing system. In *Parallel Processing in Neural Systems and Computers* (R. Eckmiller, G. Hartmann, and G. Hauske, Eds.), pp.307–310. Elsevier, Amsterdam, 1990.
[36] Y. Akiyama *et al.* Combinatorial optimization with Gaussian machines. In *Proceedings of the IEEE First IJCNN*, Washington, DC, Vol. 1, pp. 533–540, 1989.
[37] M. Kajiura *et al.* Solving large scale puzzles with neural networks. In *Proceedings IEEE TAI'89*, pp. 562–569, 1989.
[38] Y. Takefuji and K.-C. Lee. A parallel algorithm for tiling problems. *IEEE Trans. Neural Networks* 1:143–145, 1990.
[39] J. L. Johnson. A neural network approach to the 3-satisfiability problem. *J. Parallel and Distributed Process.* 6:435–449, 1989.
[40] Y. Takefuji and K. C. Lee. Neural network computing for knight's tour problems. *Neurocomputing* 4:249–254, 1992.

[41] H. N. Schaller. Design of neurocomputer architectures for large-scale constraint satisfaction problems. *Neurocomputing* 8:315–339, 1995.
[42] H. N. Schaller. Entwicklung hochgradig paralleler Rechnerarchitekturen zur Lösung diskreter Belegungsprobleme. Dissertation, Technical University of Munich, 1994.
[43] B. Nadel. Representation selection for constraint satisfaction: a case study. *IEEE Expert* 5:16–25, 1990.
[44] V. N. Rao and V. Kumar. On the efficiency of parallel backtracking. *IEEE Trans. Parallel Distributed Systems* 4:427–437, 1993.
[45] V. Plesser *et al.* Ein Workstation-LAN als verteiltes System zum parallelen Lösen von kombinatorischen Suchproblemen. *Informationstechnik*, 273–279, 1992.
[46] A. Miller. Optimierung eines Backtrackingverfahrens zur Lösung von Belegungsproblemen. Diploma Thesis, Lehrstuhl für Datenverarbeitung, Technical University of Munich, 1992.
[47] H. S. Stone and J. M. Stone. Efficient search techniques—an empirical study of the N-queen problem. *IBM J. Res. Develop.* 31:464–474, 1987.
[48] P. J. M. van Laarhofen and E. H. L. Aarts. *Simulated Annealing: Theory and Applications*. Kluwer Academic, Dordrecht, 1987.
[49] A. Cichocki and R. Unbehauen. *Neural Networks for Optimization and Signal Processing*. Teubner/Wiley, Stuttgart/Chichester, 1993.
[50] A. Homaifar *et al.* The N-queens problem and genetic algorithms. In *Proceedings IEEE SOUTHEASTCON'92*, pp. 262–267, 1992.
[51] R. Sosic and J. Gu. A polynomial time algorithm for the N-queens problem. *SIGART Bull.* 1:7–11, 1990.
[52] R. Sosic and J. Gu. 3,000,000 queens in less than one minute. *SIGART Bull.* 2:22–24, 1991.
[53] J. J. Hopfield. Neurons with graded response have collective computational properties like those of two-state neurons. *Proc. Nat. Acad. Sci.* 81:3088–3092, 1984.
[54] J. J. Hopfield and D. W. Tank. "Neural" computation of decisions in optimization problems. *Biol. Cybernet.* 52:141–152, 1985.
[55] J. J. Hopfield and D. W. Tank. Computing with neural circuits: a model. *Science* 233:625–633, 1986.
[56] J. J. Hopfield. Collective computation, content-addressable memory, and optimization problems. In *Complexity in Information Theory* (Y. Abu-Mostafa, Ed.), pp. 99–113. Springer-Verlag, New York.
[57] D. W. Tank and J. J. Hopfield. Collective computation in neuronlike circuits. *Scientific American* 62–70, 1987.
[58] B. Müller and J. Reinhardt. *Neural Networks—An Introduction*. Springer-Verlag, Berlin, 1990.
[59] G. A. Tagliarini and E. Page. Solving constraint satisfaction problems with neural networks. In *Proceedings of the First ICNN*, San Diego, Vol. 3, pp. 741–747, 1987.
[60] O. Shagrir. A neural net with self-inhibiting units for the N-queens problem. *Internat. J. Neural Systems* 3:249–252, 1992.
[61] A. Johannet *et al.* Specification and implementation of a digital Hopfield-type associative memory with on-chip training. *IEEE Trans. Neural Networks* 3:529–539, 1992.
[62] A. Kuh and B. W. Dickinson. Information capacity of associative memories. *IEEE Trans. Inform. Theory* 35:59–68, 1989.
[63] H. J. Sussmann. On the number of memories that can be prefectly stored in a neural net with Hebb weights. *IEEE Trans. Inform. Theory* 35:174–178, 1989.
[64] M. D. Johnston and H.-M. Adorf. Learning in stochastic neural networks for constraint satisfaction problems. In *Proceedings of the NASA Conference on Space Telerobotics*, Pasadena, 1989.
[65] G. A. Tagliarini and E. W. Page. Learning in systematically designed networks. In *Proceedings of the IEEE First IJCNN*, Washington, DC, Vol. 1, pp. 497–502, 1989.

[66] 80170NX, electrically trainable analog neural network, experimental data sheet. Intel Corporation, 1991.
[67] E. W. Page and G. A. Tagliarini. Algorithm development for neural networks. In *Proceedings IEEE SPIE: High Speed Computing*, Vol. 880, pp. 11–19, 1988.
[68] G. A. Tagliarini. Undesirable equilibria in systematically designed neural networks. In *Proceedings of the IEEE Southeastcon Region 3 Conference*, Columbia, SC, pp. 63–67, 1989.
[69] G. A. Tagliarini and E. W. Page. A neural-network solution to the concentrator assignment problem. In *First IEEE Conference NIPS*, Denver, pp. 775–782. Amer. Inst. of Phys., New York, 1987.
[70] G. A. Tagliarini *et al.* Optimization using neural networks. *IEEE Trans. Comput.* 40:1347–1358, 1991.
[71] T. Nakagawa and H. Kitagawa. SDNN: an $O(1)$ parallel processing with strictly digital neural networks for combinatorial optimization. In *Artificial Neural Networks* (T. Kohonen, K. Mäkisara, O. Simula, and J. Kangas, Eds.), pp. 1181–1184. Elsevier, Amsterdam, 1991.
[72] T. Nakagawa *et al.* SDNN: a computation model for strictly digital neural networks and its application. In *Proceedings of the Fifth AAAIC'89*. ACM/SIGART, Dayton, OH, 1989.
[73] T. Nakagawa *et al.* SDNN-3: a simple processor architecture for $O(1)$ parallel processing in combinatorial optimization with strictly digital neural networks. In *IJCNN'91*, Singapore, pp. 2444–2449, 1991.
[74] T. Nakagawa *et al.* Strictly digital neurocomputer based on a paradigm of constraint set programming for solving combinatorial optimization problems. *ICNN'93*, San Francisco, pp. 1086–1091, 1993.
[75] T. Nakagawa and K. Murakami. Evaluation of virtual slack-neurons for solving optimization problems in circuit design using neural networks based on the between-l-and-k-out-of-n design rule. In *WCNN'93*, Portland, OR, pp. 122–125, 1993.
[76] K. Murakami *et al.* A high-speed and low-cost parallel convergence in coloring map problems with virtual slack-neurons. In *WCNN'93*, Portland, OR, Vol. 4, pp. 421–424, 1993.
[77] Y. Takefuji and K.-C. Lee. A super-parallel sorting algorithm based on neural networks. *IEEE Trans. Circuits Systems* 37:1425–1429, 1990.
[78] Y. Takefuji and K.-C. Lee. A near-optimum parallel planarization algorithm. *Science* 245:1221–1223, 1989.
[79] Y. Takefuji and K. C. Lee. An artificial hysteresis binary neuron: a model suppressing the oscillatory behaviors of neural dynamics. *Biol. Cybernet.* 64:141–152, 1991.
[80] Y. Takefuji *et al.* An artificial maximum neural network: a winner-take-all neuron model forcing the state of the system in a solution domain. *Biol. Cybernet.* 67:243–251, 1992.
[81] K. C. Lee *et al.* A parallel improvement algorithm for the bipartite subgraph problem. *IEEE Trans. Neural Networks* 3:139–145, 1992.
[82] H.-M. Adorf. Connectionism and neural networks. In *Knowledge Based Systems in Astronomy* (F. Murtagh and A. Heck, Eds.), pp. 215–245. Springer-Verlag, Heidelberg, 1989.
[83] H.-M. Adorf and M. D. Johnston. A discrete stochastic neural network algorithm for contraint satisfaction problems. In *Proceedings of the International Conference on Neural Networks*, San Diego, pp. 917–924, 1990.
[84] M. D. Johnston and H.-M. Adorf. Scheduling with neural networks—the case of the Hubble Space Telescope. *Comput. Oper. Res.* 19:209–240, 1992.
[85] J. Lorenz. Vergleich verschiedener Neuronenmodelle zur Lösung kombinatorischer Suchprobleme. Diploma Thesis, Lehrstuhl für Datenverarbeitung, Technical University of Munich, 1993.
[86] S. Minton *et al.* Minimizing conflicts: a heuristic repair method for constraint satisfaction problems. *Artificial Intelligence* 58:161–205, 1992.
[87] Y. Takefuji and H. Szu. Design of parallel distributed Cauchy machines. In *Proceedings of the First IJCNN*, Washington, DC, Vol. 1, pp. 529–532, 1989.

[88] G. Bilbro *et al.* Optimization by mean field annealing. In *Advances in Neural Information Processing Systems* (D. Touretzky, Ed.), Vol. 1, pp. 91–98. Morgan Kaufmann, San Mateo, CA, 1989.

[89] B. Hellstrom and L. N. Kanal. Asymmetric mean-field neural networks for multiprocessor scheduling. *Neural Networks* 5:671–686, 1992.

[90] E. Majani *et al.* On the K-winners-take-all network. In *Advances in Neural Information Processing Systems* (D. Touretzky, Ed.), Vol. 1, pp. 634–642. Morgan Kaufmann, San Mateo, CA, 1989.

[91] R. Erlanson and Y. Abu-Mostafa. Analog neural networks as decoders. In *Advances in Neural Information Processing Systems* (R. Lippmann, Ed.), Vol. 3, pp. 585–588. Morgan Kaufmann, San Mateo, CA, 1991.

[92] D. S. Touretzky. Analyzing the energy landscapes of ditributed winner-take–all networks. In *Advances in Neural Information Processing Systems* (D. Touretzky, Ed.), Vol. 1, pp. 626–633. Morgan Kaufmann, San Mateo, CA, 1989.

[93] H. N. Schaller. Problem solving by global optimization. In *Proceedings IJCNN'93*, Nagoya, Japan, pp. 1481–1484, 1993.

[94] H. N. Schaller. On the problem of systematically designing energy functions for neural expert systems based on combinatorial optimization networks. In *Proceedings Neuro-Nîmes'92*, Nanterre Cedex, Vol. 2, pp. 648–653, 1992.

[95] H. N. Schaller and K. Ehrenberger. Defining the attractor of a recurrent neural network by Boolean expressions. In *Proceedings ICANN'93* (S. Gielen and B. Kappen, Eds.), pp. 712–715. Springer-Verlag, London, 1993.

[96] M. Angermayer. Entwicklung eines heuristischen Minimierungsverfahrens für eine Klasse neuronaler Strukturen. Diploma Thesis, Lehrstuhl für Datenverarbeitung, Technical University of Munich, 1992.

[97] J. Nievergelt. Das Zahlenkreuz—Eiger-Nordwand des parallelen Rechnens? In *Informatik-Spectrum*, Vol. 13, pp. 344–346. Springer-Verlag, Berlin, 1990.

[98] B. Abramson and M. Yung. Divide and conquer under global constraints: a solution to the N-queens problem. *J. Parallel Distributed Process.* 6:649–662, 1989.

[99] B. Bernhardsson. Explicit solutions to the N-queens problem for all N. *SIGART Bull.* 2:7, 1991.

[100] A. C. Chen and C.-L. Wu. A parallel execution model of logic programs. *IEEE Trans. Parallel Distributed Systems* 2:79–92, 1991.

[101] J. Mandziuk and B. Macukow. A neural network to solve the N-queens problem. *Biol. Cybernet.* 66:375–379, 1992.

Dominant Neuron Techniques

Jar-Ferr Yang
Department of Electrical Engineering
National Cheng Kung University
Tainan, Taiwan, Republic of China

Chi-Ming Chen
Department of Electrical Engineering
Kao Yuan College of Technology
and Commerce
Luchu, Kaohsiung, Republic of China

I. INTRODUCTION

Neural networks in various realization platforms have become a very popular research field for cognitive science, neurobiology, computer science, signal processing, and system modeling. Various applications can be implicitly related to the construction of neural processing models. To acquire a specific neural processing model, the associated training process plays an important role in optimally updating the connection weights in neural networks. In terms of their corresponding training algorithms, neural networks can generally be categorized as fixed-weight networks, supervised networks, and unsupervised networks.

Fixed-weight networks with predetermined connection weights are mostly directed at association information processes. The autoassociation retrieves the complete pattern while partial information of the desired pattern is given. The cross-association retrieves a corresponding pattern defined in one metric space when another pattern is given from a different metric space. The weights in associative neural networks are usually determined from either the autocorrelation or the cross-correlation formulation. There is no learning required for fixed-weight networks.

In supervised training networks, the learning processes involve pairs of input and output patterns, which represent the assistance of the teacher. The most popular supervised network is the well-known multilayer back-propagation network, which is more suitable for applications involving a large number of classes for more complex separating clusters. Many classification applications, such as pattern and speech recognition, use supervised learning networks because their input and output patterns are one to one. When the relationship between the input and the output is unclear, unsupervised training networks should be introduced to overcome best-match problems. Thus, the unsupervised network adapts the weights and verifies the result based exclusively on the input patterns. Without benefit of any teacher, the networks learn to adapt with their experiences gathered through the previous trained results.

In the preceding neural networks, the dominant neuron techniques to select the neuron (neurons), which has (have) the maximum initial activation (activations), are usually involved in many applications. In recognition problems, for example, we need to identify which one (ones) has the highest matched scores. With back-propagation models, trained multilayer neural networks can only provide the maximum tendency for the inputs, which are not the exact training patterns. Determination of one or more maximum activated neurons becomes important for supervised neural networks. During unsupervised learning processes, the best-matched candidate (or candidates) should be identified for the later reference patterns. Of course, trained unsupervised neural networks also require the same dominant neuron techniques to pick the best-matched pattern(s). Hence, the exploration of dominant neuron techniques has gained a lot of attention recently.

In this chapter, we shall provide an integrated and intensive investigation of the fundamental issues of the design and analysis of unsupervised learning neural networks for resolving the dominant neuron (or neurons), which has the maximum preference. The explorations of the dominant neuron and K neurons can be related to the techniques for solving the winner-take-all (WTA) and K-winners-take-all (KWTA) problems, respectively. Well-known neural networks such as Grossberg's competitive learning [1], adaptive resonance theory [2], fuzzy associative memory [3], learning vector quantizer [4], and their various versions [5] all require a WTA neural network. In classification applications, the WTA neural networks can apply to the error correction system, fuzzy associative memory, Gaussian classifiers, nearest-match content-addressable memory (CAM) [6], and so on. Recently, winner-take-all networks have been widely applied to signal processing and have become the fundamental building block of many complex systems [7, 8].

Consider the inputs, $X_1, X_2, \ldots,$ and X_M, which are assumed in the range of $[X_{\min}, X_{\max}]$, where M denotes the number of inputs. Furthermore, $X_{\min}$ and $X_{\max}$ represent the minimum and the maximum values of all possible inputs, respectively. If the inputs are arranged in ascending order of magnitude, which satisfy

$$X_{\langle 1\rangle} \leq X_{\langle 2\rangle} \leq \cdots \leq X_{\langle M\rangle}, \tag{1}$$

where $X_{\langle m\rangle}$ represents the mth value increasingly ordered from the minimum, the content of $\langle m\rangle$ carries the original index of the mth ordered input. Thus, the maximum activation of the inputs is denoted by $X_{\langle M\rangle}$, the second maximum activation is expressed by $X_{\langle M-1\rangle}, \ldots$, and the minimum activation, of course, is represented by $X_{\langle 1\rangle}$.

The *winner-take-all* (WTA) neural network should process the M inputs to converge such that the M outputs become

$$Z_{\langle m\rangle} = \begin{cases} X_{\max}, & \text{for } m = M, \\ X_{\min}, & \text{for } m \neq M. \end{cases} \tag{2}$$

In other words, the neuron initially most activated will gradually dominate and become maximally activated, while the other competitive neurons die out in this so-called WTA neural system.

Generally, the *K-winners-take-all* (KWTA) neural network performs a selection of the K competitors whose activations are larger than the remaining $(M-K)$ ones. When K is equal to one, the KWTA network achieves the winner-take-all (WTA) process, which can verify the neuron with the maximum activation. Hence, we may treat the KWTA network as a generalization of the WTA network. The K-winners-take-all neural network, therefore, should make M neurons converge respectively to

$$Z_{\langle m\rangle} = \begin{cases} X_{\max}, & \text{for } m = (M-K+1), (M-K+2), \ldots, M, \\ X_{\min}, & \text{for } m = 1, 2, \ldots, (M-K). \end{cases} \tag{3}$$

For example, if $X_1 = 0.1$, $X_2 = 0.9$, $X_3 = 0.05$, $X_4 = 0.2$, and $X_5 = 0.5$ are the inputs of the WTA and the 2WTA (i.e., $K = 2$) neural networks, then the maximum value of these inputs can be represented by $X_{\langle 5\rangle} = 0.9$ and the minimum is initially expressed by $X_{\langle 1\rangle} = 0.01$. It is obvious that the order indices related to their corresponding original indices are $\langle 5\rangle = 2$, $\langle 4\rangle = 5$, $\langle 3\rangle = 4$, $\langle 2\rangle = 1$, and $\langle 1\rangle = 3$ in this case. Equation (2) shows that the WTA neural network should finally output $Z_2 = 1$ and $Z_1 = Z_3 = Z_4 = Z_5 = 0$ if the global maximum and the global minimum values of the WTA and KWTA networks are 1 and 0, respectively. Similarly, Eq. (3) shows that the final outputs of the two-winners-take-all network will be $Z_2 = Z_5 = 1$ and $Z_1 = Z_3 = Z_4 = 0$.

In general, WTA and KWTA neural networks can be categorized into two types. The first type is attained by continuous active networks, whose competitive processes are asynchronously converged until the dynamic states are stabilized. Usually, the WTA competitive mechanisms embedded in continuous neural networks can be characterized by differential equations. The second category, which is performed by iterative procedures, is treated as an iterative updating WTA process controlled by a synchronous system clock. The WTA competitive mechanisms for iterative neural networks are designated in difference equations.

In this chapter, we shall discuss and analyze the well-known WTA and KWTA neural networks. In Section II, we first introduce continuous WTA neural networks. In Section III, iterative WTA neural networks are discussed. The convergence analyses of the recently developed iterative networks are also provided. Simulation examples associated with the analyses are given for verifying the convergence behaviors of the reviewed iterative WTA networks. In Section IV, continuous and iterative KWTA networks are examined and analyzed.

II. CONTINUOUS WINNER-TAKE-ALL NEURAL NETWORKS

Continuous WTA neural networks are the simplest competitive processes. Each competitive process can generally be characterized by differential equations as

$$\frac{dZ_i}{dt} = F_i(Z_1, Z_2, \ldots, Z_M, X_i), \tag{4}$$

where

$$\frac{\partial F_i}{\partial Z_j} \leq 0, \quad \text{for } j \neq i. \tag{5}$$

The criterion stated in Eq. (5) explicitly reveals that all the neurons with a competitive tendency inhibit the other neurons in the network. Grossberg [9] introduced a continuous-time WTA mechanism, which satisfies the differential equation as follows:

$$\frac{dZ_i}{dt} = -AZ_i + (B - Z_i)f(Z_i) - Z_i \sum_{k \neq i} f(Z_k) + X_i, \tag{6}$$

where Z_i denotes the activation of the ith neuron and X_i is the ith input. In Eq. (6), A and B are positive constants with the requirement that $B > Z_i$ to assure the positive self-excitation. And $f(Z)$ is a neuron function to be selected properly for causing the network to perform the WTA mechanism. Thus, the neuron function is with nonlinear and nondecreasing characteristics. It is almost impossible to directly obtain a closed-form solution from Eq. (6). However, Grossberg [9] gave several comprehensive discussions about the convergence behavior of competitive mechanisms. The WTA behaviors expressed in Eq. (6) are the exponential decay through the term $-AZ_i$; shunting self-excitation through $(B-Z_i)f(Z_i)$; shunting inhibition of other units through $Z_i \sum_{k\neq i} f(Z_k)$; and externally applied inputs through X_i. Each neuron of the network competes with the others by sending positive signals to itself and negative signals to all its neighbors of the network. The convergence of the preceding competitive processes has been verified in detail by Grossberg. With the existence of an equilibrium state, we can subsequently verify the WTA behavior by checking the order-preserving condition

$$\frac{dZ_i(t)}{dt} \geq \frac{dZ_j(t)}{dt} \quad \text{for } Z_i(t-1) > Z_j(t-1), \tag{7}$$

which should be self-possessed in any WTA competitive learning process.

Feldman and Ballard [10] offered another WTA mechanism in which each unit would set itself to zero if it knew of a higher input to another competitor. Koch and Ullman [11] suggested two similar WTA mechanisms. In the first mechanism, the activity of the ith unit was assumed to satisfy the following differential equation:

$$\frac{dZ_i}{dt} = Z_i\left(X_i - \sum_{j=1}^{M} X_j Z_j\right). \tag{8}$$

It is obvious that the differential equation stated in Eq. (8) also satisfies both the competitive requirements shown in Eq. (5) and the order-preserving condition depicted in Eq. (7). With the initial condition $\sum_j Z_j(0) = 1$, we can easily prove that the activation of the ith unit is given by

$$Z_i(t) = \frac{Z_i(0)\exp(X_i t)}{\sum_j Z_j(0)\exp(X_j t)}. \tag{9}$$

Of course, we can verify that $\sum_j Z_j(t) = 1$. By inspecting Eq. (9), we can see immediately that if X_i is the maximum among all X_j, the exponential function $\exp(X_i t)$ will gradually dominate the remaining exponential functions $\exp(X_j t)$. Thus, the corresponding Z_i will tend asymptotically to 1, when all other Z_j's decay to 0. Koch and Ullman admitted the preceding WTA process was biologically improbable and suggested a second mechanism consisting of units arranged in a treelike structure. Each unit at a branch could select the winner of, say, two competing units at subbranches (or at the "leaves" of the tree where input values are presented) and faithfully transmit the value of the winner to the next (smaller) level in the tree. A parallel tree structure would help keep track of the path of the winner from the leaves to the root of the tree so that the winner could be identified.

Yuille and Grzywacz [12] designed another WTA mechanism. In their network, the activity of the ith unit is governed by the following differential equation:

$$\tau\frac{dZ_i}{dt} = -Z_i + X_i F\left(\sum_{i\neq j} Z_j\right), \tag{10}$$

where the first and last terms correspond to time decay and lateral shunting inhibition, respectively. In Eq. (10), $F(\cdot)$ is a monotone decreasing function, which can be an exponential, step, hyperbolic, or rational function. We can choose that the function $F(\cdot)$ contains the inhibition, which is implemented by

$$F\left(\sum_{i\neq j} Z_j\right) = \exp\left(-\lambda\sum_{i\neq j} Z_j\right). \tag{11}$$

We can also solve the differential equation stated in Eq. (10) to obtain

$$\begin{aligned} Z_i(t) &= \exp\left(\frac{-t}{\tau}\right) + X_i \exp\left(-\lambda\sum_{i\neq j} Z_j(t)\right) \\ &= \exp\left(\frac{-t}{\tau}\right) + X_i \frac{\exp(-\lambda\sum_j Z_j(t))}{\exp(-\lambda Z_i(t))} \end{aligned} \tag{12}$$

where λ is a constant. According to Eq. (12), if X_i is the maximum among all X_j, the corresponding Z_i will asymptotically tend to 1, when all other Z_j's decay to 0.

Lazzaro *et al.* [13] have designed and fabricated a series of compact, completely functional complementary metal oxide semiconductor (CMOS) integrated circuits that realize the WTA function, using the full analog nature. This circuit has been used successfully as a component in several very large scale integration (VLSI) sensory systems that perform auditory localization and visual stereopsis. Another dynamic circuit, which was proposed by Perfetti [14] as shown in Fig. 1, makes use of M competing units (M is the number of different inputs from among which the network must find the largest). It consists of a single-ended operational (OP) amplifier operating as an integrator. The inverting input of the OP amp is connected to its own inverted output through a resistor of conductance G_0 and to the output of the remaining $M-1$ amplifiers through resistors of conductance G.

The configuration of the circuit that performs a continuous WTA function is stated as follows. The first pulse moves the switches to position 1, feeding the input V_i into the capacitor C as a sample-and-hold shown in Fig. 1. The second pulse moves the switches to position 2. Then the state of the network, represented by the voltages v_{ci} across the capacitors, evolves continuously through the state space from the initial conditions $v_{ci}(0) = V_i$ for $i = 1, 2, \ldots, M$. At equilibrium, the value of $V_{\text{out}}^{(i)}$ relative to the maximum V_i, is at a high saturation level, while all others are at zero level.

Applying the Kirchhoff current law (KCL) at the node connected with the inverting input of the OP amplifier in Fig. 1 allows us to obtain the state equations

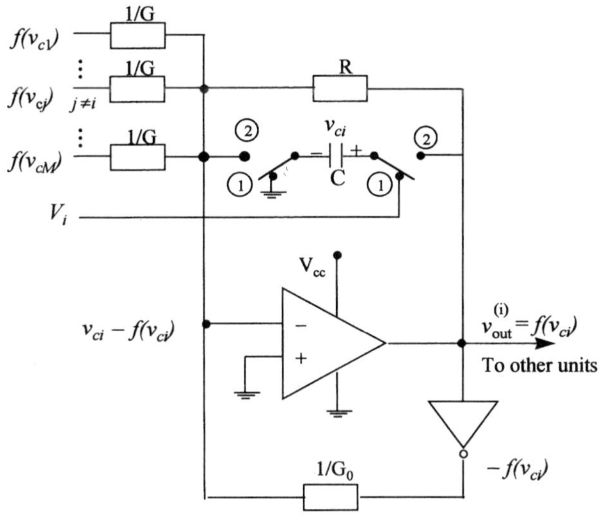

Figure 1 Basic structure of the differential WTA network.

of the network as

$$C\frac{dv_{ci}}{dt} = -\frac{v_{ci}}{R} + G_0[2f(v_{ci}) - v_{ci}] - \sum_{j=1, j\neq i}^{M} G[f(v_{cj}) - f(v_{ci}) + v_{ci}]$$

$$\text{for } i = 1, 2, \ldots, M. \quad (13)$$

As the dynamic behavior is completed, in other words, the right-hand side of Eq. (13) is equal to zero, the maximum output will converge to saturation voltage in the OP amp. And other outputs will be inhibited to zero, if the values of the conductances satisfy the following design constrains:

$$G + G_0 > 1/R \quad (14)$$

and

$$(N - 1)G - G_0 > -1/R. \quad (15)$$

The detailed discussions of the preceding network, are complex (see [14] for further investigation). However, we can simply apply the similar results in solving the differential equation depicted in Eq. (6) to verify the convergence of the network.

For other continuous WTA networks, Coultrip *et al.* [15] suggested a WTA mechanism consisting of several excitatory neurons jointly innervating and receiving feedback from a common inhibitory interneuron. Ermentrout [16] designed a very simple network of excitatory cells coupled via a single inhibitory neuron for which the time constant is allowed to vary. The basic model is assumed to mimic a small piece of cortex where the ratio of excitatory pyramidal cells to inhibitory interneurons is large. Lemmon and Kumar [17] provided an input–output description to predict the active neurons which resulted in a dynamic for laterally inhibited neural networks. Pankove *et al.* [18] suggested an optically controlled WTA circuits based on the *pnpn* structure that can be used in the optical implementation of a competitive network. Implementation of a WTA network suitable for content-addressable-memory application was presented by Johnson and Jalaleddine [19]. Seiler and Nossek [20] presented an implementation of winner-take-all behavior in inputless cellular neural networks. Although this approach basically requires a fully interconnected network, a simplified structure with only linear architectural complexity exists. The design and implementation of a high-precision VLSI winner-take-all circuit that can be arranged to process 1024 inputs was presented by Choi and Sheu [21]. The cascade configuration can be used to significantly increase the competition resolution and maintain high-speed operation for a large-scale network. The circuit can be easily extended to a large scale of over 1000 competitive cell for real-world applications. The prototype chip has been fabricated in a 2-μm CMOS technology and successfully connected to construct a 200-input winner-take-all circuit.

III. ITERATIVE WINNER-TAKE-ALL NEURAL NETWORKS

In this section, we shall introduce well-known iterative WTA neural networks and discuss the principles of their competitive processes. In iterative WTA neural networks, the neurons are synchronously updated to achieve competition systematically. The realization of iterative WTA networks could be either in a recursive style with single-layer competitions or in direct feedforward fashion with multiple-layer competitions.

A. Pair-Compared Competition

The pair-compared neural network (PACNET) shown in Fig. 2 is an eight-input multilayer feedforward structure composed of pair comparison subnets to pick a maximum in a pair-by-pair hierarchy. The PACNET is recognized as the simplest structure of multiple-layer competition. The basic element in the PACNET is the

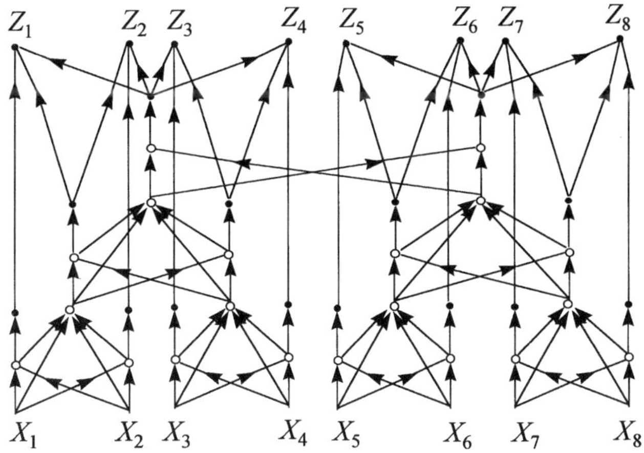

Figure 2 Complete eight-input PACNET.

comparison subnet [22], which could be treated as the smallest WTA network for two-input competition. The detailed network configuration of the comparison subnet is shown in Fig. 3. There are three outputs in this basic pair comparison subnet: two WTA outputs and the maximum output. The WTA outputs Z_1 and Z_2 can be respectively obtained by two hard limiters as

$$Z_1 = f_h(X_1 - X_2) \tag{16}$$

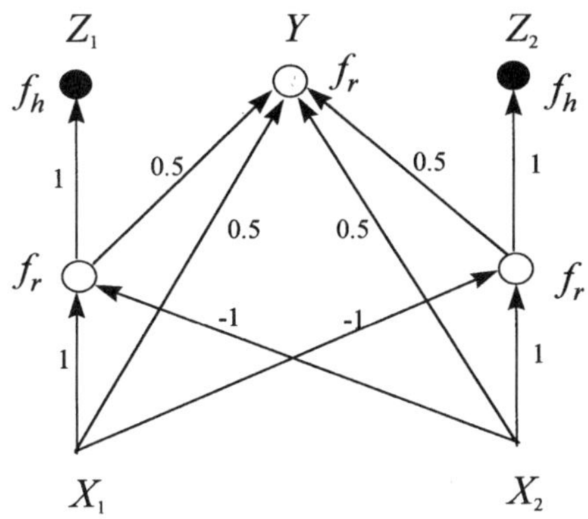

Figure 3 Basic pair-comparison subnet.

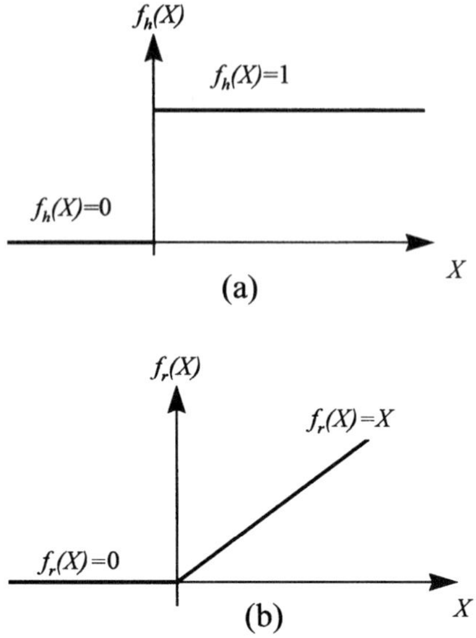

Figure 4 (a) Transfer characteristics of hard-limiter function; (b) transfer characteristics of ramp function.

and

$$Z_2 = f_h(X_2 - X_1), \tag{17}$$

where $f_h(\cdot)$ is the hard-limiter function, whose transfer characteristics, shown in Fig. 4a, perform

$$f_h(X) = \begin{cases} 1, & \text{for } X \geq 0, \\ 0, & \text{for } X < 0. \end{cases} \tag{18}$$

The maximum output of the pair comparison subnet is expressed by

$$Y = 0.5 f_r(X_1 - X_2) + 0.5 f_r(X_2 - X_1) + 0.5 X_1 + 0.5 X_2, \tag{19}$$

where $f_r(\cdot)$ is the ramp threshold function whose transfer characteristics, shown in Fig. 4b are

$$f_r(X) = \begin{cases} X, & \text{for } X \geq 0, \\ 0, & \text{for } X < 0. \end{cases} \tag{20}$$

When $X_1 > X_2$, the WTA outputs become $Z_1 = 1$ and $Z_2 = 0$ and the maximum output produces $Y = 0.5(X_1 - X_2) + 0 + 0.5X_1 + 0.5X_2 = X_1$. When $X_1 < X_2$, it is obvious that $Z_1 = 0$, $Z_2 = 1$, and $Y = X_2$. Each subnet not only determines the dominant input but also feeds the maximum value of the two inputs X_1 and X_2 forward to the next layer for further comparisons.

The PACNET for M inputs constructed in pairwise comparisons requires $\lfloor \log_2 M \rfloor$ layers, where $\lfloor \cdot \rfloor$ denotes the nearest integer which is greater than the argument. In Fig. 2, the PACNET must be constructed into a three-layer structure network for resolving eight-input comparisons. Because the PACNET requires $\lfloor \log_2 M \rfloor$ layers to achieve a WTA process for M inputs, it needs exactly $\lfloor \log_2 M \rfloor$ iterative cycles to complete a competitive process. The convergence of the PACNET is obviously assured, however, with layers which are too complicated to be utilized for performing the WTA process in hardware implementation.

B. Fixed Mutually Inhibited Competition

The MAXNET [22] shown in Fig. 5 is a one-layer neural network with a feedback structure whose design mimics the heavy use of lateral inhibition of the human brain. This mutually inhibited M-competitor WTA network consists of M neurons and M^2 connection weights. The connection weight from node j to node i in the MAXNET is fixed as

$$W_{ij} = \begin{cases} 1, & \text{for } i = j, \\ -\varepsilon, & \text{for } i \neq j,\ 1 \leq i, j \leq M. \end{cases} \tag{21}$$

The weight from each node to itself is set to 1 and the weights with a fixed value $-\varepsilon$ between distinct nodes perform the inhibition to the others. The characteris-

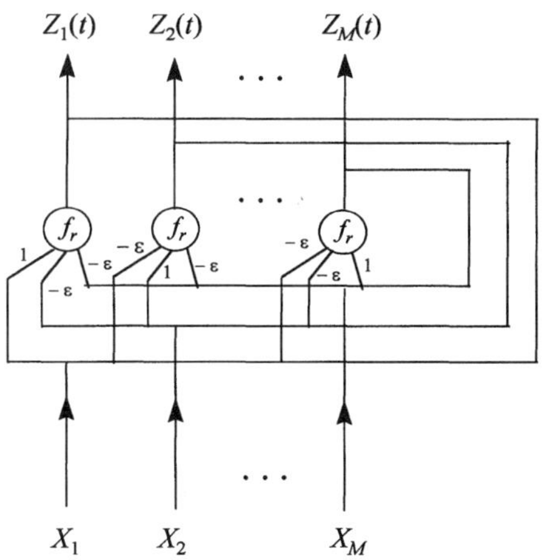

Figure 5 Configuration of the MAXNET.

tics of neurons in the MAXNET could be sigmoid, ramp, or any threshold function that has a monotonic increasing nature. The monotonic increasing nature preserves the original orders of the inputs and the threshold vitality ensures the dominant neuron (neurons) remain active and disables the inactive ones. In this chapter, without loss of generality, we assume that the following iterative WTA networks use the neurons, which are with the ramp threshold function described in Eq. (20). Hence, in the MAXNET, the activation of the ith neuron after the tth iteration becomes

$$\begin{aligned} Z_i(t+1) &= Z_i(t) - \varepsilon(Z_1(t) + \cdots + Z_{i-1}(t) + Z_{i+1}(t) + \cdots + Z_M(t)) \\ &= Z_i(t) - \varepsilon \sum_{j=1, j\neq i}^{M} Z_j(t) \end{aligned} \tag{22}$$

for $i = 1, 2, \ldots, M$. To ensure the winning neuron remains active (i.e., $Z_{\langle M\rangle}(t+1) > 0$), the selection of ε should satisfy

$$Z_{\langle M\rangle}(t) - \varepsilon \sum_{j=1, j\neq\langle M\rangle}^{M} Z_j(t) > 0 \tag{23}$$

for all n. Because $Z_{\langle M\rangle}(t) > Z_j(t)$ for $j \neq \langle M\rangle$, we can develop another inequality

$$\begin{aligned} Z_{\langle M\rangle}(t) - \varepsilon \sum_{j=1, j\neq\langle M\rangle}^{M} Z_j(t) &> Z_{\langle M\rangle}(t) - \varepsilon \sum_{j=1, j\neq\langle M\rangle}^{M} Z_{\langle M\rangle}(t) \\ &= \{1 - \varepsilon(M-1)\} Z_{\langle M\rangle}(t). \end{aligned} \tag{24}$$

If we choose

$$1 - \varepsilon(M-1) > 0, \tag{25}$$

we can ensure that the inequality stated in Eq. (23) will hold for all cases. Equation (22) shows that all the neurons will be inhibited by the others in each iteration. In other words, the activations of neurons will be gradually decreased. With the bound condition of ε stated in Eq. (25), the winning neuron will never die out. Thus, the convergence of the MAXNET will be confirmed if $\varepsilon < 1/(M-1)$. Hence, Lippmann [22] claimed that the mutual inhibition ε should be less than $1/M$ in order to ensure the convergence of the MAXNET. By observing Eq. (22), the mutual inhibition embedded in the MAXNET can be expressed as the iterative thresholding as

$$Z_i(t+1) = (1+\varepsilon) Z_i(t) - \varepsilon \sum_{j=1}^{M} Z_j(t). \tag{26}$$

Because $(1 + \varepsilon)$ is a magnifying factor in the WTA processing, we can ignore this factor without altering the final result. Thus, the equivalent thresholding in the MAXNET is given by

$$Z_i(t+1) = Z_i(t) - \frac{\varepsilon}{1+\varepsilon}\sum_{j=1}^{M} Z_j(t) = Z_i(t) - T_{\text{MAX}}(t), \tag{27}$$

where

$$T_{\text{MAX}}(t) = \frac{\varepsilon}{1+\varepsilon}\sum_{j=1}^{M} Z_j(t). \tag{28}$$

Note that this threshold function is time-varying but constant for all neurons.

To explore the convergence behavior of the MAXNET, we can examine the difference between the first largest (i.e., the winner) activation, $Z_{\langle M\rangle}(t)$ and the second largest one, $Z_{\langle M-1\rangle}(t)$. If the difference between initial activations of these two neurons is defined as $Z_{\langle M\rangle}(1) - Z_{\langle M-1\rangle}(1) = \Delta$, from Eq. (26), we can obtain the difference between the first and second maximum activations in each iteration as

$$\begin{aligned} Z_{\langle M\rangle}(t) - Z_{\langle M-1\rangle}(t) &= (1+\varepsilon)\big\{Z_{\langle M\rangle}(t-1) - Z_{\langle M-1\rangle}(t-1)\big\} \\ &= (1+\varepsilon)^2\big\{Z_{\langle M\rangle}(t-2) - Z_{\langle M-1\rangle}(t-2)\big\} \\ &= (1+\varepsilon)^t\,\Delta. \end{aligned} \tag{29}$$

For a large number of competitors, M, we have very small ε, resulting in

$$Z_{\langle M\rangle}(t) - Z_{\langle M-1\rangle}(t) = \Delta(1+\varepsilon)^t \approx \Delta(1+t\varepsilon). \tag{30}$$

The difference between the first two maximum activations increases linearly with the number of iterations. We can say that the MAXNET is converged when the difference is greater than a certain threshold, say $\Delta_{\max}$. Then the number of iterations required for the convergence becomes

$$t_{\text{MAXNET}} > \frac{\Delta_{\max} - \Delta}{\Delta\varepsilon} = \kappa\frac{1}{\varepsilon}, \tag{31}$$

where $\kappa = (\Delta_{\max} - \Delta)/\Delta$. It is obvious that κ depends on the initial difference Δ. However, the initial difference depends on not only the distribution of activations but also the number of input competitors. Usually, the larger M, the number of inputs, causes the smaller Δ, the initial difference. Once these two factors are fixed, the convergence speed will be determined by ε. For the MAXNET [22], the mutual inhibition ε is suggested to be no larger than $1/M$. If we choose $\varepsilon \approx 1/M$, the number of iterations required for the convergence is then

$$t_{\text{MAXNET}} > \kappa M. \tag{32}$$

For a small number of competitors, the convergence of the MAXNET is linearly proportional to M. If the number of competitors is large or several possible activations are close to each other, the convergence speed becomes even slower. However, the simple configuration of the net is the favorable advantage of the MAXNET.

C. Dynamic Mutual-Inhibition Competition

The major problem of the MAXNET is its slow convergence rate if the values of the competitors are nearly the same or the number of inputs is large. To overcome these problems, Yen and Chang [24] designed a dynamic WTA scheme, in which the strength of the mutual inhibitions between each processing unit is adaptively updated. The dynamic MAXNET, which is also a one-layer neural network with feedback structure, utilizes the dynamic connection weight [24],

$$W_{ij} = \begin{cases} 1, & i = j, \\ -\dfrac{1}{M(t)}, & i \neq j, 1 \leq i, j \leq M, \end{cases} \tag{33}$$

between node i and node j, where $M(t)$ is the number of active neurons at the tth iteration. The adaptation of the dynamic MAXNET tends to increase the effects of mutual inhibitions in the iteration process. As shown in Eq. (27), the equivalent threshold function of the improved MAXNET (IMAXNET) becomes

$$T_{\text{IMAX}}(t) = \frac{\varepsilon(t)}{1+\varepsilon(t)} \sum_{j=1}^{M} Z_j(t), \tag{34}$$

where $\varepsilon(t) = 1/M(t)$.

Similar to the MAXNET, the difference between the first and second maximum activations of the improved MAXNET is

$$\begin{aligned} Z_{\langle M\rangle}(t) - Z_{\langle M-1\rangle}(t) &= \big(1+\varepsilon(t)\big)\big(Z_{\langle M\rangle}(t-1) - Z_{\langle M-1\rangle}(t-1)\big) \\ &= \Delta \prod_{k=1}^{n} \big(1+\varepsilon(t)\big). \end{aligned} \tag{35}$$

Because $M(t) \leq M$, we know that $\varepsilon(t) > \varepsilon$. Comparing Eq. (28) to Eq. (34), the threshold function in the improved MAXNET is larger than that in the MAXNET. At the same time, we can also show that the difference stated in Eq. (35) is larger than that depicted in Eq. (29). Thus, we can prove that the speed of convergence of the improved MAXNET is faster than that of the MAXNET [24].

D. Mean-Threshold Mutual-Inhibition Competition

When we increase the strength of inhibition, the number of iterations in the dynamic MAXNET network will be decreased. The GEMNET proposed in [25] was first introduced from the basic idea of the *ge*neral *m*ean-based WTA neural *net*work. It is well known that the maximum is always greater than the mean of any subset of activations in all competitive neurons. So, the decision of the winner is processed by the following simple mean-based WTA axiom:

> *Inhibit the processing elements whose activations are less than the mean of activated neurons until the stable state is reached.*

Because the maximum is always greater than the mean value when the activated neurons are more than one, a portion of neurons after each mean-based WTA process will be inhibited if the activations are less than the mean value of the active activations. The threshold function for the GEMNET, of course, is expressed as

$$T_{\mathrm{MEAN}}(t) = \frac{1}{M(t)} \sum_{j=1}^{M} Z_j(t), \tag{36}$$

where $M(t)$ is the number of active neurons. The whole process will continue iterating until only one neuron, which contains the maximum activation, remains activated. When only one activated neuron remains, the maximum value and the mean value will be the same. Under the guideline of the mean-based WTA axiom, no neuron will be further inhibited for a single active neuron; therefore, the GEMNET reaches the desired stable state. After the simplification, the connection weight $W_{ij}(t)$ in the GEMNET becomes

$$W_{ij} = \begin{cases} \gamma, & \text{for } i = j, \\ -\dfrac{\gamma}{M(t) - 1}, & \text{for } i \neq j, 1 \leq i, j \leq M, \end{cases} \tag{37}$$

where $\gamma (1 \leq \gamma)$ acts as a compensation factor. The compensation factor which varies with distributions of the activations helps to preserve the maximum activation constant. The detailed configuration of the GEMNET is depicted in Fig. 6. It has been verified that the suggested WTA neural network on average requires fewer than $\mathrm{Log}_2(M)$ iterations to complete a WTA process for uniform, normal, and peak-uniform distributed inputs, where M is the number of competitors. Comparing Eq. (36) to Eqs. (28) and (34), it is obvious that the threshold function in the GEMNET is larger than those in the MAXNET and improved MAXNET. Thus, after n iterations, the difference between the maximum and the second maximum of the GEMNET is larger than that of the MAXNET [22] and the improved

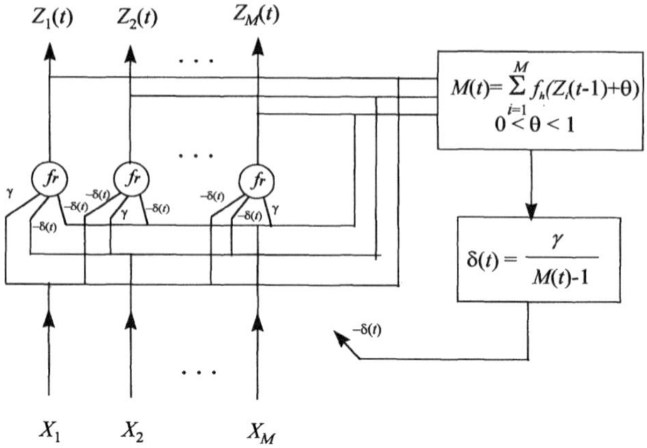

Figure 6 Configuration of the GEMNET.

MAXNET [24]. In other words, the convergence speed of the GEMNET is faster than that of the MAXNET and the improved MAXNET. Suter and Kabrisky [26] suggested a WTA network which requires different types of nodal computation elements and no nodal switches. This WTA network under the concept of the statistical mean is similar to the GEMNET network with $\mathrm{Log}_2(M)$ iterations for completing the WTA process.

E. Highest-Threshold Mutual-Inhibition Competition

If we can increase the level of inhibition, the convergence speed of the GEMNET will be further improved. By using an acceleration factor, the GEMNET can be extended to the higher-order-based neural network (HOSNET) [27, 28]. The basic concept of the HOSNET is evolved from the mean-base threshold. By raising the threshold up to the average of the second maximum, the mutual-inhibition process will be finished in just one iteration in expectation. Thus, the threshold function in HOSNET is designed to become

$$T_{\mathrm{HOS}}(t) = E[Z_{\langle M-1\rangle}]. \tag{38}$$

The connection weight between node i and node j of the HOSNET is given by

$$W_{ij}(t) = \begin{cases} \gamma, & \text{for } i = j, \\ -\dfrac{\gamma\beta(t)}{M(t)-\beta(t)}, & \text{for } i \neq j,\ 1 \leq i, j \leq M, \end{cases} \tag{39}$$

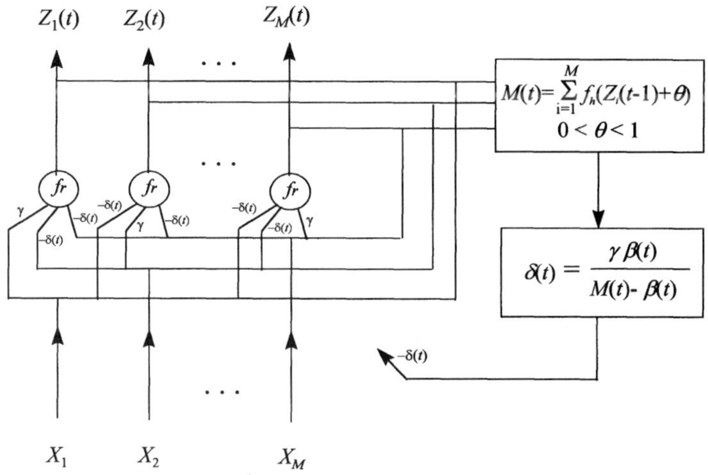

Figure 7 Configuration of the HOSNET.

where $M(t)$ and $\beta(t)$ denote the number of active neurons and the proposed acceleration factor in the HOSNET, respectively. The HOSNET is depicted in Fig. 7. If $\beta(t) = 1$, the HOSNET is identical to the GEMNET. How to determine a better acceleration factor to increase the convergence speed becomes an important issue [28]. The acceleration factor can be designed dynamically to optimally increase the convergence speed. However, $\beta(t)$ is suggested as

$$\beta(t) = \begin{cases} \delta(t), & \text{for } M(t) > 1, \\ 1, & \text{for } M(t) = 0, \end{cases} \tag{40}$$

where the selection of the optimal acceleration factor $\delta(t)$, which is developed in [28], should be greater than one in order to improve the convergence speed. The convergence speed of the HOSNET is higher than that of the GEMNET for the case of large competitors.

F. Dynamic Thresholding Competition

In the preceding mutual-inhibition WTA neural networks, the activations of neurons are varied during competition processes once the original inputs, $\{X_1, X_2, \ldots, X_M\}$ are fed into the networks. After the completion of the WTA process, one and only one neuron will survive, and the remaining neurons will die out completely. The preceding WTA neural networks do not require storing the original inputs.

By storing the original inputs, $\{X_1, X_2, \ldots, X_M\}$, the SELECTRON [29] is proposed to physically select a maximum threshold to result in a WTA neural process. The self-excited connection weight in the SELECTRON is designed as

$$W_{ii}(t) = \begin{cases} g(X_i), & \text{for } Z_i(t) = 0, \\ \dfrac{M(t)-1}{M(t)}, & \text{for } Z_i(t) = 1, \end{cases} \tag{41}$$

where $g(X)$ denotes the function where $g(X) = -1$ for $X \geq 0$ and $g(X) = 1$ for $X < 0$. The mutual-inhibition connection weight between node i and node j $(i \neq j)$ is given by

$$W_{ij}(t) = \begin{cases} 0, & \text{for } Z_i(t) = 0 \text{ or } Z_j(t) = 0, \\ -\dfrac{1}{M(t)}, & \text{otherwise.} \end{cases} \tag{42}$$

In Eqs. (35) and (36), the WTA output

$$\begin{aligned} Z_i(t) &= f_h\left(\sum_{j=1}^{M} + W_{ij}(t-1)X_j + I_i\right) \\ &= f_h\left(X_i - \frac{1}{M(t-1)} \sum_{j \in \text{active}} X_j + I_i\right) \end{aligned} \tag{43}$$

represents the hard-limited output of the output node i at the kth iteration and $I_i \leq 0$. It is obvious that $Z_i(t)$ is an indication of the action or nonaction of the ith output neuron. The number of active output neurons can be obtained by $M(t) = \sum_{j=1}^{M} Z_j(t)$. The dynamic threshold is obtained from the average of the original inputs corresponding to the active output neurons. In the dynamic thresholding process, Eq. (43) shows that the SELECTRON gradually raises the threshold as

$$T_{\text{SEL}}(t-1) = \frac{1}{M(t-1)} \sum_{j \in \text{active}} X_j, \tag{44}$$

which represents the average of the largest $M(t-1)$ activations. The SELECTRON converges if $M(t+1) = M(t)$. The number of iterations required for convergence is bounded above by M. The average number of iterations is not greater than $\text{Log}_2(M)$ for large competitors. The convergence speed of the SELECTRON network is similar to that of the GEMNET network; however, the SELECTRON can select the minimum neuron for any data sets, if we modify some of functions and parameters.

Table I

Average Number of Iterations Required for Completion of a WTA Process (Uniformly Distributed Inputs in [0, 1])

WTA networks	10 inputs	50 inputs	100 inputs	500 inputs	1000 inputs	2000 inputs	3000 inputs	4000 inputs
MAXNET	11.89	72.51	145.3	645.4	1000*	2000*	3000*	4000*
IMAXNET	4.54	7.01	8.05	10.12	11.35	12.13	12.73	13.17
GEMNET	3.12	5.39	6.38	8.76	9.76	10.78	11.35	11.66
HOSNET	2.97	3.18	3.30	3.51	3.67	3.80	3.89	3.94

*Not converged at this point.

G. Simulation Results

The inputs with uniform, normal, and peak-uniform distributions are randomly generated by Monte Carlo simulations to evaluate the WTA behavior of the MAXNET, improved MAXNET, GEMNET, and HOSNET. After 1000 independent runs, Tables I–III show the average number of iterations required for complete convergence using these four WTA nets. The *complete convergence* is defined so that all of the activations except the winner are inhibited to zero by the WTA nets. For the uniform and the normal distributions, the MAXNET cannot converge within M iteration cycles where M is the number of inputs. However, the MAXNET does converge within reasonable iteration cycles for peak-uniform distribution if M is less than 50. When M is greater than 1000, the MAXNET cannot converge for the reasons stated in the previous section. The MAXNET requires the largest number of iterations to converge, whereas the HOSNET with the

Table II

Average Number of Iterations Required for Completion of a WTA Process (Normally Distributed Inputs with $N[0, 1]$)

WTA networks	10 inputs	50 inputs	100 inputs	500 inputs	1000 inputs	2000 inputs	3000 inputs	4000 inputs
MAXNET	8.60	50.16	106.6	515.3	1066	2000*	3000*	4000*
IMAXNET	3.30	5.27	5.84	7.66	8.52	9.10	9.51	9.68
GEMNET	2.12	3.94	4.44	6.28	6.90	7.60	8.04	8.42
HOSNET ($\beta = 3$)	1.30	3.86	4.09	3.69	4.41	4.84	4.92	4.83

*Not converged at this point.

Table III

Average Number of Iterations Required for Completion of a WTA Process (One Peak Value at 0.9 and the Rest Uniformly Distributed in [0, 0.5])

WTA networks	10 inputs	50 inputs	100 inputs	500 inputs	1000 inputs	2000 inputs	3000 inputs	4000 inputs
MAXNET	3.15	11.29	19.42	46.68	67.72	97.93	122.6	143.8
IMAXNET	2.34	3.71	4.01	5.05	6.04	6.32	6.89	7.03
GEMNET	2.02	3.16	4.00	5.00	5.92	6.00	6.34	7.00
HOSNET ($\beta = 1.5$)	1.31	2.00	2.24	3.00	3.02	3.79	4.00	4.00

optimal acceleration factor requires the least. In other words, the HOSNET shows the fastest WTA behavior among the four nets in the cases of uniformly, normally, and peak uniformly distributed inputs. The simulation results meet the preceding statistical convergence analyses of the four iterative WTA neural networks.

IV. K-WINNERS-TAKE-ALL NEURAL NETWORKS

The K-winners-take-all (KWTA) neural network performs a selection of the K competitors whose activations are larger than the remaining $(M - K)$ ones. When K is equal to one, the KWTA network achieves the well-known winner-take-all (WTA) process, which can verify the neuron with the maximum activation. When $K = (M - 1)$, we can select the minimum activation if the indication of the minimum output is reversed. Thus, the KWTA network is a generalization of the WTA network. The KWTA neural networks can be categorized into two types. The first type of KWTA network uses dynamically mutual inhibition in a one-layer structure. The second category of KWTA network, which adopts dynamic threshold in decaying variations, is also built in a one-layer competitive architecture.

A. Continuous K-Winners-Take-All Competition

Majani *et al.* [30] suggested a KWTA mechanism, which uses the continuous Hopfield model of dynamics for reliable K-winners convergence. In the continuous Hopfield K-winners-take-all (CHKWTA) network [30], the convergence behavior can be characterized by the differential equations as

$$C\frac{dZ_i}{dt} = -\lambda Z_i + (a+1)f_s(Z_i) - \left(\sum_{j=1}^{M} f_s(Z_j) - E\right) \qquad \text{for } i = 1, 2, \ldots, M, \tag{45}$$

where $\lambda = M-1+a$ $(a < 1)$, $E = 2K-M$, and $f_s(Z_i)$ is a sigmoid function that varies between -1 and $+1$. Thus, the CHKWTA network possesses the weight between node i and node j given by

$$w_{ij} = \begin{cases} 1, & i = j, \\ a, & i \neq j \text{ and } 1 \leq i, j \leq M \end{cases} \tag{46}$$

if E is treated as the external input. It has been verified in [30] that the stable equilibrium state of the CHKWTA network converges to K positive numbers in the positions of the K largest inputs and $(M-K)$ negative numbers in the remaining inputs.

B. Interactive Activation *K*-Winners-Take-All Competition

Wolfe *et al.* [31] proposed and analyzed a special class of mutually inhibitory networks to complete a KWTA network by using interactive activations. The interactive activation K-winners-take-all (IAKWTA) network [31] suggests a special class of mutually inhibitory network and provides parameters for reliable K-winners performance. In the IAKWTA network, the weight between node i and node j is given by

$$w_{ij} = \begin{cases} 0, & i = j, \\ w, & i \neq j \text{ and } 1 \leq i, j \leq M. \end{cases} \tag{47}$$

where $w < 0$, which is usually chosen as -1. The network dynamics are modeled by using interactive activation to update the units as

$$Z_i(t+1) = Z_i(t) + \gamma \Delta Z_i(t), \tag{48}$$

$$\Delta Z_i(t) = \text{Net}_i(t)[X_{\max} - Z_i(t)] \qquad \text{if } \text{Net}_i(t) \geq 0, \tag{49}$$

$$\Delta Z_i(t) = \text{Net}_i(t)[Z_i(t) - X_{\min}] \qquad \text{if } \text{Net}_i(t) \leq 0, \tag{50}$$

$$\text{Net}_i(t) = \sum_{j=1}^{M} Z_j w_{ij} + E_i, \tag{51}$$

where E_i is the (constant) external input to unit i and $\gamma > 0$ is the step size. In the preceding equations, the selection of $X_{\min}$ is constrained by $X_{\min} =$

$-K/(M-K)X_{\max}$. The IAKWTA network converges more rapidly if the step size γ is larger. However, the order preservation of $Z_i(t)$ and $Z_j(t)$ should be constrained by the condition that $\gamma < 1/[-w(X_{\max}+\Sigma)]$, where Σ denotes the sum of all the activations except $Z_i(t)$ and $Z_j(t)$. In other words, the step size should be chosen small enough to ensure stability. Large competitors always cause Σ to increase and result in a small selectable step size. Thus, the convergence speed of the IAKWTA network will become sluggish for large competitors.

When the parameters of the variable $X_{\min}/X_{\max}$ ratio and connection weights w are properly chosen, the IAKWTA network [31] has been proven to be a dual of the CHKWTA network. The CHKWTA network should be similar in convergence rate to the IAKWTA network. The convergence speed of the CHKWTA network is also slow in very large number of the input cases.

C. Coarse-Fine Mutual-Inhibition *K*-Winners-Take-All Competition

Yen and Chang [32] extended the mutual-inhibition concept of the MAXNET to achieve the KWTA process. The MAXNET or other mutually inhibited WTA neural networks gradually inhibit the smallest-activation neurons, then the second smallest, etc. Hence, the number of winners (active neurons), $M(t)$ will be successively decreased in each mutual-inhibition process. By continuous observation of $M(t)$, the mutual-inhibition neural network exhibits the same convergence behavior as the MAXNET if $M(t)$ is greater than K. Thus, the coarse-fine KWTA (CFKWTA) [32] network possesses a weight between node i and node j as

$$W_{ij}(t) = \begin{cases} 1, & i = j, \\ \varepsilon(t), & i \neq j \text{ and } 1 \leq i, j \leq M, \end{cases} \tag{52}$$

in the coarse search stage with normal WTA processes. These normal WTA processes will be completed if $M(t)$ is exactly equal to K. In the final normal WTA process, the overinhibition could result in $M(t)$ less than K. Once the overinhibition stage is reached, the CFKWT network should retrieve the previous state and change into the fine search stage by choosing a much smaller $\varepsilon(t)$ to ensure the convergence. Of course, the development of coarse-fine mutual-inhibition rules can be further explored by using various WTA competitions.

D. Dynamic Threshold-Search *K*-Winners-Take-All Competition

Departing from typical mutual inhibition, the KWTA network can be built in a one-layer competitive architecture, which possesses dynamic threshold with decaying variations. Figure 8 shows the configuration of the dynamic KWTA net-

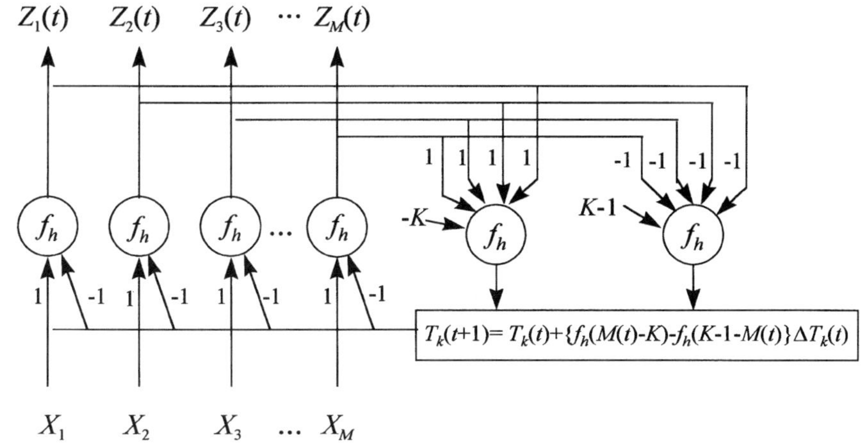

Figure 8 Dynamic KWTA (DKWTA) network.

work (DKWTA) in [33]. If K winners are desired, the DKWTA network progressively provides a proper threshold, $T_k(t)$, which should gradually approach and finally fall into the range between $X_{\langle K-1\rangle}$ and $X_{\langle K\rangle}$. The dynamic threshold search algorithm suggested in [33] is given by

$$T_k(t+1) = T_k(t) + \big\{f_h\big(M(t) - K\big) - f_h\big(K - 1 - M(t)\big)\big\}\Delta T_k(t), \tag{53}$$

where the threshold search step is designed as a sequence of decaying functions given by

$$\Delta T_k(t) = (0.5)^t \delta \tag{54}$$

for $t = 1, 2, 3, \ldots$. Once the threshold $T_k(t)$ is in the range between the $(K-1)$th and the Kth maximum input values, that is, $X_{\langle K-1\rangle} < T_k(t) \le X_{\langle K\rangle}$, the outputs of the hard limiters exactly provide the KWTA results. The DKWTA network in this case will provide K active neurons and $M - K$ inhibited ones.

Because the dynamic threshold $T_k(t)$ can be updated by the bisection concept, the DKWTA network can achieve high convergence speed. Thus, the DKWTA network needs

$$i > \mathrm{Log}_2\left[\frac{1}{(X_{\langle K\rangle} - X_{\langle K-1\rangle})}\right] \tag{55}$$

iterations to achieve convergence.

If the inputs are randomly selected from the other distribution, the convergence highly depends on the expected resolution of $[\mu_{K-1:M}, \mu_{K:M}]$. The DKWTA network requires

$$i > \mathrm{Log}_2\left[\frac{1}{(\mu_{K:M} - \mu_{K-1:M})}\right] \tag{56}$$

iterations for convergence, where $\mu_{m:M}$ denotes the expectation of the mth-order statistics in M inputs. Generally, the uniformly distributed inputs are more difficult for the KWTA process than the others. Conservatively, we may say that the convergence speed of the proposed DKWTA network is logarithmic.

In general, mutual inhibition possesses convergence rate on the order of M, where M is the number of competitors. Thus, the mutual-inhibition KWTA networks become infeasible for a very large number of competitors. Therefore, the DKWTA network can conquer the problem for suitably large inputs.

E. Simulation Results

The competitors uniformly distributed in [0, 1] are randomly generated by Monte Carlo experiments to evaluate the KWTA behavior of the CHKWTA, IAKWTA, and DKWTA networks. For the DKWTA [33] network, we select $T_k(0) = 0.5$ and $\delta = 0.5$. In the IAKWTA network [31], we choose $X_{\min} = -0.428571$, $X_{\max} = +1$, $\gamma = 0.02$, and $E_i = 0$. For fair comparison, we use the finite-difference equation to realize the CHKWTA network iteratively. Then we pick $\alpha = 0.5$ and $C = 50$ for the CHKWTA network [30]. All simulation results in Table IV depict the averages of 1000 independent runs.

Table IV shows the average number of iterations required for completion of the convergence of 3WTA using the preceding KWTA networks. The *completion of convergence* for the IAKWTA network is defined as obtaining K neurons with maximum activation values which are greater than the threshold, $(X_{\min} + X_{\max})/2$. Similarly, the threshold of the CHKWTA network is also chosen as $(X_{\min} + X_{\max})/2$. However, the DKWTA network must attain K neurons with $X_{\max}$ and $(M - K)$ neurons with $X_{\min}$ for the completion of convergence.

Table IV

Average Number of Iterations Required for Completion of a 3WTA Process

KWTA networks	10 inputs	20 inputs	30 inputs	40 inputs	50 inputs	100 inputs
IAKWTA	41.20	160.75	310.49	427.90	457.51	956.08
CHKWTA	101.84	218.49	315.37	443.73	585.10	1051.30
DKWTA	3.57	4.37	5.07	5.48	5.88	6.85

From Table IV, the convergence speeds of the IAKWTA and CHKWTA networks are nearly proportional to the number of competitors. However, the DKWTA network achieves logarithmic convergence, which is the same as the theoretical result depicted in Eq. (56). Therefore, the DKWTA network with simulation results shows faster KWTA behavior than the IAKWTA and CHKWTA networks.

V. CONCLUSIONS

Many neural networks need a dominant neuron technique to select the neuron (neurons) which has (have) the maximum initial activation (activations) for various applications. In this chapter, we have investigated the design and analysis of WTA and KWTA neural networks for resolving the dominant neuron (neurons) which has (have) the maximum preference. We first introduced some continuous and iterative WTA neural networks. Theoretically, the relationship between continuous and iterative WTA neural networks is similar to the relationship between continuous-time and discrete-time systems. For example, we can readily transfer the continuous WTA neural networks to the iterative ones by replacing dZ_i/dt with $Z_i(t) - Z_i(t-1)$. The development of continuous WTA networks is more concentrated on the issues of stability and order preserving. For iterative WTA neural networks, the mutual-inhibition processes, however, are mostly focused on the convergence speed of the network. The MAXNET with fixed connection weights has the advantage of simplest structure but suffers from slowest convergence. Provided with the dynamics in their connections, the latest iterative WTA neural networks, such as the GEMNET and SELECTRON, can speed the linear convergence up to logarithmic convergence. By extending the concept of mean-based thresholding, which is used in the GEMNET, a faster iterative WTA network can be designed by raising the threshold, which is higher than the mean. An increase in the level of threshold is equivalent to an increase in the strength of the mutual inhibition in iterative WTA networks. Beyond mean-based thresholding, the extra control mechanism to detect the possible overinhibition should be figured along with the dynamic connection weights. Usually, this control mechanism becomes possible if the overinhibition detection associated with the adaptive procedure is properly designed. The continuous competitive mechanism for K-winners-take-all neural networks is similar to that for WTA neural networks. However, the design of system parameters for the KWTA is much more difficult than that for WTA networks. As to the iterative KWTA networks, we also found that the update procedure is also more complex than that of iterative WTA networks although the concept of mutual inhibition is still obviously used. From simulation results, the DKWTA network is better than the IAKWTA and CHWTA networks. The DKWTA network uses the bisection concept to gradually locate the proper threshold such that the K winners will maintain active. Because the mutual inhibition and the thresholding of inputs are equivalent for iterative competitions, we can conclude that the fastest dominant neural networks are conceptually to find a proper threshold function which is, at best, $X_{\langle M-1\rangle}$ for WTA and $X_{\langle M-k\rangle}$ for KWTA if the stability of the network can be resolved.

REFERENCES

[1] S. Grossberg. Adaptive pattern classification and universal recording, I. Parallel development and coding of neural feature detectors. *Biol. Cybernet.* 23:121–134, 1976.

[2] G. Carpenter and S. Grossberg. Adaptive resonance theory: stable self-organization of neural recognition codes in response to arbitrary lists of input patterns. In *Eighth Annual Conference of the Cognitive Science Society*, pp. 45–62. Erlbaum Hillsdale, NJ, 1986.

[3] B. Kosko. Fuzzy associative memories. In *Fuzzy Expert System* (A. Kandel, Ed.). Addison-Wesley, Reading, MA, 1987.

[4] T. Kohonen. Automatic formation of topological maps in a self-organizing system. In *Proceedings of the Second Scandinavian Conference on Image Analysis* (E. Oja and O. Simula, Eds.), pp. 214–220, 1981.

[5] T. Kohonen. *Self-Organization and Associative Memory* 3rd ed. Springer-Verlag, New York, 1989.

[6] S. Grossberg. Nonlinear neural networks: principles, mechanisms, and architectures. *Neural Networks* 1:17–61, 1988.

[7] A. K. Krishnamurthy *et al.* Neural networks for vector quantization of speech and images. *IEEE J. Select. Areas Comm.* 8:1449–1457, 1990.

[8] T. Kohonen. Self-organized formation of topologically correct feature maps. *Biol. Cybernet.* 43:59–69, 1982.

[9] S. Grossberg. Contour enhancement, short term memory, and constancies in reverberating neural networks. *Stud. Appl. Math.* 52:213–257, 1973.

[10] J. A. Feldman and D. H. Ballard. Connectionist models and their properties. *Cognitive Sci.* 6: 205–254, 1982.

[11] D. Koch and S. Ullman. Selecting one among the many: a simple network implementing shifts in selective visual attention. *Human Neurobiol.* 4:219–227, 1985.

[12] A. L. Yuille and N. M. Grzywacz. A winner-take-all mechanism based on presynaptic inhibition feedback. *Neural Comput.* 1:334–337, 1989.

[13] J. P. Lazzaro, S. Ryckebusch, M. A. Mahowald, and C. A. Mead. Winner-take-all networks of $O(N)$ complexity. Computer Science Department, California Institute of Technology, Technical Report Caltech-Cs-TR-21-8, 1989.

[14] R. Perfetti. Winner-take-all circuit for neurocomputing applications. *IEE Proc.* 137:353–359, 1990.

[15] R. Coultrip, R. Granger, and G. Lynch. A cortical model of winner-take-all competition via lateral inhibition. *Neural Networks* 5:47–54, 1992.

[16] B. Ermentrout. Complex dynamics in winner-take-all neural nets with slow inhibition. *Neural Networks* 5:415–431, 1992.

[17] M. Lemmon and B. V. Kumar. Emulating the dynamics for a class of laterally inhibited neural networks. *Neural Networks* 2:193–214, 1989.

[18] J. Pankove, C. Radehaus, and K. Wagner. Winner-take-all neural net with memory. *Electron. Lett.* 26:349–350, 1990.

[19] L. G. Johnson and S. M. S. Jalaleddine. MOS implementation of winner-take-all network with application to content-addressable memory. *Electron. Lett.* 27:957–958, 1991.

[20] G. Seiler and J. A. Nossek. Winner-take-all cellular neural networks. *IEEE Trans. Circuits Systems II Analog Digital Signal Process* 40:184–190, 1993.

[21] J. Choi and B. J. Sheu. A high-precision VLSI winner-take-all circuit for self-organizing neural networks. *IEEE J. Solid-State Circuits* 28:576–583, 1993.

[22] R. P. Lippmann. An introduction to computing with neural nets. *IEEE ASSP Mag.*, 1987.

[23] H. K. Kwan. One-layer feedforward neural network fast maximum/minimum determination. *Electron. Lett.* 1583–1585, 1992.

[24] J. C. Yen and S. Chang. Improved winner-take-all neural network. *Electron. Lett.* 662–664, 1992.

[25] J. F. Yang, C. M. Chen, W. C. Wang, and J. Y. Lee. A general mean based iteration winner-take-all neural network. *IEEE Trans. Neural Networks* 6:14–24, 1995.

[26] B. W. Suter and M. Kabrisky. On a magnitude preserving iterative Maxnet algorithm. *Neural Comput.* 4:224–233, 1992.

[27] J. F. Yang and C. M. Chen. An improved general mean based iterative winner-take-all neural network. In *1994 International Symposium on Artificial Neural Networks*, pp. 429–434.

[28] J. F. Yang and C. M. Chen. Winner-take-all neural networks based on higher order statistics. *IEEE Trans. Neural Networks*. To appear.

[29] J. C. Yen, F. J. Chang, and S. Chang. A new winners-take-all architecture in artificial neural networks. *IEEE Trans. Neural Networks* 5:838–843, 1994.

[30] E. Majani, R. Erlanson, and Y. Abu-Mostafa. On the K-winners-take-all network. *Advances in Neural Information Processing Systems* (D. S. Touretzky, Ed.), Vol. 1, pp. 634–642. Morgan Kaufmann, Los Altos, CA, 1989.

[31] W. J. Wolfe *et al.* K-winners network. *IEEE Trans. Neural Networks* 2:310–315, 1991.

[32] J. C. Yen and S. Chang. A new first-k-winners neural network. *ISANN'93* pp. D-01–D-06, 1993.

[33] J. F. Yang and C. M. Chen. A dynamic K-winners-take-all neural network. *IEEE Trans. System Man Cybernet. B* 6:523–526, 1997.

CMAC-Based Techniques for Adaptive Learning Control

Chun-Shin Lin
Department of Electrical Engineering
University of Missouri-Columbia
Columbia, Missouri 65211

Ching-Tsan Chiang
Department of Electrical Engineering
University of Missouri-Columbia
Columbia, Missouri 65211

Hyongsuk Kim
Department of Control and Instrumentation Engineering
Chonbuk National University
Republic of Korea

I. INTRODUCTION

In this chapter, we shall introduce the cerebellar model articulation controller (CMAC) and CMAC-based techniques, which are often used in learning control applications. The CMAC was first developed by Albus in the mid-1970s for robot manipulator control and functional approximation [1, 2]. It is an efficient table-lookup technique. Its most attractive characteristic is that learning always converges to the result with a least square error [3, 4] and the convergence is fast. The CMAC technique did not receive much attention until the mid-1980s when researchers started developing strong interest in neural networks. The CMAC is now considered one type of neural network with major applications in learning control.

In Section II, we will first give a brief introduction to neurocontrol. Neural networks are often used in modeling plants or inverse plants, implementing adaptive

Optimization Techniques

controllers, learning to evaluate control performance, and learning nonlinear control techniques. The CMAC and other improved CMAC-based techniques, which are adequate for such usage, are introduced in Sections III to V. Section VI gives a conclusion.

II. NEURAL NETWORKS FOR LEARNING CONTROL

Adaptive control techniques are often developed for control problems with nonlinear dynamic processes and uncertainty. Conventional methods are usually based on the assumption that the plant can be linearized at an operating point and an adaptive linear controller can be developed. Many of today's complex control problems, however, involve *highly* nonlinear dynamic processes and require the use of general nonlinear models for better control performance. Many control engineers and researchers feel that neural networks are good candidates for such usage. Most research in this area considered the use of multilayer neural networks (MNNs) [5–7] for necessary modeling in learning control. The choice of using MNNs was probably because this type of neural network was among those first introduced and could well perform functional approximation. After the MNN and its associated error back-propagation learning algorithm were introduced, many other single-layer neural nets that use basis functions have been extensively studied. These neural networks include radial basis function networks [8–10], functional-link networks [11, 12], polynomial neural networks [13, 14], CMACs, etc. All these possess the capability of learning general nonlinear static mapping and are capable of handling nonlinearity and adapting to uncertainty. Later in Sections III to V, we shall introduce the CMAC and some CMAC-based schemes. Before that, we shall give a brief introduction to neurocontrol techniques, to which CMAC-related structures can be applied. In learning control, neural networks can be used for necessary system identification, inverse plant implementation, etc. Although the theory and methodology in linear control has been well developed, neural networks are not likely to provide an advantage in the control of linear systems. The promising area is on *nonlinear* system control.

A. NONLINEAR CONTROLLER: IDENTIFICATION OF INVERSE PLANT AND ITS USAGE

The most direct usage of a neural network in learning control is the identification of an inverse plant. The identified inverse plant model can be used as a controller. There are two typical ways to develop an inverse plant. The first one

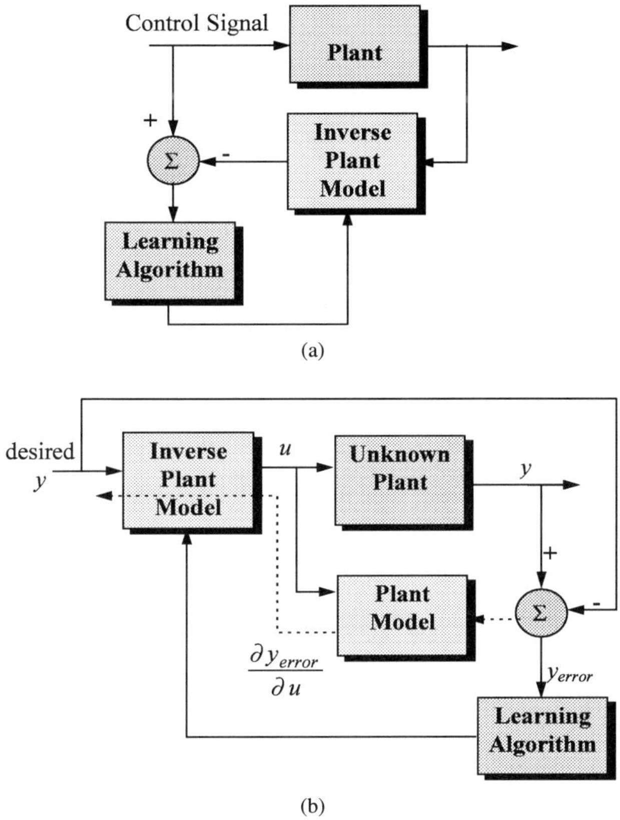

Figure 1 Identification of an inverse plant: (a) direct method for inverse plant identification; (b) specialized method for inverse plant identification.

is the direct method as shown in Fig. 1a. Using this method, the control signal is applied to the plant and the plant output is collected. What the inverse plant should implement is the mapping from the plant output to the plant input. The input–output pairs are used by the neural network learning algorithm for adjusting neural network parameters in order to implement the mapping. Upon learning, the inverse plant can be used for control. The desired plant output is provided as the input to the neurocontroller, and the output of the controller is used to control the plant. The cascade of the inverse plant model and the plant gives an identity transformation.

Another way to develop the inverse plant model relies on the use of a plant model. This specialized learning technique is illustrated in Fig. 1b. The plant should be identified first. With the plant model, the partial derivative $\partial y_{\text{error}}/\partial u$ can be obtained. This partial derivative indicates the direction that u should be changed to reduce the control output error. This piece of information can be used for adjusting the parameters in the inverse plant model. Neural networks can be used for modeling both the plant and the inverse plant. The specialized approach is based on the error back propagation through the plant.

B. Model Reference Adaptive Controller

The model reference adaptive controller (MRAC) uses a reference model and tries to control the plant to have its output asymptotically track the reference model's output. The reference model is a linear model representing a stable system. The neural network can be used as the controller. The structure is shown in Fig. 2. The training is related to the identification of the inverse plant. However, the controller is trained to generate the same output from the reference model, instead of the command input.

C. Learning a Sequence of Control Actions by Back Propagation through Time

There are control problems that require a sequence of control actions to achieve specific goals. One example is the truck backer-upper problem [15] shown in Fig. 3a. The goal is to steer the truck backward to have the trailer at the duck

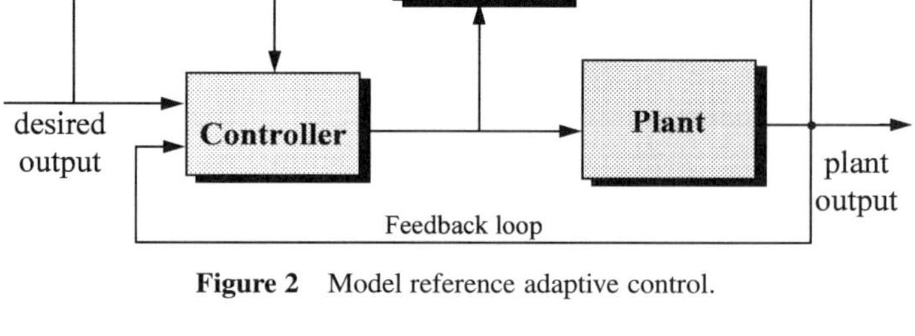

Figure 2 Model reference adaptive control.

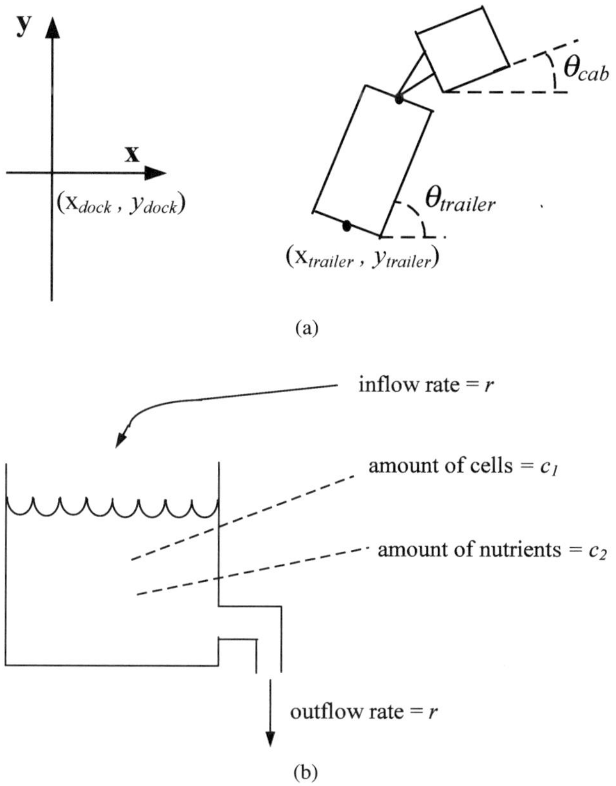

Figure 3 Examples whose controllers can be developed by BTT: (a) truck backer-upper problem [15]; (b) bioreactor problem [16].

for loading and unloading. The accomplishment of the goal requires a sequence of control actions. Another example is the control of a bioreactor [16] (see Fig. 3b). The tank contains water and cells. The inflow liquid contains nutrients. The cells consume nutrients and produce more cells. The goal is to maintain the amount of cells at a desired level by controlling the inflow and outflow rate. If the amount of cells is away from the desired level, a sequence of control actions is needed to bring the amount back to the correct level. In these two examples, several steps are needed to bring the system to the desired state. Back propagation through time (BTT) [17] can be used to learn the control sequence.

The BTT technique requires a plant model, for which a neural network can be used. It also requires a learning controller. The learning process starts from an initial state. The neurocontroller generates a control action, which will change the state of the plant. The output of the plant will be sent back to the input of the neurocontroller and the next control action will be generated. The control process continues until the system fails or a specified time period elapses. The BTT learning algorithm will then use the sequence of control actions and the final plant output to adjust the neurocontroller. The BTT learning algorithm starts from the last action and the final result. The difference between the final output and the desired one is back-propagated through the plant model and then through the controller. Back propagation through the plant gives information on how the control signal should be altered to reduce the error at the plant output. Back propagation through the controller has two functions. One function is to give the information on how the previous output at time $k - 1$ can be altered in order to reduce the error at the plant output at time k. The other function is to give information on how the parameters of the neurocontroller can be adjusted to improve the quality of the control action. The former information is then further brought to the output end of the plant for another iteration of back propagation and adjustment. The process described is the same as that shown in Fig. 4, in which the entire back propagation through the plant and the neurocontroller is unfolded along the time axis. While the plant model is assumed well trained, it will not be further adjusted during the BTT learning. Only the necessary adjustments for the neurocontroller will be memorized and accumulated. The total adjustment will be made at the end of the BTT learning procedure.

Narendra [18] once demonstrated, using a nonlinear system example, that a neurocontroller developed by the BTT can be more robust than a linear controller developed through linearization and analysis. His results show that the neurocontroller can control the system to bring the state from a farther point back to the operating point.

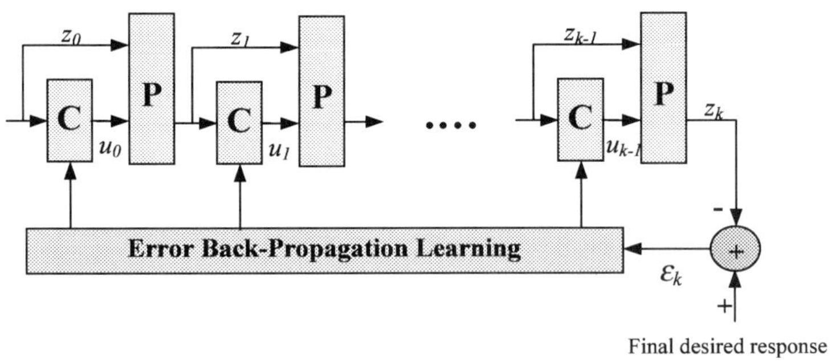

Figure 4 Unfolding along the time axis for explanation of BTT (after [15]).

D. Neural Networks for Adaptive Critic Learning

Adaptive critic learning (ACL) [19–21] is one kind of reinforcement learning. The potential usage of this scheme is on problems that have a delayed reward/penalty. The immediate reinforcement signal is not available from the environment and must be generated by the learning structure itself. Because of the need in learning to provide the internal reinforcement, the scheme is called "adaptive critic." As shown in Fig. 5, the ACL structure consists of two main modules, a control module and an evaluation module. These two modules both require learning capability and can be implemented using neural networks. The evaluation module learns to provide an adequate internal reinforcement signal, which is used for adjusting the controller. The necessity of an evaluation module is due to the need of an immediate evaluation of a control action. Barto *et al.* [19] first introduced the technique using the cart–pole system as an example. The cart–pole example has a pole attached onto a cart with a hinge. The cart is allowed to move on a limited-length track in order to keep the pole balanced. It is assumed that, at the very beginning, the control system does not know how to evaluate a state. The capability to tell if a state is good (and how good) is developed through learning. The only signal available to develop the evaluation capability is the "control failure" signal, which is delayed. The failure happens when the pole angle measured from the vertical is greater than a specific angle (e.g., 12°) or the cart hits either end of the track. In the learning control, if the state is getting better, earlier states are credited; otherwise, they are discredited. Although the reward/penalty is the result of a time series of actions, the previous actions will all be responsible for the current result. However, the earlier actions are assigned less responsibility. If the state is getting better, actions are reinforced.

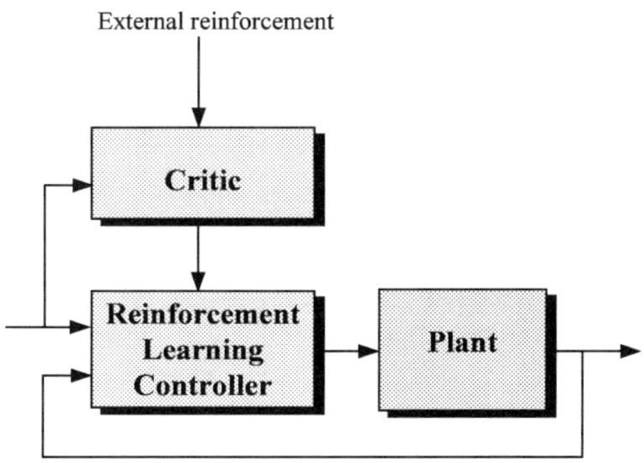

Figure 5 Adaptive critic learning.

III. CONVENTIONAL CEREBELLAR MODEL ARTICULATION CONTROLLER

Neurocontrol techniques have been introduced in the previous section. In this section, we shall introduce the CMAC, which is one of the most promising neural network structures for learning control problems. The most important advantage of the CMAC neural network is the fast learning speed and excellent convergence property. The structure is easy to implement.

A. SCHEME

The CMAC is basically a table-lookup technique. Figure 6 shows Albus's CMAC model, which imitates the function of a cerebellum. The input space is quantized into discrete states. Several memory locations associated with a discrete state are used to store information for that state. The output for the state is the sum of data retrieved from the associated memory locations. Retrieved data are added together as the output.

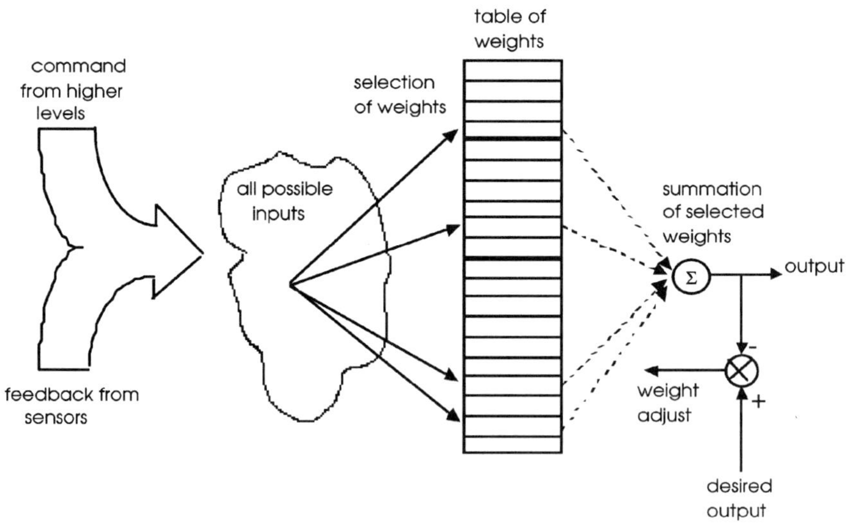

Figure 6 CMAC model.

The technique basically includes three mapping relations:

$\mathbf{S} \to \mathbf{I}$ Quantized state to intermediate variables

$\mathbf{I} \to \mathbf{P}$ Intermediate variables to physical memory addresses

$\mathbf{P} \to \mathbf{Y}$ Memory addresses to CMAC output

where **S** is a state input vector, **I** is a set of intermediate variables, **P** is a set of physical memory addresses, and **Y** is the output vector.

Figure 7 is used for further explanation. This example has two input variables, s_1 and s_2. Each input variable is quantized into several discrete regions, called *blocks*. For instance, s_1 can be divided into A, B, C, D, and E and s_2 divided into a, b, c, d, and e. A, B, C, D, etc. are intermediate variables. Areas formed by quantized regions, labeled as Aa, Ab, Ba, etc., are called *hypercubes*. Each hypercube is assigned a memory location (or a vector for the vector output case).

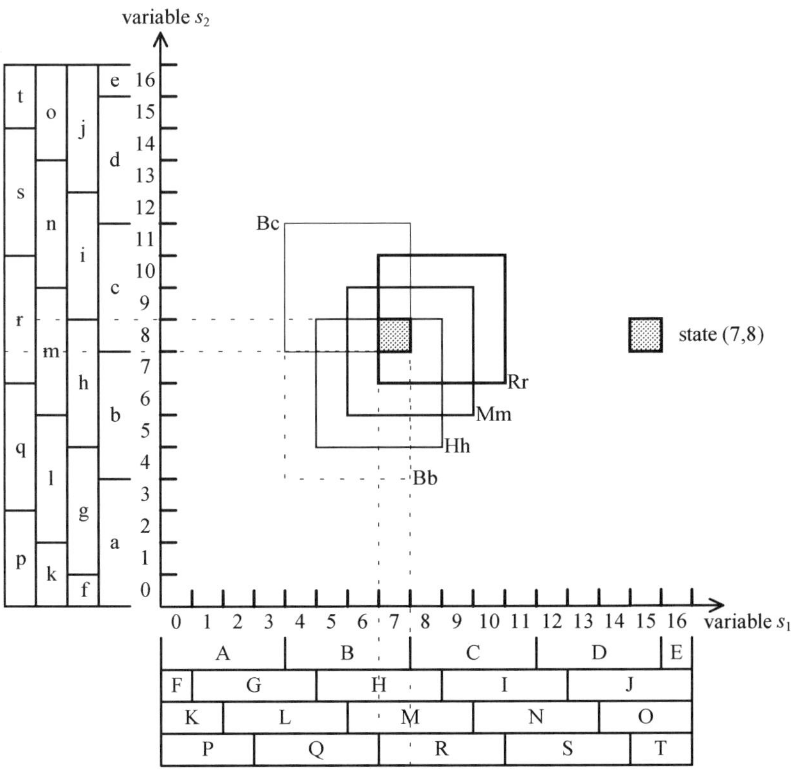

Figure 7 Block division of CMAC for a two-variable example.

If the quantization for each variable is shifted by one small interval (called an *element*), different hypercubes will be obtained. F, G, H, I, J for s_1 and f, g, h, i, j for s_2 are shifted regions. Fh, Fi, Hg, etc. are new hypercubes from the shifted regions. In most CMAC schemes, no hypercube is formed by the combination of different layers such as "A, B, C, D, E" and "f, g, h, i, j." With this kind of quantization and hypercube composition, each state is covered by N_e different hypercubes, where N_e is the number of elements in a complete block. The set of names for blocks that cover the quantized state is the set **I**. In Fig. 7, for the state (7, 8), the set **I** is $\{Bc\ Hh\ Mm\ Rr\}$. The set **P** consists of addresses assigned to the N_e hypercubes, which are Bc, Hh, Mm, and Rr. The mapping from **I** to **P** can be in a one-to-one correspondence if the memory size is not a concern. It can also be many-to-one if the entire information for quantized states does not need to be stored. Hashing, which is a random mapping, is used in many-to-one mapping. The CMAC scheme can be viewed as a technique with the basis function that is constant in each hypercube.

Information for a quantized state is distributively stored in N_e memory locations. The output for a state is obtained as the sum of stored contents for hypercubes covering the state. Assume that M is the memory size. Then, using the CMAC technique, a stored datum can be mathematically expressed as

$$y(\mathbf{s}) = \mathbf{a}^T(\mathbf{s})\mathbf{w} = \sum_{j=1}^{M} a_j(\mathbf{s})w_j, \tag{1}$$

where $\mathbf{s}$ is a specific state, $\mathbf{w} = [w_1, w_2, \ldots, w_{pm}]^T$ is the vector of memory contents, and $\mathbf{a}(\mathbf{s})$ is a memory element selection vector that has N_e 1's. Note that the element $a_j(\mathbf{s})$ of the vector $\mathbf{a}(\mathbf{s})$ is 1 if memory location j is used by one of the hypercubes that cover the state $\mathbf{s}$. In learning, w_j's are the parameters to be adjusted. Although the information can be expressed in the form in Eq. (1), it is actually retrieved from just a small number (N_e) of memory locations allocated to the hypercubes covering the state.

The conventional CMAC uses iterative learning to create information in the memory. For a given sample, the updating rule can be expressed as

$$\mathbf{w}_{\text{new}} = \mathbf{w}_{\text{old}} + \frac{\alpha}{N_e}\mathbf{a}(\mathbf{s})\big(\hat{y}(\mathbf{s}) - \mathbf{a}^T(\mathbf{s})\mathbf{w}_{\text{old}}\big), \tag{2}$$

where α is a learning rate, $\hat{y}(\mathbf{s})$ is the target function value, and $\hat{y}(\mathbf{s}) - \mathbf{a}^T(\mathbf{s})\mathbf{w}_{\text{old}}$ is the error for this training sample. Premultiplication by $\mathbf{a}(\mathbf{s})$ distributes the error to those memory elements used by this sample.

B. Application Example of Cerebellar Model Articulation Controller

In Section II, the general development of a neurocontroller was discussed. In this section, an advanced application of the CMAC to robotic manipulator control studied by Miller [22] will be introduced. The problem was to have a four-joint manipulator learn to track an object on a conveyor. A video camera was mounted on the wrist of the manipulator as a sensor. The object used in Miller's experiment was a plastic disposable razor. The centroid of the razor in the video image coordinates was (X, Y) and the size of its handle was Z. The orientation of the razor in the image was denoted by R. These four image parameters were obtained by the video camera. The goal of the control was to drive joints to (1) keep the object image at the center, (2) keep the handle of the razor parallel to the horizontal video axis (i.e., X), and (3) keep the size of the object image a constant (implying a constant altitude of the video camera). The learning control system assumed very little knowledge on the kinematics and inverse dynamics of the robot manipulator. The conveyor speed and object orientation were also assumed unknown.

Figure 8 shows the block diagram of the learning controller. There are two CMAC memory blocks in this diagram; the forward-model CMAC estimates the object position and the inverse-model CMAC generates the feedforward control voltages for joints. The voltages generated by the inverse-model CMAC are feedforward components to be added with those from the fixed-gain controller. This CMAC represents an inverse model for the system being controlled. Its output should compensate the highly nonlinear system property. Another CMAC is used

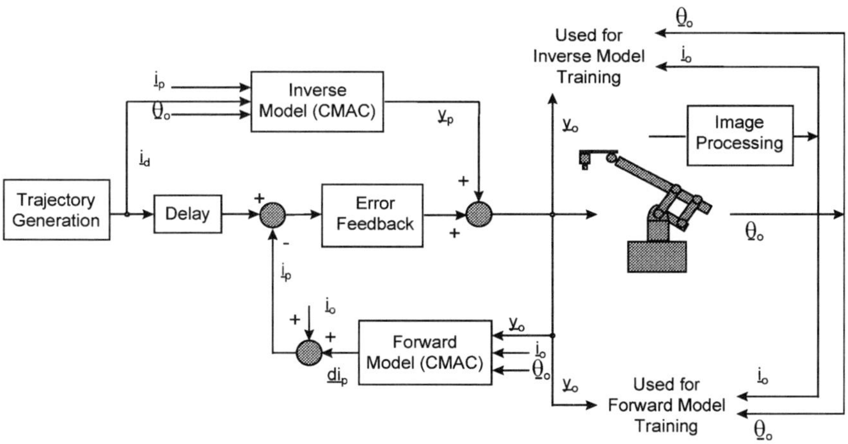

Figure 8 Block diagram of the control structure [22].

to estimate the current object location. It represents a plant model. Its output gives the estimated changes on the image parameters between two image acquisitions. The estimation of the object position is necessary because of the 280-ms delay in image processing. The object will move some significant distance during this processing period. A simple fixed-gain feedback controller is included in this structure to make sure that the end effector of the manipulator can move roughly above the conveyor for necessary learning.

The inverse model CMAC needs to generate $\mathbf{v}_p$ for object tracking. It has the following inputs and outputs:

12 input components:	4 joint positions (θ_0) 4 object image parameters [$\mathbf{i}_p = (X_p, Y_p, Z_p, R_p)$] from another CMAC 4 desired image parameters [$\mathbf{i}_d = (X_d, Y_d, Z_d, R_d)$] from the trajectory generator [note that $\mathbf{i}_d - \mathbf{i}_p$ gives the desired change of image parameters, i.e., $d\mathbf{i} = (dX, dY, dZ, dR)$ for the next control period]
4 outputs:	control voltages for four joints

To be able to generate adequate control signals, this CMAC needs to learn the nonlinear relationship between the preceding inputs and outputs over particular regions of the system state space.

In the control structure, the fixed-gain error-feedback controller is in parallel with the inverse-model CMAC. The error terms are computed as the difference between the desired image parameters and the estimated values of the present image parameters. At the end of each control cycle, the drive signals from the fixed-gain controller will be added to those from the CMAC for control usage. The fixed-gain control portion is described as follows:

if$(-2950 < \theta 5 < 2950)$
{$\mathbf{v1} = 10^*(\mathrm{X_d} - \mathrm{X_p})$
$\mathbf{v2} = -10^*(\mathrm{Y_d} - \mathrm{Y_p})$}
else if $(\theta 5 > 2950)$
{$\mathbf{v1} = -10^*(\mathrm{Y_d} - \mathrm{Y_p})$
$\mathbf{v2} = -10^*(\mathrm{X_d} - \mathrm{X_p})$}
else
{$\mathbf{v1} = 10^*(\mathrm{Y_d} - \mathrm{Y_p})$
$\mathbf{v2} = 10^*(\mathrm{X_d} - \mathrm{X_p})$}
$\mathbf{v3} = -10^*(\mathrm{Z_d} - \mathrm{Z_p})$
$\mathbf{v5} = 10^*\mathrm{R_p}$

The image parameters, which specify the object location and orientation, are supposed to be from the image system. The time required for image processing causes the location information being delayed. In order to have accurate "current" parameters, estimation is necessary. Another CMAC is used to learn the nonlinear relationship between the commanded voltages and the changes in image parameters. This CMAC has the following inputs and outputs:

12 input components:	4 joint positions (θ_0) 4 object image parameters from the video camera $[\mathbf{i}_0 = (X_0, Y_0, Z_0, R_0)]$ 4 control voltages ($\mathbf{v}_0$) for four joints
4 outputs:	4 estimated image parameter changes $[d\mathbf{i}_p = (dX_p, dY_p, dZ_p, dR_p)]$

To train the two CMACs, θ_0, $\mathbf{i}_0$ $d\mathbf{i}_0$, and $\mathbf{v}_0$ obtained from the past control cycles are used. The forward-model CMAC can be updated by

$$\Delta \mathbf{w}_2 = \beta\big(d\mathbf{i}_0 - \mathbf{f}_2(\theta_0, \mathbf{i}_0, \mathbf{v}_0)\big), \tag{3}$$

where β is a learning rate, $\mathbf{f}_2$ is the output of this CMAC, and $d\mathbf{i}_0$ is the *actual* change used as the target value.

The inverse-model CMAC, which provides feedforward voltages, can be updated by

$$\Delta \mathbf{w}_1 = \beta\big(\mathbf{v}_0 - \mathbf{f}_1(\theta_0, \mathbf{i}_0, d\mathbf{i}_0)\big), \tag{4}$$

where $\mathbf{f}_1$ is the output of this CMAC using θ_0, $\mathbf{i}_0$, and the actual change $d\mathbf{i}_0$ as inputs, and $\mathbf{v}_0$ is the vector of actual control voltages.

In Miller's experiment, the image-processing time was 280 ms and the control cycle was selected to be 350 ms. The control procedure is summarized below.

1. Obtain θ_0. Take the object image.
2. Process the image to obtain the object image parameters $\mathbf{i}_0$.
3. θ_0, $\mathbf{i}_0$, and $\mathbf{v}_0$ are provided to the image CMAC to obtain the estimated image parameter change $d\mathbf{i}_p$.
4. θ_0, $\mathbf{i}_0 + d\mathbf{i}_p$, and $\mathbf{i}_d$ are provided to the feedforward control CMAC to retrieve $\mathbf{v}_p$.
5. $\mathbf{v}_0 = \mathbf{v}_p +$ (feedback control voltage) is applied. Go to step 1.

IV. ADVANCED CEREBELLAR MODEL ARTICULATION CONTROLLER-BASED TECHNIQUES

The CMAC scheme has very attractive properties on learning convergence and speed, and is useful for learning control. Because the CMAC is a table-lookup technique, a model implemented by the structure cannot provide a derivative of its output. This creates the difficulty and inconvenience in learning that requires derivative information. One example that the derivative is needed is in back propagation through a model, which is used in BTT and some other learning techniques. This section further introduces two modified schemes that can better provide derivative information.

A. Cerebellar Model Articulation Controller with Weighted Regression

In this section, we introduce a scheme that integrates locally weighted regression (LWR) with the CMAC addressing technique. The technique is referred to as LWR-CMAC [23, 24]. Derivatives exist except on the boundary of quantized regions. With the use of the CMAC addressing technique, data points in the input space are efficiently organized. Only those in the local area are used for regression. This limits the computational load. Because of the use of regression, the approximate function is rather smooth and precise, and the function derivative can be retrieved. Compared with the conventional CMAC, the LWR-CMAC requires the same size of memory, has a similar learning speed, but provides output differentiability and more precise output. Compared with the typical weighted regression technique, this scheme offers an efficient way to organize and utilize the collected information.

The CMAC addressing technique is adopted for systematically selecting the set of neighboring points. The same method of hypercube decomposition for CMAC is used but the output is computed in a different way. The LWR-CMAC intends to have the target function value at the hypercube center stored in the memory allocated to that hypercube. To retrieve information for a given input $\mathbf{s}$, the hypercubes covering $\mathbf{s}$ are first identified. A local regression model is formed using the data stored for these hypercubes. The output for $\mathbf{s}$ is computed from this model. Construction of the local regression model and the computation of the function value for a given $\mathbf{s}$ are introduced in the following section.

1. Local Regression Model and Output Computation

In discussing information retrieval, we assume that correct data have been stored. The weighted regression technique assigns different weights for data points at different distances. Given an input point **s**, the weight for a hypercube depends on the distance from **s** to the center of the hypercube. Small modification on Eq. (1) will describe the change. The element $a_j(\mathbf{s})$, which is either 0 or 1 in Eq. (1), will be redefined as a function of the distance between the input and the center of hypercube j as

$$a_j(\mathbf{s}) = \begin{cases} \exp\left(-\operatorname{dist}^2(\mathbf{s}, \mathbf{C}_j)/2H^2\right) & \text{if hypercube } j \text{ covers } \mathbf{s}, \\ 0 & \text{otherwise,} \end{cases} \tag{5}$$

where $\mathbf{C}_j$ is the center of hypercube j, dist( ) gives the distance, and H is a given constant. The centers and memory contents $(\mathbf{C}_j, w_j)$'s for hypercubes that cover **s** are used to generate the local regression model. The $\mathbf{C}_j$ and w_j are weighted before the regression is done. Let $\mathbf{x}_1, \mathbf{x}_2, \ldots, \mathbf{x}_{N_e}$ indicate the weighted centers of N_e involved hypercubes and $z_1, z_2, \ldots, z_{N_e}$ indicate the corresponding weighted memory contents. Assume that $\mathbf{x}_j$ and z_j are for hypercube k. Then they are defined mathematically as

$$\mathbf{x}_j(\mathbf{s}) = a_k(\mathbf{s}) \begin{bmatrix} \mathbf{C}_k \\ 1 \end{bmatrix} \quad \text{and} \quad z_j = a_k(\mathbf{s}) w_k. \tag{6}$$

Note that a "1" is added in the vector of the center in Eq. (6). The locally weighted regression is to determine the vector of regression coefficients for the following equation to achieve the least square error:

$$\mathbf{z} = \mathbf{X}\mathbf{b}, \tag{7}$$

where $\mathbf{z} = (z_1, z_2, \ldots, z_{N_e})^T$ and $\mathbf{X} = (\mathbf{x}_1, \mathbf{x}_2, \ldots, \mathbf{x}_{N_e})^T$. The vector **b** can be solved as

$$\mathbf{b} = (\mathbf{X}^T\mathbf{X})^{-1}\mathbf{X}^T\mathbf{z}. \tag{8}$$

To prevent possible singularity, a small positive number can be added to all diagonal elements of $\mathbf{X}^T\mathbf{X}$ in implementation. The effect of weighting is that the error for a distant point is considered less important. With **b** obtained in Eq. (8), the output $y(\mathbf{s})$ is computed as

$$y(\mathbf{s}) = \mathbf{s}^T\mathbf{b}. \tag{9}$$

The output is continuous and differentiable except at the boundary of a quantized element. The vector **b** gives the derivative information.

2. Learning Algorithm

The objective of learning is to determine the function values for the hypercube centers. However, these target values are not explicitly provided. Data available for training could be at any location in the input space. The memory contents must be developed from the available information. Given an **s**, the learning algorithm uses the current memory contents to form the local regression model and uses the model to evaluate the output for the **s**. The difference between the target value and the evaluated value is used to modify the vector **b**. Note that the new vector **b** is not to be memorized, but is used to compute a "guessed" target vector (guessed value at hypercube centers) using Eq. (7). The guessed target values are used to update memory contents. The mathematics used in the updating is provided in the following discussion.

With the retrieval procedure given in Eqs. (7)–(9), if the target value at **s** is $\hat{y}(\mathbf{s})$, the error for the input **s** will be $\mathrm{err}(\mathbf{s}) = \hat{y}(\mathbf{s}) - y(\mathbf{s})$. Note that $\hat{y}(\mathbf{s})$ is given. The rule for modifying b_i should be

$$\Delta b_i = -\frac{1}{2}\eta \frac{\partial \mathrm{err}^2(\mathbf{s})}{\partial b_i} = \eta \frac{\partial y(\mathbf{s})}{\partial b_i}\mathrm{err}(\mathbf{s}) = \eta s_i \mathrm{err}(\mathbf{s}). \tag{10}$$

With the change in Eq. (10), the new $y(\mathbf{s})$ will be

$$\sum_{i=1}^{N_v+1} (b_i + \Delta b_i)s_i = \sum_{i=1}^{N_v+1} b_i s_i + \sum_{i=1}^{N_v+1} \eta s_i^2 \mathrm{err}(\mathbf{s}), \tag{11}$$

where N_v is the number of input variables and $\mathbf{s}_{N_v+1}$ is the added constant 1. With η selected as $1/\sum s_i^2$, the error at **s** would be completely corrected. However, one may use a smaller updating rate and have the following rule for modifying the vector **b**:

$$b_i \leftarrow b_i + \alpha_b \frac{\mathrm{err}(\mathbf{s})}{\sum_{k=1}^{n} s_k^2} s_i, \tag{12}$$

where α_b is the learning rate. A value of 1 for α_b gives the fastest learning speed and does not affect the final precision. Note that no target values are available at centers of hypercubes. Thus, Eq. (7) with the new coefficient vector $\mathbf{b}_{\mathrm{new}}$ is used to compute the guessed target vector. The amount $\alpha_m(\mathbf{x}_j^T \mathbf{b}_{\mathrm{new}} - z_j)$ is added back to the corresponding memory contents. Assume that $\mathbf{x}_j$ and y_j are for hypercube k. Then w_k is updated by

$$w_k \leftarrow w_k + \alpha_m[\mathbf{x}_j^T \mathbf{b}_{\mathrm{new}} - z_j]. \tag{13}$$

The learning procedure is summarized next.

LWR-CMAC Learning Procedure

1. Initialize all memory contents (i.e., w's) to 0.
2. For a given input $\mathbf{s}$ and the target output $\hat{y}(\mathbf{s})$, find all hypercubes that cover the input $\mathbf{s}$.
3. Use Eq. (6) to obtain the weighted center vectors and the weighted memory contents for all involved hypercubes. They are $\mathbf{x}_j$ and z_j for $j = 1, \ldots, N_e$.
4. Compute $\mathbf{b} = (\mathbf{X}^T\mathbf{X})^{-1}\mathbf{X}^T\mathbf{z}$ as in Eq. (8).
5. Compute $y(\mathbf{s}) = \mathbf{s}^T\mathbf{b}$ as in Eq. (9).
6. Compute $\text{err}(\mathbf{s}) = \hat{y}(\mathbf{s}) - y(\mathbf{s})$.
7. Update b_i by $b_i \leftarrow b_i + \alpha_b(\text{err}(\mathbf{s})/\sum_{k=1}^{n} s_k^2)s_i$ as given in Eq. (12).
8. Obtain the "guessed" target value $\mathbf{x}_j^T\mathbf{b}_{\text{new}}$.
9. Update the memory content by $w_k \leftarrow w_k + \alpha_m[\mathbf{x}_j^T\mathbf{b}_{\text{new}} - z_j]$ as given in Eq. (13).
10. Go to step 2 if the learning result is not satisfactory.

Compared to CMAC, the computation in Eq. (8) is the major extra computation. In that equation, $\mathbf{X}$ is an $(N_v + 1) \times N_e$ matrix and $\mathbf{X}^T\mathbf{X}$ is an $(N_v + 1) \times (N_v + 1)$ matrix, where N_v is the number of variables. For the two-variable case, the inversion is for a 3×3 matrix. In learning, a similar computation is also required.

When the CMAC addressing technique is adopted, the number of parameters for this scheme remains the same as that of the conventional CMAC. For the N_b-block and N_e-element case, the number of parameters is the same as the number of hypercubes and is equal to $(N_b)^{Nv} \times N_e$.

3. Example: Functional Approximation

The example is to approximate the function $\hat{g}(\theta_1, \theta_2) = \sin\theta_1 \cos\theta_2$. Figures 9 and 10 show plots of $\hat{g}$, g, $\partial\hat{g}/\partial\theta_1$, and $\partial g/\partial\theta_1$ from a learning experiment with $H = 0.2$, $\alpha_b = 1$, and α_m being initialized to 1 and reduced to 0.1. Figure 9a and b shows the three-dimensional plots of $\hat{g}$ and g. Figure 10a and b shows plots of the function and its derivative versus θ_1 while θ_2 is set to 0.

B. Cerebellar Model Articulation Controller with General Basis Functions

Another approach is to use nonconstant basis functions. The cerebellar model of articulation controller (CMAC) can be viewed as a basis function network (BFN). The conventional CMAC uses local constant basis functions and has as

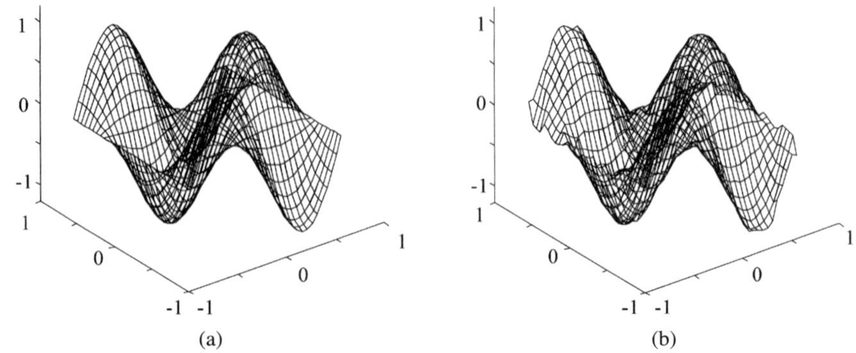

Figure 9 Three-dimensional plots of $\hat{g}$ and g. $\hat{g}(\theta_1, \theta_2) = \sin\theta_1 \cos\theta_2$. (a) Target output $\hat{g}$; (b) LWR-CMAC output g [23].

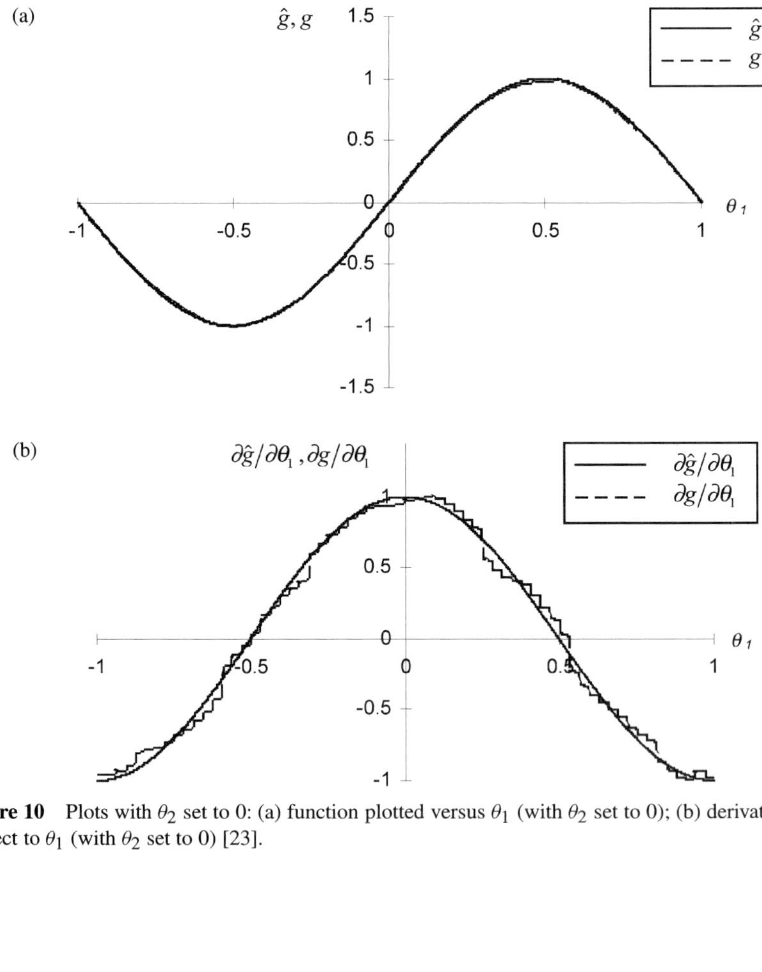

Figure 10 Plots with θ_2 set to 0: (a) function plotted versus θ_1 (with θ_2 set to 0); (b) derivative with respect to θ_1 (with θ_2 set to 0) [23].

its output a constant in each quantized state. Derivative information is not preserved. If the constant basis functions are replaced by nonconstant differentiable basis functions, the derivative can be stored into the structure as well. Spline functions [25] and Gaussian functions are possible selections as the basis functions. In this section, the generalized scheme that uses general basis functions [4] will be introduced. The conventional CMAC is a special case of the generalized technique.

1. Scheme

The generalized technique uses general basis functions and is called the GBF-CMAC. The basis function $f_i(\cdot)$ is associated with the ith hypercube. All basis functions have bounded values inside their hypercube areas. Note that if $f_i(\cdot)$ is a constant inside the area covered by the ith hypercube and zero outside, the generalized scheme becomes the conventional CMAC.

To determine an output value for a given input, a linear combination of the basis functions associated with the involved hypercubes is used. A stored datum $y(\mathbf{s})$ for the input $\mathbf{s}$ can be mathematically expressed as

$$\begin{aligned} y(\mathbf{s}) = \mathbf{a}^T(\mathbf{s})\mathbf{w}(\mathbf{s}) &= \begin{bmatrix} a_1(\mathbf{s}) & a_2(\mathbf{s}) & \cdots & a_{N_h}(\mathbf{s}) \end{bmatrix} \begin{bmatrix} w_1(\mathbf{s}) \\ w_2(\mathbf{s}) \\ \vdots \\ w_{N_h}(\mathbf{s}) \end{bmatrix} \\ &= \sum_{j=1}^{N_h} a_j(\mathbf{s}) w_j(\mathbf{s}), \end{aligned} \tag{14}$$

where $\mathbf{a}(\mathbf{s})$ is a basis function selection vector that has N_e 1's and $\mathbf{w}(\mathbf{s})$ is a vector with the ith element $w_i(\mathbf{s}) \equiv v_i f_i(\mathbf{s})$. Note that v_i is a weight to be obtained through learning.

With $w_i(\mathbf{s}) \equiv v_i f_i(\mathbf{s})$, $\mathbf{w}(\mathbf{s})$ can be rewritten as

$$\begin{aligned} \mathbf{w}(\mathbf{s}) &= \begin{bmatrix} v_1 f_1(\mathbf{s}) \\ v_2 f_2(\mathbf{s}) \\ \vdots \\ v_{N_h} f_{N_h}(\mathbf{s}) \end{bmatrix} = \begin{bmatrix} f_1(\mathbf{x_s}) & 0 & \dots & 0 \\ 0 & f_2(\mathbf{s}) & \ddots & \vdots \\ \vdots & \ddots & \ddots & 0 \\ 0 & \dots & 0 & f_{N_h}(\mathbf{s}) \end{bmatrix} \begin{bmatrix} v_1 \\ v_2 \\ \vdots \\ v_{N_h} \end{bmatrix} \\ &\equiv \mathbf{F}(\mathbf{s})\mathbf{v}. \end{aligned} \tag{15}$$

Thus, Eq. (14) becomes

$$y(\mathbf{s}) = \mathbf{a}^T(\mathbf{s})\mathbf{F}(\mathbf{s})\mathbf{v}. \tag{16}$$

Use $\hat{y}(\mathbf{s})$ to denote the target output value for the input $\mathbf{s}$. With the energy function selected as

$$E = \tfrac{1}{2}\big(\hat{y}(\mathbf{s}) - y(\mathbf{s})\big)^2 = \tfrac{1}{2}\big(\hat{y}(\mathbf{s}) - \mathbf{a}^T(\mathbf{s})\mathbf{w}(\mathbf{s})\big)^2, \tag{17}$$

the updated amount for v_k can be set equal to

$$\begin{aligned}\Delta v_k &= -\frac{\alpha}{N_e}\frac{\partial E}{\partial v_k} \\ &= -\frac{\alpha}{N_e}\frac{\partial E}{\partial w_k(\mathbf{s})}\frac{\partial w_k(\mathbf{s})}{\partial v_k} \\ &= \frac{\alpha}{N_e}\big(\hat{y}(s) - \mathbf{a}^T(\mathbf{s})\mathbf{w}(\mathbf{s})\big)a_k(\mathbf{s})f_k(\mathbf{s}),\end{aligned} \tag{18}$$

where α/N_e is the learning rate.

2. Examples of Cerebellar Model Articulation Controller with Gaussian Basis Functions

As pointed out previously, the CMAC with differentiable basis functions can provide derivative information. In this section, we examine this capability with the use of Gaussian functions. The basis function is $f_k(\mathbf{s}) \equiv \prod_{j=1}^{N_v}\phi_{kj}(s_j)$ with

$$\phi_{kj}(s_j) = \exp\left[-\frac{(s_j - \mu_{kj})^2}{\sigma_{kj}^2}\right], \tag{19}$$

where μ_{kj} is the mean and σ_{kj} is the variance. N_v is the number of variables in the target function. Consequently, $w_k(\mathbf{s}) = v_k\prod_{j=1}^{N_v}\phi_{kj}(s_j)$.

From Eq. (18), Δv_k can be further derived as

$$\Delta v_k = \frac{\alpha}{N_e}\big(\hat{y}(\mathbf{s}) - \mathbf{a}^T(\mathbf{s})\mathbf{w}(\mathbf{s})\big)a_k(\mathbf{s})\prod_{j=1}^{N_v}\phi_{kj}(s_j). \tag{20}$$

The function $\hat{g}(\theta_1, \theta_2) = \sin(\theta_1\pi)\cos(\theta_2\pi)$ used previously shows the performance of the GBF-CMAC. Figures 11 and 12 show plots of $\hat{g}$, g, $\partial\hat{g}/\partial\theta_1$, and $\partial g/\partial\theta_1$ from a learning experiment with the learning parameter α initialized to 1 and later reduced to 0.1. $\hat{g}(\mathbf{s})$ and $g(\mathbf{s})$ are the actual function (target) and the computed output from the learning module. Figure 11a and b shows the three-dimensional plots of $\hat{g}$ and g. Figure 12a and b shows plots of the function and its derivative versus θ_1 while θ_2 is set to 0.

This GBF-CMAC has good accuracy while the learning speed is very close to that of the conventional CMAC. Computation in this technique only involves the

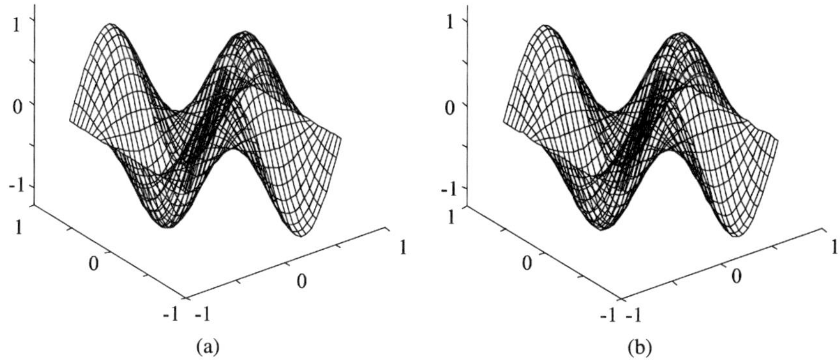

Figure 11 Three-dimensional plots of $\hat{g}(\theta_1, \theta_2) = \sin(\theta_1 \pi)\cos(\theta_2 \pi)$ and g: (a) target output $\hat{g}$; (b) GBF-CMAC output g [4].

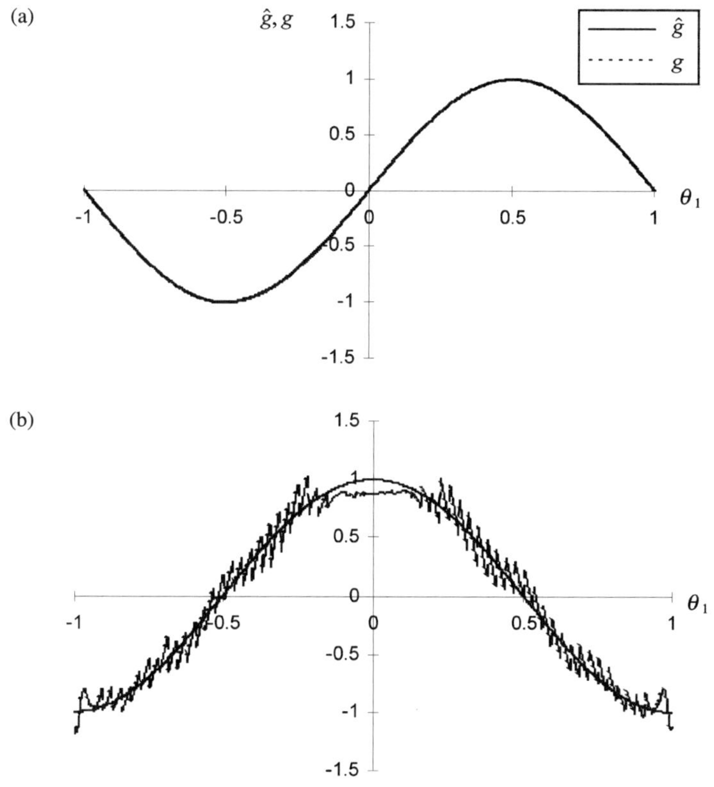

Figure 12 Plots with θ_2 set to 0: (a) function plotted versus θ_1 with θ_2 set to 0; (b) derivative with respect to θ_1 with θ_2 set to 0 [4].

basis functions used by the current state. Therefore, the computation load is much lower than that for the general global basis function networks. One advantage of this technique is that it is able to provide the output derivative information with respect to input if differentiable basis functions are used.

V. STRUCTURE COMPOSED OF SMALL CEREBELLAR MODEL ARTICULATION CONTROLLERS

In this section, we shall further discuss one structure for high-dimensional learning control. MNNs and RBFNs are widely used neural networks for static mapping. However, they both have some drawbacks and limitations. It is known that MNNs often encounter difficulty in learning. For a slightly complicated mapping, it is hard to predict how long the learning will take and whether the learning will converge to an acceptable result. Another type of neural network, RBFNs, often use the Gaussian function as the basis function. Because a Gaussian function provides fitting in a local area, the learning convergence is less of a problem compared to that in MNNs. However, the number of basis functions may become enormous for problems with high-dimensional input space. The conventional CMAC has a similar problem in high-dimensional mapping because it is a table-lookup technique. One suggestion to relax the problem is to use a structure composed of small CMACs [26].

A. NEURAL NETWORK STRUCTURE WITH SMALL CEREBELLAR MODEL ARTICULATION CONTROLLERS

Figure 13 shows one example of the structure with small CMACs. The network output ONET( ) is the sum of outputs from a set of submodules. Each CMAC in the submodule has a subset of system inputs as its inputs and each submodule implements a basis function. The number of inputs for each CMAC can be selected by the designer (two inputs in the example in Fig. 13). Let N_b be the number of blocks in the decomposition of each variable, N_e be the number of elements in each complete block, and N_v be the number of input variables. For $N_e = 8$, $N_b = 8$, and $N_v = 6$, if one uses two inputs for each CMAC in the submodules, the memory size will be 512 ($= 8 \times 8^2$) for each CMAC and 3×512 for a submodule. Using a single conventional CMAC, the memory size will be $2097152 (= 8 \times 8^6)$. For this memory size (2097152), one can have up to 1365 submodules; most problems will need just a small portion of that. Each submodule can implement a self-generated basis function. One may use the same input

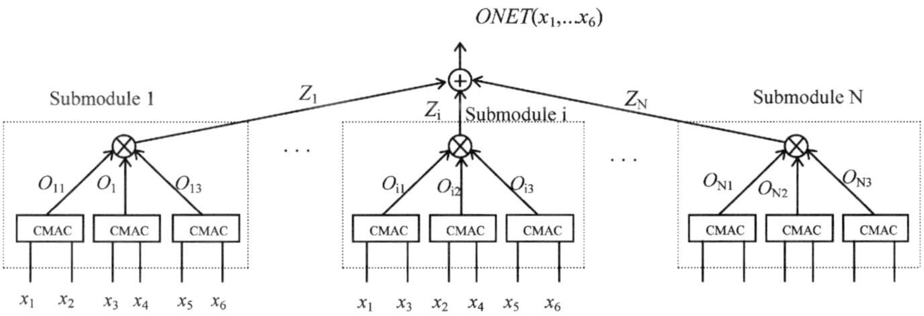

Figure 13 Small CMAC-based neural network [26].

arrangement for all submodules, that is, x_1 and x_2 as inputs of the first CMAC, x_3 and x_4 for the second CMAC, and x_5 and x_6 for the third CMAC. Such an input arrangement can be denoted by $\{(1,2),\ (3,4),\ (5,6)\}$. One can also use different kinds of combinations such as $\{(1,2),\ (3,4),\ (5,6)\}$, $\{(1,3),\ (2,4),\ (5,6)\}$, $\{(1,4),\ (2,3),\ (5,6)\}$, etc. for different submodules. The combinations can be repeated if necessary.

One question that arises is whether the structure, given enough submodules, can implement a general mapping. Let us use the structure with two-input CMACs for explanation. It is noted that a two-input CMAC can well implement a two-dimensional Gaussian function. Thus, the output of a submodule, which is a product of several two-dimensional Gaussian functions, can implement a high-dimensional Gaussian function. Consequently, the proposed structure can at least implement what a Gaussian function network can implement. The structure is more powerful than typical basis function networks because the "basis function" is self-generated and not necessary to be limited to a Gaussian or other specific type. The structure with an adequate number of blocks, suitable resolution, and enough submodules will be able to approximate any smooth function well.

B. Learning Rules

Contents of the CMAC memory must be obtained through learning. In learning, the output error can be distributed to all submodules. Let ε denote the error distributed to each submodule and Z_i indicate the output of submodule i. Then

$$Z_i = O_{i1} O_{i2} \cdots O_{im}, \tag{21}$$

where O_{ij} is the output of the jth CMAC in submodule i, and m is the number of CMACs in this submodule.

The gradient decent learning rule can be derived and used. The partial derivative of Z_i with respect to a CMAC output O_{ij} is

$$\frac{\partial Z_i}{\partial O_{ij}} = \prod_{k \neq j} O_{ik}. \tag{22}$$

The amount to be updated in the jth CMAC should be

$$\Delta O_{ij} = -\alpha \prod_{k \neq j} O_{ik} \varepsilon, \tag{23}$$

where α is the learning rate. The updated amount will be distributed to all involved hypercubes as that done for updating a conventional CMAC described in Section III.A.

The following summarizes a basic learning procedure in which the neural network structure is fixed:

Basic Learning Procedure with Predetermined Neural Network Structure

1. Initialize all CMACs with random memory contents between $-\delta$ and δ.
2. Obtain a training sample.
3. Compute the overall output for this sample and calculate the error.
4. Update all CMACs using Eq. (23).
5. If the error for the past N samples is acceptable, then stop. Otherwise, go to step 2.

This procedure uses the fixed neural network. Actually, one can increase the number of submodules during the training if it is necessary. The following summarizes the learning procedure in which the neural network is allowed to grow:

Learning with Neural Network Growing

1. Initialize the neural network with one submodule (note that one can start with more than one submodule). Select all learning parameters.
2. Initialize all CMACs of the new submodule with random memory contents between $-\delta$ and δ.
3. Obtain a training sample.
4. Compute the overall output for this sample and calculate the error.
5. Use $\Omega\%$ of the error to update the new submodule and $(100 - \Omega)/K\%$ of the error for each of the old submodules, where K is the current number of old submodules.
6. Update all CMACs using Eq. (23).
7. If the error for the past N_1 samples is acceptable, then stop.

8. If the improvement in the last N_2 samples is insignificant, then add one more submodule to the neural network and go to step 2. Otherwise, go to step 3.

The value Ω in step 5 specifies the weight in updating the memory contents in the new and old submodules. Having Ω equal to 100% disables the learning in all old submodules. The use of unequal back-propagated errors helps fully utilize the new submodule to speed up the learning. One reasonable suggestion is to have Ω equal to 50%. This makes a large updating weight in the new submodule during the training. Note that at the very beginning without any old submodule, the error should be evenly distributed to initial submodules.

C. Example: Function Approximation

This example is to approximate the following four-variable nonlinear function:

$$\hat{g}(x_1, x_2, x_3, x_4) = x_1 + \sin(x_1\pi)\cos(x_2\pi)\sin(x_3\pi)\left[\sin^2(x_4\pi) - 1\right]. \quad (24)$$

The fixed structure with three submodules is used. Each submodule consists of two two-input CMACs, and each CMAC uses eight blocks for each input variable and has eight elements in each complete block (i.e., $N_b = 8$ and $N_e = 8$). This neural network requires a memory size of 3K (# submodules $*$ # CMACs per submodule $*N_e * N_b^2$). The learning rate α was set to 0.02.

The target function and the neural network output are plotted in Fig. 14. Since there are four inputs to plot, two of them are fixed. Figure 14a shows the target function $\hat{g}_1$ with x_2 and x_3 both set to 0.25, and Fig. 15a shows the function with

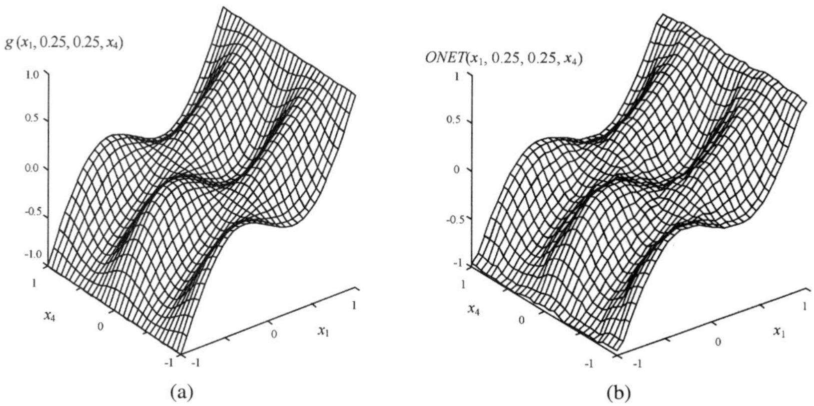

Figure 14 (a) Target function and (b) network output (with $x_2 = 0.25$ and $x_3 = 0.25$).

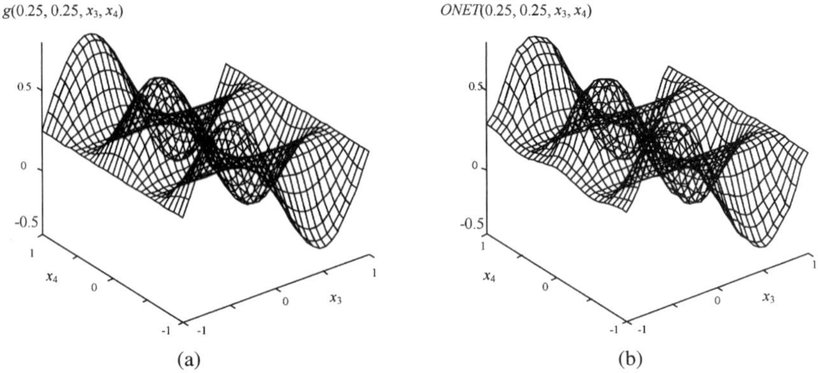

(a) (b)

Figure 15 (a) Target function and (b) network output (with $x_1 = 0.25$ and $x_2 = 0.25$).

x_1 and x_2 both set to 0.25. The corresponding plots for the output of the trained neural network are given in Figs. 14b and 15b.

VI. CONCLUSIONS

The capability for modeling nonlinear functions/relations makes the applications of neural networks to nonlinear system learning control promising. Neural networks are often used to model plants and inverse plants. In many learning control methods, a neural network may be trained and used to provide some derivative information. This happens in back propagation through the plant model. Multilayer neural networks have been the most popular type of structure referred to in the neurocontrol literature. One often encountered problem is the difficulty on learning, owing to possible local minima in the used energy function. The long training time is often unacceptable in an application. This chapter focuses on another type of neural network structure, which is believed to have advantages over the preceding issues. Four different types of CMAC-related techniques have been introduced in this chapter. The first one is the conventional CMAC. It has very good learning speed and its learning always converges. One disadvantage is that the derivative information, which is needed in many learning control techniques, is not directly available. The second one is the integration of CMAC with weighted regression. The derivative information is created with the use of regression in this scheme. The third one uses the general basis function to replace the constant basis used in the conventional CMAC and can also generate derivative information. The last scheme aims at applications to learning control problems with a high-dimensional input space. It is composed of small CMACs and has a sum-of-product structure. Although the CMAC is a table-lookup technique, the use of small CMACs releases the possible requirement of large memory space.

REFERENCES

[1] J. S. Albus. A new approach to manipulator control: the cerebellar model articulation controller (CMAC). *J. Dynamic Systems Measurement Control* 220–227, 1975.

[2] J. S. Albus. Data storage in the cerebellar model articulation controller (CMAC). *J. Dynamic Systems Measurement Control* 220–227, 1975.

[3] P. C. Parks and J. Militzer. Convergence properties of associative memory storage for learning control system. *Automat. Remote Control* 50:254–286, 1989.

[4] C. T. Chiang and C. S. Lin. CMAC with general basis functions. *Neural Networks* 9:1199–1211, 1996.

[5] D. E. Rumelhart, G. E. Hinton, and R. J. Williams. Learning internal representation by error propagation. In *Parallel Distributed Processing: Exploration in the Microstructure of Cognition* (D. E. Rumelhart and J. L. McClelland, Eds.), Vol. 1. MIT Press, Cambridge, MA, 1986.

[6] R. Hecht-Nielsen. Theory of the back-propagation neural network. In *Proceedings of the International Joint Conference on Neural Networks*, Vol. 1, pp. 593–611, 1989.

[7] K. Hornik, M. Stinchcombe, and H. White. Multi-layer feedforward networks are universal approximators. *Neural Networks* 2:359–366, 1989.

[8] S. Lee and R. M. Kil. A Gaussian potential function network with hierarchically self-organizing learning. *Neural Networks* 4:207–224, 1991.

[9] Q. Zhang and A. Benveniste. Wavelet network. *IEEE Trans. Neural Networks* 3:889–898, 1992.

[10] T. D. Sangers. A tree-structure adaptive network for function approximation in high-dimensional space. *IEEE Trans. Neural Networks* 2:285–293, 1991.

[11] Y. H Pao, G. H. Park, and D. J. Sobajic. Learning and generalization characteristics of the random vector functional-link net. *Neurocomputing* 6:163–180, 1994.

[12] B. Igelnik and Y.-H. Pao. Stochastic choice of basis functions in adaptive function approximation and the functional-link net. *IEEE Trans. Neural Networks* 6, 1995.

[13] C. L. Giles and T. Maxwell. Learning, invariance, and generalization in a high-order neural network. *Appl. Optics* 26:4972–4978, 1987.

[14] J. A. Feldman and D. H. Ballard. Connectionist models and their properties. *Cognitive Sci.* 6:205–254, 1982.

[15] D. N. Nguyen and B. Widrow. Neural networks for self-learning control systems. *IEEE Control Mag.* 18–23, 1990.

[16] L. H. Ungar. A bioreactor benchmark for adaptive network-based process control. In *Neural Networks for Control* (W. T. Miller III, R. S. Sutton, and P. J. Werbos, Eds.), pp. 388–402. MIT Press, Cambridge, MA, 1990.

[17] P. J. Werbos. Backpropagation through time: what it does and how to do it. *Proc. IEEE* 78:1550–1560, 1990.

[18] A. S. Levin and K. S. Narendra. Control of nonlinear dynamical systems using neural networks: controllability and stabilization. *IEEE Trans. Neural Networks* 4:192–206, 1993.

[19] A. G. Barto, R. S. Sutton, and C. W. Anderson. Neuronlike adaptive elements that can solve difficult learning control problems. *IEEE Trans. Systems Man Cybernet.* 13:834–846, 1983.

[20] C. W. Anderson. Learning to control an inverted pendulum using neural networks. *IEEE Control Mag.* 31–37, 1989.

[21] C. S. Lin and H. Kim. CMAC-based adaptive critic self-learning control. *IEEE Trans. Neural Networks* 2:530–533, 1991.

[22] W. T. Miller III. Real-time application of neural networks for sensor-based control of robots with vision. *IEEE Trans. Systems Man Cybernet.* 19:825–831, 1989.

[23] C. S. Lin and C. T. Chiang. Integration of CMAC and weighted regression for efficient learning and output differentiability. *IEEE Trans. Systems, Man, and Cybernetics*, to appear.

[24] C. T. Chiang. CMAC addressing technique based learning structure. Ph.D. Dissertation, Department of Electrical and Computer Engineering, University of Missouri-Columbia, Columbia, Missouri, 1994.

[25] S. H. Lane, D. A. Handelman, and J. J. Gelfand. Theory and development of high-order CMAC neural networks. *IEEE Control Mag.* 23–30, 1992.

[26] C. S. Lin and C. K. Li. A new neural network structure composed of small CMACs. *Proceedings of the 1996 IEEE International Conference on Neural Networks*, Washington, DC, Vol. 3, pp. 1777–1783, 1996.

Information Dynamics and Neural Techniques for Data Analysis

Gustavo Deco
Corporate Research and Development
Siemens AG
81739 Munich, Germany

I. INTRODUCTION

One of the most essential problems in the fields of neural networks and nonlinear dynamics is the extraction and characterization of the statistical structure underlying an observed set of data. In the context of neural networks, the problem is posed as the data-based learning of a parametric form of the statistical dependences behind the data. In this parametric formulation, the goal is to model the observed process. On the other hand, an *a priori* requirement for the extraction of statistical structures is the detection of their existence and their characterization. For time series, for example, it is useful to know if the dynamics that originate the observed values are stationary or nonstationary, if the time series is deterministic or stochastic, and to be able to distinguish between white noise, colored noise, Markov processes, and chaotic and nonchaotic determinism. The detection and characterization of such dependences should therefore be previously performed in a nonparametric fashion in order to be able to model the process *a posteriori* in a parametric form. The basic problem is of statistical nature and therefore information theory offers the ideal theoretical framework for a mathematical formulation.

The aim of this chapter is to review a detailed and unifying formulation of the theory of parametric and nonparametric statistical structure extraction based on an information-theoretic approach developed by the author in the last years. The formulation presented in this chapter establishes a consistent theoretical framework for the problem of discovering knowledge behind empirical data. This approach offers a great potential for achieving optimal solutions of complex real-world problems, as illustrated in the different sections of this chapter by simulations and experiments. The experiments utilize biological data such as electroencephalogram (EEG) signals, financial data such as the Germann Stock Index (DAX) time series, and physical data such as sunspot data.

The chapter is organized as follows. Section II, concentrates on the problem of *parametric extraction of the statistical structure* underlying the measured data. We present the solution of this problem in the framework of information theory and the theory of neural networks. Feature extraction is one of the principal goals of unsupervised learning. In biological systems, this is the first step of the cognitive mechanism that enables the processing of higher-order cognitive functions. Therefore, the parametric extraction of a statistical structure in the data can be posed as an information-theoretic approach to the problem of unsupervised learning.

The concept of feature extraction is defined as *independent component analysis* (ICA) where independence is formulated in the statistical sense. An information-theoretic-based formulation of ICA is presented for the case of arbitrary input probability distributions and arbitrary, possibly nonlinear, input–output transformations. We apply this architecture on the problem of modeling time series by learning statistical correlations between the past and present elements of the series in an unsupervised fashion.

We also discuss in this parametric framework the problem of learning and generalization from examples by using neural networks and frameworks from statistics. In the statistical approach, an ensemble of neural networks is used to address the problem of generalization from a finite number of noisy training examples. The ensemble treatment of neural networks assumes that the final model is built by an integration of singular models weighted with the appropriate probability distribution. *Gibbs' distribution* is obtained from the *maximum entropy principle* or alternatively by imposing the equivalence of the minimum error and the maximum likelihood criteria for training the network. This theory is used to obtain a general formulation of a statistical approach to unsupervised learning. The duality between supervised and unsupervised learning is also discussed.

Section III handles the problem of *nonparametric statistical structure extraction* from an information-theoretic point of view. We restrict ourselves to the presentation of time series but the theory is valid in general.

We present a nonparametric formalism for the extraction of statistical structures. We introduce a nonparametric cumulant-based statistical approach for detecting linear and nonlinear statistical dependences in nonstationary time series. The statistical dependence is detected by measuring the predictability which tests the null hypothesis of statistical independence, expressed in Fourier space, by the surrogate method. Therefore, the predictability is defined as a higher-order cumulant-based significance discriminating between the original data and a set of scrambled surrogate data which corresponds to the null hypothesis of a noncausal relationship between past and present, that is, of a random process with statistical properties equivalent to those of the original time series. In this formulation linear and nonlinear as well as Gaussian and non-Gaussian temporal dependences can be detected in time series. The theory introduced herein is illustrated by artificial and real-world stochastic as well as chaotic time series.

In Section IV, we extend the nonparametric and test-statistics-based theory presented in Section III to the concept of *information flow* in such a way that we cannot only answer the question about the existence of structure but also characterize the dynamics that originate the temporal structure. This section introduces an information-theoretic-based concept for the characterization of the information flow in systems in the framework of symbolic dynamics. The information flow characterizes the loss of information about the initial conditions, that is, the decay of statistical correlations (i.e., nonlinear and non-Gaussian) between the entire past and a point p steps into the future as a function of p. In the case where the partition that generates the symbolic dynamics is finite, the information loss is measured by a mutual information. The profiles in p of the mutual information describe the short- and long-range forecasting possibilities for the given partition resolution. The information loss provides us with a mean measure of sensitivity on the initial conditions which is the main characteristic of chaotic, dynamical systems. Therefore, it is more relevant to study the evolution of information loss for the cases of infinitesimal partitions which characterize the intrinsic behavior of the dynamics on an extremely fine scale. In this case, the information flow is characterized by the evolution of a conditional entropy which generalizes the Kolmogorov–Sinai entropy to the case of observing the uncertainty more than one step ahead into the future. This definition gives a rigorous definition of chaotic systems in the language of information theory. A dynamical system is chaotic if the generalized Kolmogorov–Sinai entropy for a point p steps into the future increases linearly with p which is equivalent to say that the conditional mutual information is always constant and equal to the Kolmogorov–Sinai entropy.

II. STATISTICAL STRUCTURE EXTRACTION: PARAMETRIC FORMULATION BY UNSUPERVISED NEURAL LEARNING

Parametric extraction of statistical dependencies in a given data set can be formulated as an unsupervised learning task. In biological systems it is the first step of the cognitive mechanism that enables the processing of higher-order cognitive functions. The idea of the nervous system and brain being regulated by an economy principle has been well known since the publication of the pioneer work of Zipf [1] and the ideas of Attneave [2] about information processing in visual perception. In the neural network society, these ideas were introduced by the important papers of Barlow [3, 4], where the author presented the connectionist model of unsupervised learning under the perspective of *redundancy reduction*. Barlow describes the process of cognition as a preprocessing of the sensorial information performed by the nervous system in order to extract the statistically relevant and independent features of the inputs without losing information. This means that the brain should statistically decorrelate the extracted information. As a learning strategy, Barlow formulated the principle of redundancy reduction. The term redundancy refers to the statistical dependence between the components involved, and therefore the principle of learning by redundancy reduction tries to find a transformation such that the transformed components are as statistically independent as possible. This kind of learning is called *factorial learning*. In the binary case, factorial learning leads to *factorial codes*, that is, codes with no redundancy [5].

This section formulates the problem of *independent component analysis* (ICA) as a search for an information-preserving mapping (linear or nonlinear) which results in statistically independent output components, that is, as a factorial learning paradigm. We present an information-theoretic-based formulation of ICA for the case of arbitrary input probability distributions and arbitrary, possibly nonlinear, input–output transformations. Two general criteria the optimization of which leads to the desired solution of ICA are defined under the assumption of invertibility of the input–output map. The first criterion establishes the connection between statistical independence and the properties of the cumulant expansion of the output joint probability. The second criterion formulates a measure of statistical dependence as the mutual information between the individual elements of the output vector variable.

In the linear case, Deco and Obradovic [6] apply Barlow's principle and derive a learning rule that performs *principal component analysis* (PCA). In this chapter, PCA is derived as a linear orthogonal transformation which conserves the transmitted entropy and which minimizes the mutual information between the outputs in order to decorrelate them. This point of view is an alternative to the "infomax" principle presented by Linsker [7–9], who essentially proposes a learning rule which maximizes the mutual information between input and output layer. A most general formulation of linear ICA for the case of non-Gaussian and nonorthogonal correlated inputs was developed by Obradovic and Deco [10].

Some nonlinear extensions of ICA for statistical decorrelation of sensorial input signals were recently introduced. The information-theoretic formulation of the principle of redundancy minimization in the nonlinear and non-Gaussian case was devised for different neural architectures in the papers of Deco and Brauer [11, 12], Deco and Schürmann [13, 14] and Parra *et al.* [15, 16].

In this section, we review these approaches. First, however, we review the concepts of information theory which are required for the mathematical formulation of our theory.

A. Basic Concepts of Information Theory

The concept of entropy is introduced as a measure of the uncertainty of a random variable. Let us consider the case of discrete random variables, which will be denoted by X. The discrete random variable X takes different discrete values x from an alphabet $\aleph$. Let us define a probability $p(x)$ for all $x \in \aleph$. Then a measure of the uncertainty of the probability distribution $p(x)$ is given by the *entropy* $H(X)$ as defined by Shannon [17]:

$$H(X) = -\sum_{x\in\aleph} p(x)\log\bigl(p(x)\bigr) = E\left(\log\left(\frac{1}{p(x)}\right)\right), \tag{1}$$

where $E(\cdot)$ denotes the expectation operator. The entropy provides a measure of the sharpness of the distribution, which is nothing else than the degree of uncertainty corresponding to the random variable X. If the entropy is equal to zero, that is, $H(X) = 0$, the variable X describes a deterministic process. In other words, zero entropy implies that there is an absolute certainty that only one outcome of X is possible. On the other hand, the maximum value of $H(X)$ is reached when the distribution $p(x)$ is uniform, that is, when the uncertainty about the random variable X is maximal.

Let X, Y be a pair of random variables over the discrete alphabets $\aleph$ and $\Im$, respectively. The joint probability will be denoted by $p(x, y)$ and the conditional probability of y for a given outcome x by $p(y|x)$. Then the joint $H(X, Y)$ entropy is defined as

$$H(X,Y) = -\sum_{x\in\aleph}\sum_{y\in\Im} p(x,y)\log\bigl(p(x,y)\bigr) = E\left(\log\left(\frac{1}{p(x,y)}\right)\right). \tag{2}$$

The conditional entropy $H(Y|X)$:

$$H(Y|X) = \sum_{x\in\aleph} p(x)H(Y|X \equiv x) = -\sum_{x\in\aleph}\sum_{y\in\Im} p(x,y)\log\bigl(p(y|x)\bigr) \tag{3}$$

is defined as the average of the degree of uncertainty about Y over all concrete outcomes of X.

An important problem often encountered in statistics is to define a measure of the difference between two distributions. The Kullback–Leibler entropy $K(p, q)$, also called the relative entropy or cross-entropy, is a measure of the "distance" between two distributions $p(x)$ and $q(x)$, and it is defined as (see Kullback and Leibler [18])

$$K(p,q) = \sum_{x \in \aleph} p(x) \log\left(\frac{p(x)}{q(x)}\right). \tag{4}$$

The relative entropy is not a true distance because it is not symmetric, that is, $K(p,q) \neq K(q,p)$. However, both expressions can be interpreted as quasi-distances which are always positive and equal to zero if and only if $p(x) = q(x)$.

To measure the statistical independence between two random variables X and Y with associated probability distributions $p(x)$ and $p(y)$, respectively, it is useful to introduce the notion of *mutual information* $I(X;Y)$. The latter is defined as the Kullback–Leibler distance between the joint probability and the factorized ones, and it is equal to zero if and only if X and Y are independent. Following Shannon [17], the mutual information between X and Y is defined as

$$I(X;Y) = K\big(p(x,y), p(x)p(y)\big) = \sum_{x \in \aleph} \sum_{y \in \Im} p(x,y) \log\left(\frac{p(x,y)}{p(x)p(y)}\right). \tag{5}$$

The mutual information is symmetric, that is, $I(X;Y) = I(Y;X)$ and $I(X;X) = H(X)$. Thus, the mutual information is a measure of the amount of information that Y conveys about X (or vice versa); that is, it provides a measure of the statistical correlation between X and Y. When X and Y are defined as the input and output, respectively, of a stochastic channel, $I(X;Y)$ is the amount of transmitted information in the stochastic channel.

The definition of entropy is extended to the case of a continuous random variable X described by a probability density function $f(x)$. The entropy in the continuous case is defined as (see Cover [19])

$$h(X) = -\int_A dx\, f(x) \log\big(f(x)\big), \tag{6}$$

where A is the support of the continuous variable x. The relation between the discrete definition of entropy and the previous definition is clearly understandable if the density function $f(x)$ is Riemann integrable:

$$H\big(X^D\big) + \log(D) \to h(X), \qquad D \to 0, \tag{7}$$

where X^D is a discrete random variable generated by partitioning the continuous random variable into X bins of length D. Equation (7) implies that the entropy of an n-bit quantization of a continuous random variable X is approximately $h(X) + n$.

The Kullback–Leibler entropy of two density distributions $f(x)$ and $g(x)$ can be defined in a similar fashion [19] as

$$K(f, g) = \int dx\, f(x) \log\left(\frac{f(x)}{g(x)}\right), \tag{8}$$

whereas the mutual information between two continuous random variables X and Y with associated joint density distribution $f(x, y)$ is given by

$$I(X; Y) = \int dx\, dy\, f(x, y) \log\left(\frac{f(x, y)}{f(x) f(y)}\right). \tag{9}$$

The properties of $K(f, g)$ and $I(X; Y)$ are the same as in the discrete case. In particular, in the limit $D \to 0$ the mutual information in the discretizations converges to the mutual information in the continuous distributions, that is,

$$I\left(X^D; Y^D\right) \to I(X; Y), \qquad D \to 0. \tag{10}$$

Owing to the fact that the theorems and inequalities about entropy hold in general for both discrete and continuous variables, we will not distinguish between these two cases and we will use the notation introduced for the discrete case as the common nomenclature. Note that the only difference is that in the continuous case the entropy can be negative. Fortunately, this fact usually has no influence on the interpretation and on the use of the relations involving relative entropy and mutual information.

B. Independent Component Analysis

1. Definition

Let $\mathbf{x}$ be a random vector of dimension n with the joint probability density function $p(\mathbf{x})$ whose covariance matrix is nonsingular. Furthermore, let $F(x)$ be a bijective square map which maps $\mathbf{x}$ into the random vector $\mathbf{y}$ with probability density function $p(\mathbf{y})$.

ICA is an input–output bijective map from F the d-dimensional input vector $\mathbf{x}$ to the d-dimensional output vector $\mathbf{y}$ such that the output components with joint probability

$$p(\mathbf{y}) = p(y_1 \cdots y_d), \qquad \mathbf{y} = F(\mathbf{x}) \tag{11}$$

are "as independent as possible" according to an appropriate measure. In the special case where the complete independence of the output components is achieved, the following holds:

$$p(y_1 \cdots y_d) = p(y_1) \cdots p(y_d). \tag{12}$$

If the input vector $\mathbf{x}$ is jointly Gaussian and the input–output map is linear, the corresponding linear ICA is equivalent to the problem of finding a nonsingular matrix M which diagonalizes the output covariance matrix Q_y:

$$\mathbf{y} = M\mathbf{x}, \tag{13}$$

$$Q_y = MQ_xM^T = D, \tag{14}$$

where Q_x is the input covariance matrix and D is a nonnegative diagonal matrix with real-valued entries [20]. In addition, if the matrix M is a rotation, that is, an orthogonal matrix with determinant equal to one, then linear Gaussian ICA coincides with the well-known PCA.

According to the definition, ICA implies a search for the map F which maximizes an appropriate measure of the statistical independence of the output components. Hence, by establishing a measure of statistical independence, ICA will be posed herein as an optimization problem with respect to the function F. The latter can, in general, be parameterized as a neural network whose architecture and learning mechanism guarantee required properties such as bijectivity.

The transmitted information from the input to the output of a differentiable map F is defined by the output entropy [21]

$$H(\mathbf{y}) \leq H(\mathbf{x}) + \int d\mathbf{x}\, p(\mathbf{x}) \ln\left(\left|\det\left(\frac{\partial F}{\partial \mathbf{x}}\right)\right|\right), \tag{15}$$

where $\partial F/\partial \mathbf{x}$ stands for the Jacobian of the input–output transformation. The inequality becomes an equality if and only if the map F is bijective. In the case of a linear, invertible mapping, the entropy relationship takes the following form:

$$H(\mathbf{y}) = H(\mathbf{x}) + \ln\big(|\det(M)|\big). \tag{16}$$

The entropy-preserving property will be used to simplify an appropriate measure of statistical independence. Conservation of of the input entropy is assured if the input–output transformation is bijective and *conserves the volume*, that is, if its Jacobian matrix has a determinant equal to one

$$\det\left(\frac{\partial F}{\partial \mathbf{x}}\right) = 1. \tag{17}$$

According to Eq. (17), the inequality in (15) is then reduced to

$$H(\mathbf{y}) = H(\mathbf{x}). \tag{18}$$

2. General Criteria for Independent Component Analysis

The two criteria derived in this section are based on suitable properties of joint and marginal output probability density functions and are general in the sense that they are valid for arbitrary non-Gaussian distributions. Nevertheless, because, in general, arbitrary distributions are not known explicitly, the introduced measures of statistical dependence have to be estimated from the available data. The first criterion for evaluation of the statistical dependence among output components is based on the properties of a cumulant expansion of the joint probability density function $p(\mathbf{y})$.

a. Cumulant Expansion-Based Criterion for Independent Component Analysis

To formulate the cumulant expansion which is a common tool in statistical analysis, the Fourier transform of the output distribution is needed:

$$\phi(\mathbf{w}) = \int d\mathbf{y} \exp\bigl(i(\mathbf{w} \cdot \mathbf{y})\bigr) \cdot p(\mathbf{y}), \tag{19}$$

$$\phi(w_i) = \int dy_i \exp\bigl(i(w_i \cdot y_i)\bigr) \cdot p(y_i). \tag{20}$$

Now the cumulant expansion of a distribution [22] is defined as follows:

$$\phi(\mathbf{w}) = \exp\left(\sum_{n=1}^{\infty} \frac{i^n}{n!} \sum_{i_1,\ldots,i_n}^{d} K_{i_i,\ldots,i_n} w_{i_1} \cdots w_{i_n}\right), \tag{21}$$

$$\phi(w_i) = \exp\left(\sum_{n=1}^{\infty} \frac{i^n}{n!} K_i^{(n)} w_i^n\right). \tag{22}$$

The general cumulants $K_i^{(n)}$ and $K_{i_i,\ldots,i_n}$ used in Eqs. (21) and (22) are defined in Gardiner [22] and will be defined herein as needed.

In Fourier space, the tatistical independence condition [22] is specified as

$$\phi(\mathbf{w}) = \prod_i \phi(w_i), \tag{23}$$

which is equivalent to

$$\ln\bigl(\phi(\mathbf{w})\bigr) = \sum_i \ln\bigl(\phi(w_i)\bigr). \tag{24}$$

The condition in (24) combined with the cumulant expansions in Eqs. (21) and (22) can be used to formulate the following requirement for statistical independence:

$$\sum_{n=1}^{\infty} \frac{i^n}{n!} \sum_{i_i,\ldots,i_n}^{d} K_{i_1,\ldots,i_n} w_{i_1} \cdots w_{i_n} = \sum_i^d \left(\sum_{n=1}^{\infty} \frac{i^n}{n!} K_i^{(n)} w_i^n\right). \tag{25}$$

The multidimensional cumulants are defined in Gardiner [22] as the appropriate functions of multidimensional higher-order moments $C_{i,\ldots,j} = \int d\mathbf{y}p(\mathbf{y})y_i \cdots y_j$, and the one-dimensional cumulants as a function of the one-dimensional higher-order moments $C_i^{(n)} = \int dy_i p(y_i) y_i^n$.

The cumulants of order higher than 1 are bias independent. Hence, the transformation $\mathbf{y}' = \mathbf{y} - \langle \mathbf{y} \rangle$ can be performed to simplify the cumulant calculations. Equation (25) implies that the statistical independence test requires the evaluation of cumulants of all orders. This is practically impossible and an appropriate approximation is necessary. If cumulants decay at higher orders, as is often the case, we can obtain a good approximation by retaining only the first few cumulants. By neglecting the cumulants of order higher than 4, the following "fourth"-order independence condition is derived:

$$-\frac{1}{2}\sum_{i,j} w_i w_j \{C_{ij} - C_i^{(2)}\delta_{ij}\} - \frac{i}{6}\sum_{i,j,k} w_i w_j w_k \{C_{ijk} - C_i^{(3)}\delta_{ijk}\}$$
$$+\frac{1}{24}\sum_{i,j,k,l} w_i w_j w_k w_l \Big\{(C_{ijkl} - 3C_{ij}C_{kl}) - \big(C_i^{(4)} - 3\big(C_i^{(2)}\big)^2\big)\delta_{ijkl}\Big\} = 0. \tag{26}$$

In Eq. (26), $\delta_{i,\ldots,j}$ denotes Kronecker's delta. Because Eq. (26) should be satisfied for all $\mathbf{w}$, all coefficients in each summation must be zero. This yields the following conditions:

$$C_{ij} = 0, \quad \text{if } i \neq j, \tag{27}$$
$$C_{ijk} = 0, \quad \text{if } i \neq j \vee i \neq k, \tag{28}$$
$$C_{ijkl} = 0, \quad \text{if } \{i \neq j \vee i \neq k \vee i \neq l\} \wedge \neg L, \tag{29}$$
$$C_{iijj} - C_{ii}C_{jj} = 0, \quad \text{if } i \neq j. \tag{30}$$

In (29), L is the logical expression

$$L = \big\{(i = j \wedge k = l \wedge j \neq k) \vee (i = k \wedge j = l \wedge i \neq j) \vee (i = l \wedge j = k \wedge i \neq j)\big\}, \tag{31}$$

which excludes the cases considered in Eq. (30). The conditions of independence at the "fourth" order defined in Eqs. (27)–(30) can be achieved by minimizing the cost function

$$E = \alpha \sum_{i<j} C_{ij}^2 + \beta \sum_{i<j\leq k} C_{ijk}^2 + \gamma \sum_{i<j\leq k\leq l} C_{ijkl}^2 + \delta \sum_{i<j} \big(C_{iijj} - C_{ii}C_{jj}\big)^2 \tag{32}$$

over the invertible maps F where $\alpha, \beta, \gamma, \delta$ are the inverses of the number of elements in each summation, respectively. Hence, the cost function in (32) represents a suitable measure of statistical dependence up to the cumulant order 4 of an arbitrary, possibly non-Gaussian output distribution. The numerical complexity of evaluating the expression in (32) is proportional to the output dimension and might be high. Although in the general case no simplification is possible, the special case of linear transformations offers alternative cost functions of much lower computational complexity.

The second criterion for ICA introduced herein is based on the evaluation of the mutual information at the output of the map F.

b. Mutual Information as Criterion for Independent Component Analyses

The mutual information between the output components, defined as

$$R = I(y_1; \ldots; y_{i-1}; y_i) = \sum_{i=1}^{n} I(y_1, \ldots, y_{i-1}; y_i) = \sum_{j=1}^{n} H(y_j) - H(\mathbf{y}), \quad (33)$$

is a measure of the statistical correlations between the components of the outputs. In fact, statistical independence is equivalent to

$$R = \sum_{j=1}^{n} H(y_j) - H(\mathbf{y}) = 0. \quad (34)$$

The latter implies that in order to minimize the redundancy at the output, the mutual information between the different components of the output vector should be minimized with respect to F. On the other hand, it was shown earlier that restricting the input–output map to be bijective and volume preserving guarantees that the joint entropy at the output is equal to the joint entropy at the input, that is, $H(\mathbf{y}) = H(\mathbf{x})$. Hence, the minimization of R reduces to the minimization of

$$I_1 = \sum_{j=1}^{n} H(y_j). \quad (35)$$

The restriction to invertible, volume-preserving maps has simplified the optimization of the statistical independence measure in (34) because the evaluation of the joint output entropy is unnecessary. Nevertheless, the evaluation of individual entropies in (35) is still nontrivial in the case of arbitrary probability distributions. The estimation of the individual entropies $H(y_j)$ from the data is based on the appropriate approximation of the corresponding probability densities $p(y_j)$.

According to the second Gibbs theorem, the entropy of an arbitrary distribution is bounded from above by the entropy of a Gaussian distribution with the same variance as the original one. Hence, the cost function in (35) is bounded from above by the scaled sum of the logarithm of the output variances, that is, the entropies of Gaussian distributions. If the variance of each component is denoted by σ_j^2, then

$$\text{minimization}(I_1) \equiv \text{minimization}\left(\sum_j \ln(\sigma_j^2)\right). \tag{36}$$

The minimum of this upper bound is achieved when the output covariance matrix is diagonalized. A potential problem with the cost function in (36) stems from the fact that the entropy of a continuous variable can be negative and even infinite in magnitude if a variable is deterministic. Hence, it is worth obtaining a somewhat different upper bound which does not exhibit such problems. Let $G(y_i)$ be a Gaussian distribution with mean equal to that of the real distribution $p(y_i)$ but with unit variance. Hence, the distribution $G(y_i)$ is not equal to the best Gaussian approximation of the real output distribution. The Kullback–Leibler distance between $p(y_i)$ and $G(y_i)$ is defined as

$$\begin{aligned} p &\leq \int dy_i\, p(y_i) \ln\left(\frac{p(y_i)}{G(y_i)}\right) \\ &= -H(y_i) - \int dy_i\, p(y_i) \ln\big(G(y_i)\big) \\ &= -H(y_i) + \frac{1}{2}\ln(2\pi) + \frac{\sigma_i^2}{2}. \end{aligned} \tag{37}$$

Hence, an alternative upper bound (modulo a constant) for I_1 is defined as

$$\text{minimization}(I_1) \equiv \text{minimization}\left(\sum_j \sigma_j^2\right). \tag{38}$$

With a variational approach, it can be shown that a spherical Gaussian distribution minimizes the sum of variances under the constraint of constant entropy transmission (see Deco and Obradovic [23]).

The advantage of using the previously defined upper bounds stems from their low numerical complexity because the estimate of the real entropies is substituted by the estimation of variances. Although there are several examples where the minimization of an upper bound leads to satisfactory results, there are other cases where the approximation of an arbitrary distribution by a Gaussian distribution yields poor results. One way to improve over the pure Gaussian approximation is to carry out the Edgeworth expansion [24] of the given probability distribution around its best Gaussian approximation. The upper bound in (36) can be seen as resulting from the Edgeworth expansion where only the first term is taken into account.

The Edgeworth expansion of a scalar variable with zero mean and unit variance around its best Gaussian approximation N_w is a natural candidate for the cumulant-based evaluation of the unknown probability density function $p(w)$. Writing explicitly the terms corresponding to the Cramer–Edgeworth polynomials [25] up to order 4, the Edgeworth expansion is defined as [24]

$$\begin{aligned}\frac{p_w(s)}{N_w(s)} &= 1 + \frac{1}{3!}K^{(3)}h_3(s) + \frac{1}{4!}K^{(4)}h_4(s) + \frac{10}{6!}\big[K^{(3)}\big]^2 h_6(s) \\ &\quad + \frac{1}{5!}K^{(5)}h_5(s) + \frac{35}{7!}K^{(3)}K^{(4)}h_7(s) + \frac{280}{9!}\big[K^{(3)}\big]^3 h_9(s) \\ &\quad + \frac{1}{6!}K^{(6)}h_6(s) + \frac{56}{8!}K^{(3)}K^{(5)}h_8(s) + \frac{35}{8!}\big[K^{(4)}\big]^2 h_8(s) \\ &\quad + \frac{2100}{10!}\big[K^{(3)}\big]^2 K^{(4)}h_{10}(s) + \frac{15400}{12!}\big[K^{(3)}\big]^4 h_{12}(s) + \mathrm{Rem}(s) \\ &= 1 + F_4(s) + \mathrm{Rem}(s). \end{aligned} \tag{39}$$

Hence, the function F_4 contains all the cumulants corresponding to Cramer–Edgeworth polynomials up to order 4, whereas Rem is the remainder function. In the expansion in (39), $K^{(i)}$ is the cumulant of order i of the corresponding normalized scalar variable and $h_j(s)$ denotes the Hermite polynomial of degree j. The accuracy of the "nth-order" approximation of the unknown density function $p(w)$ based on taking only the n first elements from the Edgeworth series expansion is proportional to n.

The Edgeworth expansion of density functions presented in (39) can be used to express the entropy of a normalized scalar variable w:

$$H(w) = -\int_{-\infty}^{\infty} ds\, p_w(s)\ln\big(p_w(s)\big) \tag{40}$$

as a function of cumulants by rearranging the expression in (40) as a function of the ratio between $p(w)$ and its best Gaussian approximation $N_w(s)$:

$$\begin{aligned} H(w) &= -\int_{-\infty}^{\infty} ds\, \frac{p_w(s)}{N_w(s)} N_w(s)\ln\left(\frac{p_w(s)}{N_w(s)}\right) - \int_{-\infty}^{\infty} ds\, p_w(s)\ln\big(N_w(s)\big) \\ &= -\int_{-\infty}^{\infty} ds\, \frac{p_w(s)}{N_w(s)} N_w(s)\ln\left(\frac{p_w(s)}{N_w(s)}\right) + H_N(w). \end{aligned} \tag{41}$$

The entropy $H_N(w)$ of the best Gaussian approximation of the normalized variable in (41) is equal to $0.5(1 + \ln(2\pi))$. The first term in the expression in (41) is always positive because it corresponds to the Kullback–Leibler distance [21] between p_w and N_w:

$$\begin{aligned} \int_{-\infty}^{\infty} ds\, \frac{p_w(s)}{N_w(s)} N_w(s)\ln\left(\frac{p_w(s)}{N_w(s)}\right) &= H_N(w) - H(w) \\ &= \int_{-\infty}^{\infty} ds\, p_w(s)\ln\left(\frac{p_w(s)}{N_w(s)}\right) \geq 0. \end{aligned} \tag{42}$$

The expression in (41) can be used to derive an approximation of by keeping only a finite number of elements in the Edgeworth expansion. An extreme case corresponds to keeping only the first term of the expansion, that is, the best Gaussian approximation N_w.

C. Nonlinear Independent Component Analysis

1. Redundancy Reduction by Triangular Volume-Conserving Architectures

This section formulates a neural network architecture which satisfies the conditions required by the two ICA criteria defined in Section II.B: bijectivity and volume preservation. A single-layer triangular architecture is depicted in Fig. 1a.

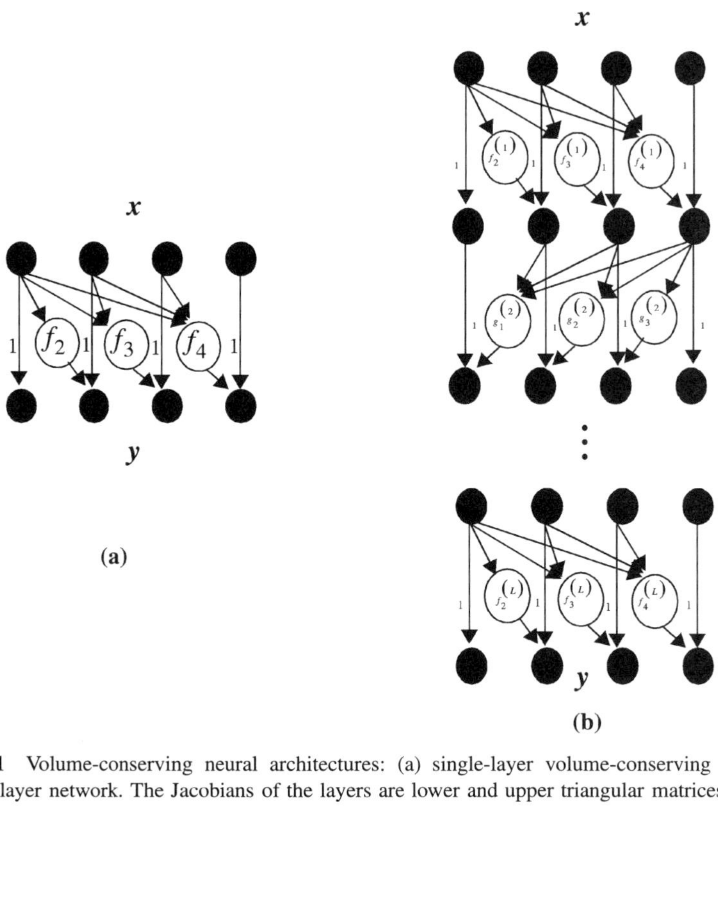

Figure 1 Volume-conserving neural architectures: (a) single-layer volume-conserving network; (b) multilayer network. The Jacobians of the layers are lower and upper triangular matrices, respectively.

The dimensions of the input and output layers are the same and equal d. The analytical formulation of the transformation defined by this architecture can be written as

$$y_i = x_i + f_i(x_1, \ldots, x_{i-1}, \mathbf{w}_i), \tag{43}$$

where $\mathbf{w}_i$ represents a set of parameters of the function f_i. Note that the network is always volume conserving; that is, its Jacobian satisfies Eq. (17) regardless of the choice of the functions f_i. In general, each f_i can be parameterized in an arbitrary way: by another neural network, by a single sigmoid neuron, by a polynomial (higher-order neurons), etc.

The Jacobian matrix of the transformation defined in Eq. (43) is upper triangular with diagonal elements all equal to one whose determinant is, therefore, equal to one. An "inverted" version of the network can be defined as

$$y_i = x_i + g_i(x_{i+1}, \ldots, x_d, \mathbf{t}_i). \tag{44}$$

This network has a lower triangular Jacobian matrix with diagonal elements all equal to one, which also guarantees volume conservation.

The vectors $\mathbf{t}_i$ represent the parameters of functions g_i. To construct a general nonlinear transformation from inputs to outputs, it is possible to build a multilayer architecture like the one shown in Fig. 1b, which consists of both networks described by Eqs. (43) and (44). Because successive application of volume-conserving transformations is also volume conserving, the multilayer architecture is also volume conserving. Once the architecture is specified, independent feature extraction can be posed as optimization of the cumulant or mutual information criterion. For the moment, we focus on the cumulant criterion defined in (32). The cost function E should be minimized in order to statistically decorrelate nonlinearly correlated non-Gaussian inputs.

The learning rule can be easily obtained by the gradient descent method over the cost functions defined in Section II.B.

2. Unsupervised Modeling of Chaotic Time Series

Modeling time series by learning from experiments can be viewed as the parametric extraction of statistical correlations between past and future values of the time series measurements. Owing to the short-term predictability of chaotic series, a thorough study of statistical correlations between components of the embedding vector yields the only way to distinguish between a purely random process and a chaotic deterministic series, eventually corrupted by colored or white noise. For short-term prediction, neural network models have been implemented using supervised learning paradigms and feedforward [26] or recurrent architectures [27]. However, the problem of extracting statistical correlations in a sensorial environment is the subject of unsupervised learning. Hence, the typically supervised problem of modeling dynamical systems is herein transformed into an unsupervised problem of independent feature extraction. Independent feature extraction is applied to extract in an unsupervised fashion the statistical correlation between the components of the embedding vector associated with a time series. A single-layer architecture is employed which attempts to extract correlations considering only the past information relative to each element of an embedding vector. The architecture is always reversible and conserves the volume and, therefore, the transmitted information [13]. In general, the environment is non-Gaussian distributed and nonlinearly correlated. The learning rule statistically decorrelates the elements of the output.

In modeling a chaotic system using observations collected from a chaotic attractor, the Takens method [28] called *phase-space reconstruction* is briefly reviewed. This method results in a d-dimensional "embedding space" in which the dynamics of the multidimensional attractor are captured. Let us assume a time series of a single (one-dimensional) measured variable from a multidimensional dynamical system. The aim of forecasting is to predict the future evolution of this variable. It has been shown that in nonlinear, deterministic, chaotic systems it is possible to determine the dynamical invariants and the geometric structure of the multivariable dynamic system from the observations of a single dynamical variable [28, 29]. Let a chaotic system be described as $\mathbf{y}(t+1) = \mathbf{g}[\mathbf{y}(t)]$. Let the observable measurement be $x(t) = f[\mathbf{y}(t)]$. The Takens theorem ensures that for an embedding $\mathbf{b}(t) = [x(t), x(t-\tau), \ldots, x(t-d\cdot\tau)]$ a map $\mathbf{b}(t+1) = \mathbf{F}[\mathbf{b}(t)]$ exists which has the same dynamical characteristics as the original system $\mathbf{y}(t)$ if the number of delays is equal to $d = 2D+1$, where D is the dimension of the strange attractor and τ is the delay. This sufficient condition may be relaxed to $d > 2D$ [28]. The theorem implies that all the coordinate-independent properties of $\mathbf{g}(\cdot)$ and $\mathbf{F}(\cdot)$ will be identical. The proper choice of d and τ is an important topic of investigation [30, 31]. The goal of unsupervised neural network modeling is to learn the map given by $\mathbf{F}(\cdot)$ by learning the statistical correlations between the time of successive elements of the embedding vector.

In the example, the architecture defined in Fig. 1a is used. The input vector is the embedding vector defined as $\mathbf{x} = [x(t-d\cdot\tau), x(t-(d-1)\tau), \ldots, x(t)]$. The learning rules were defined in the previous section. The example focuses on modeling the Mackey–Glass system. Owing to the presence of pure delay in the differential equation (45), the Mackey–Glass system formally has an infinite number of degrees of freedom but its strange attractor has finite dimension. The delay differential equation of Mackey–Glass [32] is as follows:

$$\dot{x}(t) = -bx(t) + \frac{ax(t-T)}{1+x^{10}(t-T)}, \tag{45}$$

where $a = 0.2$, $b = 0.1$, and $T = 30$. A polynomial neural network of order 2 is used in this example. The learning constant was $\eta = 0.01$, and 25000 iterations of training were performed. Five hundred training patterns corresponding to the time interval from $t = 2000$ to $t = 25000$ are generated by integrating Eq. (45). The input and output dimension is 6 and the criterion for learning is also the cumulant-expansion-based cost function in (32). The six inputs are: $x(t-50), x(t-40), x(t-30), x(t-20), x(t-10), x(t)$.

Because the neural architecture in this example is a polynomial of second order, it is obvious that it can only approximate the real dynamics of the Mackey–Glass series. Nevertheless, the embedding dimension was detected even with this second-order approximation by analyzing the weight connections or output components after training. Those weights which are negligibly small indicate statistical independence. The embedding dimension found is 4, in agreement with the results of Liebert and Schuster [30] and Liebert *et al.* [31]. Figure 2 shows the output components after the training. The output components 5 and 6 have very low variances, meaning that the network with second-order polynomial functions f_i has extracted an approximated form of the correlation between these outputs and the past four points. This also indicates that four points in the past (embedding dimension) are required to approximately model the map by the second-order

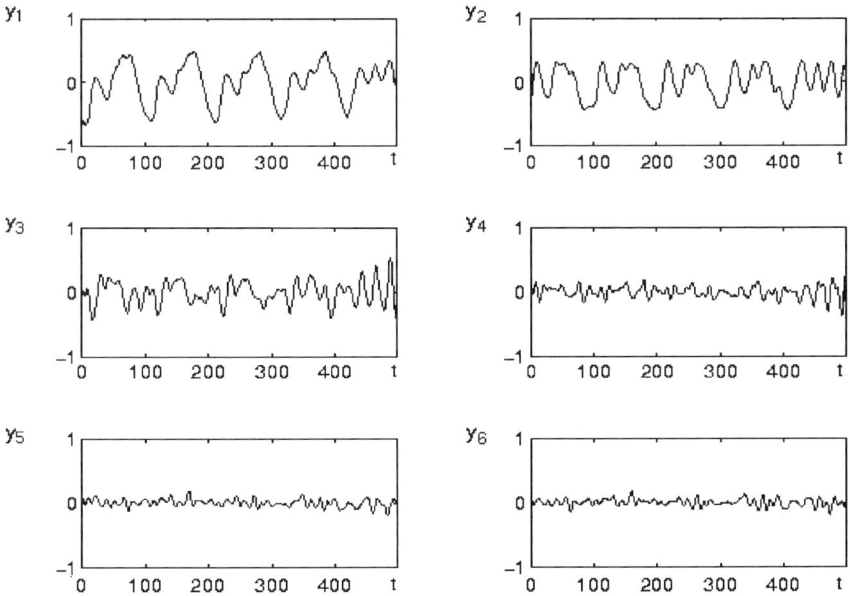

Figure 2 Outputs as a function of time of a six-input and six-output neural network trained in unsupervised fashion for extracting the decorrelation between the component of a six-dimensional embedding vector for the Mackey–Glass time series.

polynomial. The correlation cannot be totally extracted because the original series is nonpolynomial, whereas the network is a second-order polynomial. In addition, the data were generated by a differential equation and by a finite equation. This is the reason why components 5 and 6 of the output are not constant.

It is important to remember that the optimal embedding dimension is determined by the number of points in the past that are statistically correlated with the present. A strategy to measure statistical correlations is to find out how many points in the past are necessary to model, that is, to find the statistical correlations with the present. This technique is related with the ones proposed by Fraser and Swinney [33] and the works of Liebert and Schuster [30] and Liebert *et al.* [31] which formulate the minimization of the mutual information for the detection of the optimal embedding.

D. Linear Independent Component Analysis

Linear independent component analysis is defined as the search for a linear map, that is, matrix, which optimizes an appropriate measure of the statistical dependence of the output components. The conservation of information in the strict sense of Eq. (16) corresponds to matrices whose determinant is equal to one. Nevertheless, one can argue that the invertibility of the matrix is the actual requirement for the information preservation and that the determinant condition can be achieved by a simple scalar scaling of the matrix M.

One measure of the statistical dependence between the output components is defined as the average mutual information:

$$\begin{aligned} I &= \sum_{i=1}^{n} H(y_i) - H(\mathbf{y}) \\ &= \sum_{i=1}^{n} H(y_i) - H(\mathbf{x}) - \ln\big(|\det(M)|\big) \geq 0 \qquad \text{iff } |\det(M)| \neq 0. \end{aligned} \tag{46}$$

It is easy to see that the average mutual information as a measure of statistical independence is invariant to diagonal scaling as well as to the permutation of the output components. The invariance with respect to diagonal scaling can be used to reduce the number of terms in (46), which depends on the transformation M, because there always exists a scaling

$$\big[\det(M)\big]^{-1/n}, \tag{47}$$

which makes the determinant of the resulting matrix

$$M\big[\det(M)\big]^{-1/n} \tag{48}$$

equal to one. Hence, without loss of generality, it is possible to restrict the original matrix to have a unit determinant and, therefore, $H(\mathbf{y}) = H(\mathbf{x})$. Equation (46) shows that the only term which is dependent on the linear transformation M with unit determinant is the sum of entropies of the individual output elements. Consequently, linear ICA can be now formulated as the following constrained optimization problem:

$$J(M) = \min \sum_{i=1}^{n} H(y_i) \qquad \text{such that } |\det(M)|. \tag{49}$$

The cost function $J(M)$ does not impose any restriction on the probability density function of the input and output signals, but it requires estimation of the entropies corresponding to the output components. In the case where the input signal is jointly Gaussian, the entropies are simple functions of the output variances. If the matrix M is further restricted to be a rotation, the minimization of the cost function $J(M)$ results in the standard PCA [6]. When the input signal is not jointly Gaussian, the evaluation of the cost function $J(M)$ has to be based on the cumulant/moment expansion of the probability density functions $p(y_i)$ (see Deco and Obradovic [23] for a detailed discussion of the linear case).

E. Dual Ensemble Theory for Unsupervised and Supervised Learning

The problem of learning and generalization from examples using neural networks has been posed in the context of statistics [34]. In the statistical approach, an ensemble of neural networks is used to address the problem of generalization of learning from a finite number of noisy training examples. The ensemble treatment of neural networks [34] assumes that the final model is a probabilistic model built by an integration of singular models weighted with the corresponding probability distribution. Gibbs distribution is obtained from the maximum entropy principle [35] or, alternatively, by imposing the equivalence of the minimum error and the maximum-likelihood criteria for training the network [34]. Learning is defined as a maximization of the Kullback–Leibler entropy of the network distribution in parameter space, and it reduces the ensemble volume where the initial volume was fixed by the *a priori* distribution [34]. A principle similar to the principle of minimum predictive description length [36, 37] is derived in this framework by applying the maximum-likelihood approach to the problem of explaining the data by the ensemble of neural models.

This section establishes a duality between the unsupervised factorial learning and the supervised learning based on the maximum-likelihood principle and derives a common ensemble theory.

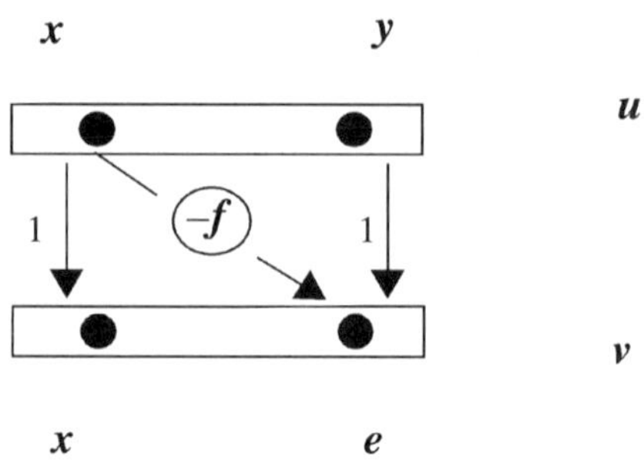

Figure 3 Unsupervised architecture for supervised learning.

Let us consider an input vector $\mathbf{u}$ for the triangular architecture of Fig. 3 as composed of two components $\mathbf{x}$ and $\mathbf{y}$ of dimensions n and m, respectively, that is, $\mathbf{u} = \{\mathbf{x}, \mathbf{y}\}$. These two vectors are related through a probability distribution $\rho(\mathbf{x}, \mathbf{y})$, and they can be regarded as the input and output of a map to be learned by supervised learning. The input vector $\mathbf{u}$ is given empirically by the set of the training data. Let us denote the output of the triangular architecture by $\mathbf{v}$ which is also composed of two vectors such that $\mathbf{v} = \{\mathbf{x}, \mathbf{e}\}$. The network output component $\mathbf{e}$ is defined by $\mathbf{e} = \mathbf{y} - \mathbf{f}(\mathbf{x}, \mathbf{w})$, where for supervised learning $\mathbf{e}$ is the error and $\mathbf{w}$ is the parameter vector that describes the input–output model $\mathbf{f}$.

The maximum-likelihood principle for supervised learning requires $\mathbf{w}$ to be chosen so that the empirical likelihood

$$L_e = \frac{1}{P} \sum_{i=1}^{P} \ln\left(p\left(\mathbf{y}^{(i)} \middle| \mathbf{x}^{(i)}, \mathbf{w}\right)\right) \tag{50}$$

is maximal. In Eq. (50), the conditional probability $p(\mathbf{y}^{(i)}|\mathbf{x}^{(i)}, \mathbf{w})$ should be regarded as a measure of the compatibility of the pairs $(\mathbf{x}^{(i)}, \mathbf{y}^{(i)})$. On the other hand, the goal of unsupervised learning is redundancy minimization. The architecture of Fig. 3 can only minimize the redundancy inherent in the relation between the vectors $\mathbf{x}$ and $\mathbf{y}$; that is, it aims to extract the correlations between these vectors which is the goal of supervised learning. In fact, unsupervised learning minimizes the redundancy at the output components given by

$$R = \sum_{j=1}^{n} H(x_j) + \sum_{j=1}^{m} H(e_j) - H(\mathbf{v}). \tag{51}$$

Using the fact that the entropy is conserved, that is,

$$H(\mathbf{v}) = H(\mathbf{u}) = \text{constant}, \tag{52}$$

and assuming that the distribution of **x** is stationary, that is,

$$\sum_{j=1}^{n} H(x_j) = \text{constant}, \tag{53}$$

minimization of redundancy is reduced to minimization of the term $\sum_{j=1}^{M} H(e_j)$. Owing to the fact that

$$H(\mathbf{e}) \leq \sum_{j=1}^{M} H(e_j) \tag{54}$$

and taking into account that

$$H(\mathbf{e}) = -L \tag{55}$$

because of

$$p(\mathbf{y}|\mathbf{x}, \mathbf{w}) \equiv p\big(\mathbf{y} - \mathbf{f}(\mathbf{x}) = 0|\mathbf{x}, \mathbf{w}\big) \equiv p\big(\mathbf{e} = 0|\mathbf{x}, \mathbf{w}\big) \equiv p(\mathbf{e}|\mathbf{x}, \mathbf{w}), \tag{56}$$

minimization of redundancy of the output components **v** is equivalent to maximization of the likelihood L. In other words, maximum-likelihood supervised learning is equivalent to minimizing the entropy of the error which is the goal of unsupervised reduction of redundancy. This is the dual formulation of maximum-likelihood-based supervised learning and factorial unsupervised learning.

The ensemble theory of unsupervised learning was derived in Deco and Schürmann [14]. In this framework, we can write the prediction probability of a new point as

$$\begin{aligned} P\big(\mathbf{y}|\mathbf{x}, D^{(P)}\big) &= p\big(\mathbf{e}|\mathbf{x}, D^{(P)}\big) \\ &= \int d\mathbf{w} \frac{\exp\big(\beta \sum_{i=1}^{P} \ln\big(p(\mathbf{e}^{(i)}|\mathbf{x}^{(i)}, \mathbf{w})\big)\big)}{Z(P)} p\big(\mathbf{e}|\mathbf{x}, \mathbf{w}, \beta\big), \end{aligned} \tag{57}$$

which is the maximum entropy principle applied to an ensemble of networks constrained by the fact that they should minimize the redundancy. Furthermore, assuming for each network the escort distribution

$$p\big(\mathbf{e}|\mathbf{x}, \mathbf{w}, \beta\big) = \frac{\exp(\beta \ln(p(\mathbf{e}|\mathbf{x}, \mathbf{w})))}{z}, \tag{58}$$

it is possible to obtain the following distribution of **y**:

$$\begin{aligned} &p\big(\mathbf{y}|\mathbf{x}\, D^{(P)}\big) \\ &\quad \approx \left(\frac{1}{z} \exp\left(\beta H' + \frac{1}{2} g'^{T} \left(F'^{(P+1)}\right)^{-T} g'\right)\right) \det{}^{1/2}\left(F'^{(P)} \left(F'^{(P+1)}\right)^{-1}\right) \end{aligned} \tag{59}$$

and, therefore,

$$p(\mathbf{y}|\mathbf{x}, D^{(P)}) \leq p(\mathbf{y}|\mathbf{x}, \mathbf{w}_P, \beta)\left(\frac{\det(F'^{(P)})}{\det(F'^{(P+1)})}\right)^{1/2}. \tag{60}$$

In the last two equations,

$$F'^{(P)} = -\sum_{i=1}^{P} \nabla\nabla \ln\left(p(\mathbf{e}^{(i)}|\mathbf{x}^{(i)}, \mathbf{w}_P)\right) \tag{61}$$

$$H' = \ln\left(p(\mathbf{e}|\mathbf{x}, \mathbf{w}_P)\right) \tag{62}$$

and

$$\mathbf{g}' = \nabla H'|_{\mathbf{w}_P}. \tag{63}$$

Hence, an *upper bound* for the prediction probability of a new data point given P training samples is obtained. The upper bound is essentially determined by the square root of the ratio between the determinants of the Fisher information matrices with and without the new point. The factor $p(\mathbf{y}|\mathbf{x}, \mathbf{w}_P, \beta)$ is the probability of observing the new pair given by the network with parameter vector $\mathbf{w}_P$. The negative logarithm of the second term on the right-hand side of Eq. (60) is the novelty measure $NW(P)$, that is,

$$NW(P) = -\frac{1}{2}\ln\left(\frac{\det(F'^{(P)})}{\det(F'^{(P+1)})}\right). \tag{64}$$

Hence, the statistical theory of unsupervised factorial learning with a volume-conserving network has been extended to the ensemble theory of supervised learning based on the maximum-likelihood approach.

III. STATISTICAL STRUCTURE EXTRACTION: NONPARAMETRIC FORMULATION

During this decade interest in the inverse problem of nonlinear systems has increased remarkably [38–42]. In the case of time series, this problem consists of determining the underlying dynamics of the process when only the measured data are given. In particular, knowledge of the kind of dynamics that generate the data opens the possibility of answering essential questions such as theoretical predictability and forecasting horizon in time series, that is, the possibility and reliability of modeling. Predictability can also be used as a mechanism for the detection of useful data for training parametric models such as neural networks or also for deciding when a prediction with the model is feasible. Especially in economic applications, the predictability can be used as a tool for developing investment strategies. A challenging related task is to detect the presence of chaos in real data, but, as was noticed by Theiler *et al.* [40], this problem is cumbersome because the erratic fluctuations that are observed in real data owe their variation to many influences such as chaos, nonchaotic but nonlinear determinism, linear correlations, and noise, both in the dynamics and in the measurement process. Therefore, a task that has to be solved before the final goal of extracting the dynamics of the system can be achieved is the reliable detection of statistical dependence (nonlinear and non-Gaussian in general) in the data. In the context of time series, the statistical dependence between the past points and the considered present point therefore yields a measure of the predictability of the corresponding system.

Several excellent works [41–43] concentrated especially on the previously mentioned problem of detection of nonlinear correlations in time series. In the paper of Palus [42], an information-theoretic-based measure of statistical dependence called redundancy was successfully used in combination with the surrogate method of Theiler *et al.* [40]. The same kind of information-theoretic-based redundancy in data has been formulated in the previous section in the context of neural networks for a parametric formulation of statistical structure extraction. A drawback of the redundancy measure of statistical dependence is that it is an entropy-based measure, and therefore usually box-counting algorithms should be used for the estimation of probability distributions involved in its calculation. In the case where the dimensionality involved is high, for example, when the redundancy between several points in the past and the present should be measured, very unreliable results can be obtained, owing to the difficulty of density estimations by histograms in high dimensions, especially when the number of data is small [44].

The aim of the present section is to formulate a nonparametric cumulant-based statistical approach for detecting linear and nonlinear statistical dependences in time series. The statistical dependence is detected by measuring the discriminating significance, which tests the null hypothesis of statistical independence, expressed in Fourier space, by the surrogate method. The surrogate data herein used correspond to the null hypothesis of a noncausal relationship between the past and present, that is, of a random process with statistical properties equivalent to the original time series. The formulation of statistical independence in Fourier space leads automatically and consistently to a cumulant-based measure. In this formulation, linear and nonlinear as well as Gaussian and non-Gaussian temporal dependences can be detected in time series. Contrary to box-counting-based methods for the estimation of temporal redundancy, this cumulant-based predictability offers a method for the estimation of temporal redundancy which is reliable even in the case of a small number of data. This fact permits us to avoid the assumption of stationarity and to use the slicing-window-based method for measuring the nonstationarity which can only be reliably employed when the estimation of predictability can be performed with a small number of data points.

A. Statistical Independence Measure

In this section, we present a measure of statistical independence based on the expansion of Fourier-transformed densities in higher-order cumulants. The measure derived here will be used in the context of time series for testing the null hypothesis of statistical independence of one present point from its past in order to measure the predictability of the system. Let us consider a time series $x(t)$ and let us define an embedding vector constructed from the observable x by $\mathbf{x}(t) = (x(t), x(t-\Delta), \ldots, x(t-(m-1)\Delta)) = (x_1, \ldots, x_m)$ of dimensionality m and time lag Δ [28]. The vector $\mathbf{x}(t)$ is a random variable distributed according to the probability distribution $P(\mathbf{x})$. A measure of the predictability of the present from the past is given by how strong the probability of the present [say the point $x(t)$] is conditioned by the past [i.e., by $(t-\Delta), \ldots, x(t-(m-1)\Delta)$]. In other words, the predictability is a measure of how different from zero is the subtraction

$$P(x_1) - P(x_1|x_2, \ldots, x_m). \tag{65}$$

Multiplying this quantity by $P(x_2, \ldots, x_m)$ yields

$$P(x_1)P(x_2, \ldots, x_m) - P(x_1, x_2, \ldots, x_m). \tag{66}$$

We see that the predictability is nothing else than a measure of the statistical independence between the present and the past. If this quantity is zero, then the present is independent of the past and therefore unpredictable from the past; that is, there is no statistical structure underlying the time series data, meaning that the data are just uncorrelated noise and therefore defined by a Bernoulli process. Because of this, we will measure the predictability by testing the null hypothesis H of statistical independence, that is,

$$H \equiv \{P(x_1, x_2, \ldots, x_m) = P(x_1)P(x_2, \ldots, x_m)\}. \tag{67}$$

To estimate from the data the null hypothesis H, we express H in terms of empirically measurable entities such as higher-order cumulants, which take into account the effect of non-Gaussianity and nonlinearity underlying the data. Therefore, let us define the Fourier transforms of the joint and marginal probability distributions by the following equations:

$$\phi(w_1, w_2, \ldots, w_m) = \int dx_1\, dx_2 \cdots dx_m \times \exp\left(i\left(\sum_{j=1}^{m} x_j w_j\right)\right) P(x_1, x_2, \ldots, x_m), \quad (68)$$

$$\varphi(w_1) = \int dx_1 \exp\big(i(w_1 \cdot x_1)\big) P(x_1), \quad (69)$$

$$\lambda(w_2, \ldots, w_m) = \int dx_2 \cdots dx_m \times \exp\left(i\left(\sum_{j=2}^{m} x_j w_j\right)\right) P(x_2, \ldots, x_m), \quad (70)$$

where $i = \sqrt{-1}$. The cumulant expansions are [22]

$$\phi(w_1, w_2, \ldots, w_n) = \exp\left(\sum_{n=1}^{\infty} \frac{i^n}{n!} \sum_{i_1,\ldots,i_n=1}^{\infty} K_{i_1,\ldots,i_n} w_{i_1} \cdots w_{i_n}\right), \quad (71)$$

$$\varphi(w_1) = \exp\left(\sum_{n=1}^{\infty} \frac{i^n}{n!} K_1^{(n)} w_1^n\right), \quad (72)$$

$$\lambda(w_2, \ldots, w_n) = \exp\left(\sum_{n=1}^{\infty} \frac{i^n}{n!} \sum_{i_1,\ldots,i_n=2}^{m} K_{i_1,\ldots,i_n} w_{i_1} \cdots w_{i_n}\right). \quad (73)$$

It is clear from the definition that $K_1^{(n)} = K_{i_1,\ldots,i_n}$ when $i_1 = \cdots = i_n = 1$. In Fourier space, the independence condition of Eq. (67) is given by Papoulis [21]

$$\phi(w_1, w_2, \ldots, w_m) = \varphi(w_1)\lambda(w_2, \ldots, w_m), \quad (74)$$

which is equivalent to

$$\ln\big(\phi(w_1, w_2, \ldots, w_m)\big) = \ln\big(\varphi(w_1)\big) + \ln\big(\lambda(w_2, \ldots, w_m)\big). \quad (75)$$

Substituting Eqs. (71)–(73) into Eq. (75), we find that in the case of independence the following equality should be satisfied:

$$\sum_{n=1}^{\infty} \frac{i^n}{n!} \sum_{i_1,\ldots,i_n=1}^{m} K_{i_1,\ldots,i_n} w_{i_1} \cdots w_{i_n} = \sum_{n=1}^{\infty} \frac{i^n}{n!} K_1^{(n)} w_1^n + \sum_{n=1}^{\infty} \frac{i^n}{n!} \sum_{i_1,\ldots,i_n=2}^{m} K_{i_1,\ldots,i_n} w_{i_1} \cdots w_{i_n} \quad (76)$$

and, therefore,

$$\sum_{n=1}^{\infty} \frac{i^n}{n!} \sum_{i_2,\ldots,i_n=1}^{m} (1 - \delta_{1i_2,\ldots,i_n}) K_{1i_2,\ldots,i_n} w_1 w_{i_2} \cdots w_{i_n} = 0. \tag{77}$$

The Kronecker delta $\delta_{1i_2,\ldots,i_n}$ is defined by

$$\delta_{1i_2,\ldots,i_n} = \begin{cases} 1, & \text{if } i_2 = 1 \wedge i_3 = 1 \wedge \cdots \wedge i_n = 1, \\ 0, & \text{otherwise.} \end{cases} \tag{78}$$

Owing to the fact that Eq. (74) should be satisfied for all $\mathbf{w}$, all coefficients in each summation of Eq. (77) must be zero. This means that the nondiagonal elements of all higher-order n-dimensional cumulants whose first index is 1 should be zero, that is,

$$K_{1i_2,\ldots,i_n} = 0 \qquad \text{if } 1 \neq i_2 \text{ or } 1 \neq i_3 \cdots \text{or } 1 \neq i_n. \tag{79}$$

Therefore, a measure that can be used for testing statistical independence in time series can be defined by the cost function

$$D = \sum_{n=1}^{\infty} \sum_{i_2,\ldots,i_n=1}^{m} (1 - \delta_{1i_2,\ldots,i_n}) \{K_{1i_2,\ldots,i_n}\}^2 \tag{80}$$

based on Eq. (77). The quantity D is such that $D \geq 0$. It measures the degree of statistical independence, indicating independence when it is minimal, that is, zero, and increasing statistical dependence with increasing positive value. Let us note that the past until now has been considered as the last m points and therefore m should be large enough to really test predictability of the present given the "whole" past.

Let us note that the marginal redundancy $I(x_m, \ldots, x_2; x_1)$ defined by the mutual information between the past and the present is minimal, that is, zero, if condition (77) is satisfied for all n. The redundancy measures the statistical independence (also nonlinear and non-Gaussian) and therefore it is consistent with our formulation. Furthermore, $I(x_m, \ldots, x_2; x_1)$ is the measure of the condition given by Eq. (66) in the form of a Kullback–Leibler entropy [19].

In the special case where only second-order statistics are considered, that is, only the expansion up to the second-order cumulant is employed, the trivial result that the autocorrelation coefficient measures the statistical independence in the linear and Gaussian case is obtained; that is, if

$$C_{1j} = 0 \qquad \text{if } 1 \neq j, \tag{81}$$

the present is linearly decorrelated with the past if the signal is Gaussian. It is relatively easy to show that, in the linear Gaussian case, Eq. (81) ensures that the redundancy I is minimal and equal to zero.

In conclusion, we have developed in this section a tool for measuring statistical correlations between the past and the future in time series. To give a quantitative meaning to this measure, especially when handling smaller and noisy data sets, we introduce in the next section the method of surrogates [40], which aims to detect predictability by a significance test of a null hypothesis corresponding to the assumption of statistical independence in time.

B. Statistical Test

1. Surrogate Method

In the framework of statistical testing, a hypothesis about the data can be rejected or accepted based on a measure that discriminates the statistics underlying a distribution consistent with the hypothesis from the original distribution of the data (see Breiman [45]). The hypothesis to be tested is called the *null hypothesis*. The measure that quantifies the aforementioned distinction between the statistics of the distributions is called the *discriminating statistics*. In general, it is very difficult to derive analytically the distribution of a given statistic under a given null hypothesis. The surrogate method proposes to estimate such distributions empirically by generating different versions of the data, called *surrogates*, such that they share some given properties of the original data and are, at the same time, consistent with the null hypothesis. The rejection of a null hypothesis is therefore based on the computation of the discriminating statistics for the original time series D_O and the discriminating statistics D_{S_i} for the ith surrogated time series generated under the null hypothesis.

If multiple realizations of the original time series are possible, then a Kolmogorov–Smirnov test is recommended [45]. In our case, we assume that only one realization is possible (which is a realistic assumption especially in the case of real-world examples) and therefore a one-realization test is adopted. The null hypothesis is rejected if the *significance* given by

$$S = \frac{|D_O - \mu_S|}{\sigma_S} \tag{82}$$

is greater than $\widehat{S}$ corresponding to a p value $p = \operatorname{erfc}(\widehat{S}/(\sqrt{2}))$ (p usually being 0.05 which corresponds to $\widehat{S} = 1.645$). This means that the probability of observing a significance $\widehat{S}$ or larger is p if the null hypothesis is true under the assumption of Gaussianity in the surrogate distribution (which is a realistic assumption if the number of surrogates is equal to or larger than 100).

In Eq. (82),

$$\mu_S = \frac{1}{N}\sum_{i=1}^{N} D_{S_i} \qquad \text{and} \qquad \sigma_S = \sqrt{\frac{1}{N}\sum_{i=1}^{N}(D_{S_i} - \mu_S)^2} \tag{83}$$

are the estimated mean value and standard deviation of the discriminating statistics under the surrogate distribution.

In our case, the concrete null hypothesis that we would like to test is the one corresponding to the assumption of a noncausal relationship between the past and present, that is, of statistical independence in time. In other words, we are trying to answer the most modest question, namely whether there are any dynamics at all underlying the data. The surrogate data used are generated from the original time series by mixing the temporal order, so that if originally any temporal dependence was present it is destroyed by this mixing process. The generated surrogates are scrambled surrogates and have the same statistical characteristics as the original data. The histogram of the original data is preserved but with the particularity that these data do not possess any statistical temporal correlation. The discriminating statistics used here measure the statistical correlations by means of the cumulant expansion, that is, by calculating the nondiagonal elements of the cumulants according to Eq. (80). In this section, we include cumulant terms up to fourth order.

2. Nonstationarity

If the assumption of stationarity of the observed time series cannot be accepted, the test of the existence of dynamics should be performed by the method of overlapping windows or slicing windows. In this method, the statistical test of temporal independence at time t is performed in a window of the data of length N_W, that is, by using the data from time $t - N_W\Delta$ to t. Subsequently, we slice the window into N_S time steps and perform again the statistical test by using the data from $t + N_S - N_W\Delta$ to $t + N_S$ and so on. In this case, it is useful to plot the significance S as a function of the time t where the slicing window finished. It is clear that for stationary time series the value of S will be approximately the same for all times. In other cases, this method can be used for the detection of regions in which a stationary behavior is observed or regions in which dynamics are really underlying the process. We present some interesting tests and applications on real-world examples in the next section.

3. Experimental Results

a. Testing Predictability: Artificial Time Series

In this section, we would like to demonstrate the detection of nonstationarity by using the method of slicing windows in combination with our cumulant-surrogate-based testing of statistical correlations. We use the chaotic time series

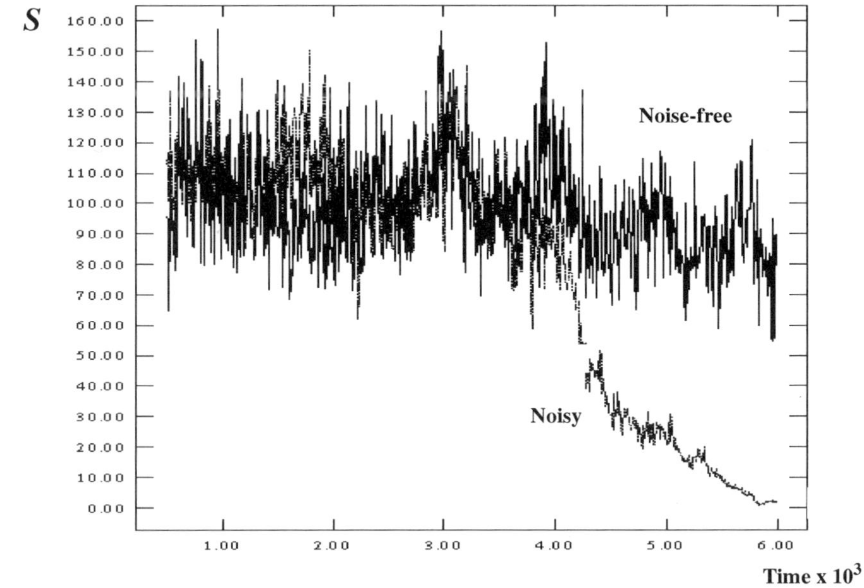

Figure 4 Significance as a function of time for the chaotic Hénon series without noise (stationary) and with additive nonstationary noise. The dimension of the embedding vector is $m = 3$ and the time lag is $\Delta = 1$. All quantities were computed for a slicing window with $N_W = 256$ and $N_S = 4$.

of Hénon [46] defined by the following iterative equation:

$$x_{t+1} = 1 - Ax_t^2 + Bx_{t-1}, \tag{84}$$

with $A = 1.4$ and $B = 0.3$. This time series contains nonlinear correlations, and it is stationary. A nonstationary variation of this time series can be implemented by adding nonstationary Gaussian noise in the following form:

$$x_{t+1} = 1 - Ax_t^2 + Bx_{t-1} + n_v v_t, \tag{85}$$

with

$$n_v = \begin{cases} 0, & \text{if } t < 1500, \\ \left(\dfrac{t - 1500}{1500}\right), & \text{if } t > 1500. \end{cases} \tag{86}$$

In this example, we calculate the significance, including cumulant terms up to fourth order. In Fig. 4, the significance for the noise-free and noisy Hénon time series is plotted as function of time for a slicing window with $N_W = 256$ and $N_S = 4$. The dimension of the embedding vector is $m = 3$ and the time lag is $\Delta = 1$. In the stationary case, an approximately horizontal evolution of the significance during time is observed indicating that the evidence of statistical correlations is strong and stationary. In the noisy case, a nonstationary evolution of the significance is observed indicating now that from $t > 1500$ on the evidence of nonlinear correlations slowly disappears until the end ($t = 3000$) where, owing to the signal-to-noise ratio, only uncorrelated Gaussian noise is observed. This clearly does not represent correlations in time, and therefore the significance is approximately zero, indicating that the null hypothesis of independence cannot be refuted.

b. Testing Predictability: Real-World Time Series

In this section, several real-world examples for the detection of temporal statistical correlations in time series are presented. In all of the examples, we calculate the significance by including terms up to fourth-order cumulants. It is important to note that we only test the existence of statistical correlations which would implicitly indicate the existence of underlying dynamics which can be stochastic, deterministic chaotic, or nonchaotic. We do not test the existence of chaos.

Sunspot Data. In this section, we test the existence of temporal statistical correlations in the yearly and monthly sunspot time series. The yearly sunspot time series has been extensively studied in the literature [40]. The monthly sunspots were recollected by the Sunspot Index Data Centre (Brussels).

In Fig. 5, the significance as a function of time for the yearly sunspot time series is shown. The dimension of the embedding vector is $m = 13$ and the time lag is $\Delta = 1$. The slicing window used had $N_W = 200$ and $N_S = 1$. A stationary and clear evidence of statistical correlations is observed which supports the usual assumption that a stable cycle of 12 years really exists.

Figure 6 shows the plot of the significance as a function of time for the monthly sunspot time series. The dimension of the embedding vector in this case is $m = 5$ and the time lag is $\Delta = 1$. All quantities were computed for a slicing window with $N_W = 300$ and $N_S = 4$. The evolution of the significance supports the existence of statistical correlations between past and future which are probably of a nonstationary nature, owing to the strong changes in the significance magnitude during the time. This would mean that, on a short-term scale, the dynamics involved are nonstationary.

Financial Data. In this section, we deal with the tick German Stock Index (DAX). The problem was to predict the 60-minute volatility of the DAX. Volatility estimates have gained major attention in financial analysis, for example, as a basis for directional market forecasts or to evaluate and hedge derivative market instruments. Intradaily stock market data are relatively difficult to predict because the

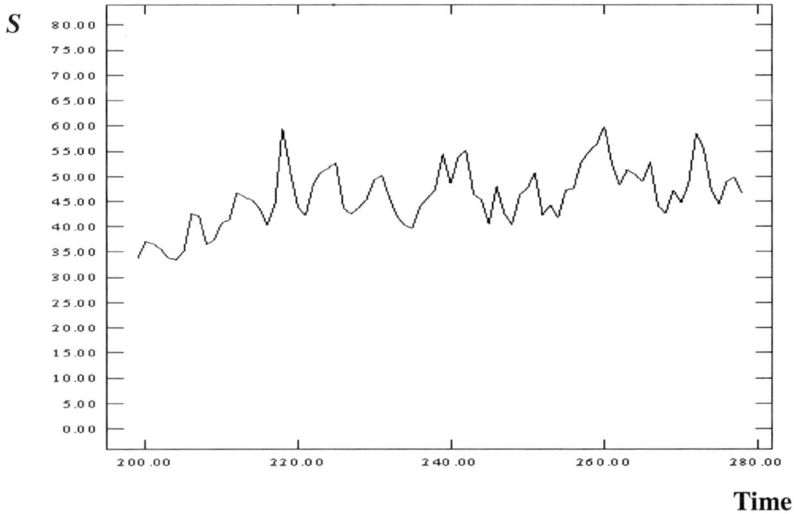

Figure 5 Significance as a function of time for the yearly sunspot time series. The dimension of the embedding vector is $m = 13$ and the time lag is $\Delta = 1$. All quantities were computed for a slicing window with $N_W = 200$ and $N_S = 1$.

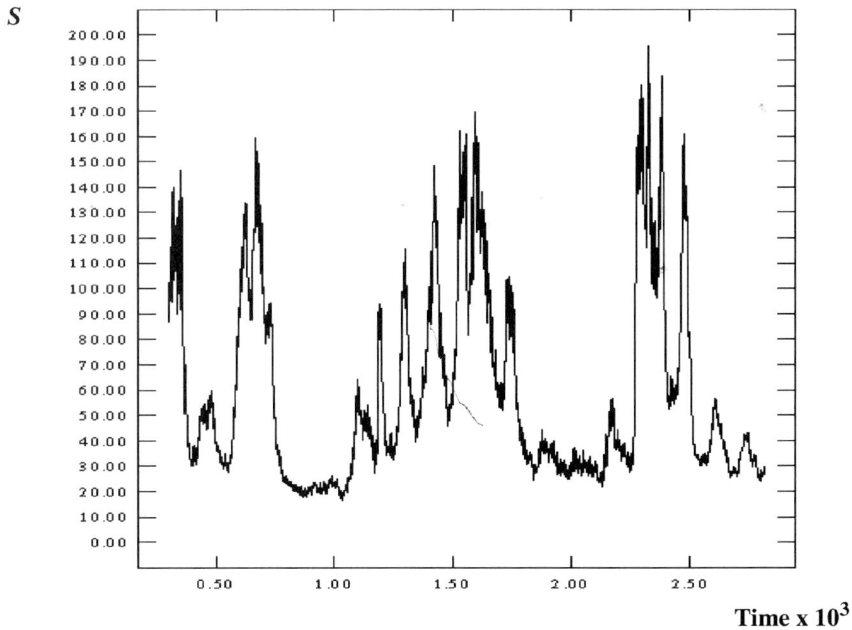

Figure 6 Significance as a function of time for the monthly sunspot time series. The dimension of the embedding vector is $m = 5$ and the time lag is $\Delta = 1$. All quantities were computed for a slicing window with $N_W = 300$ and $N_S = 4$.

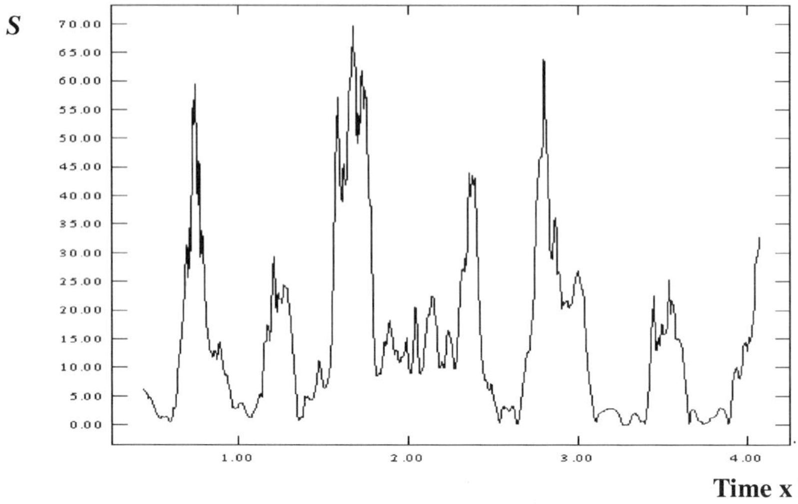

Figure 7 Significance as a function of time for the German Stock Index (DAX) time series. The dimension of the embedding vector is $m = 5$ and the time lag is $\Delta = 60$. All quantities were computed for a slicing window with $V_W = 440$ and $N_S = 8$.

of large amount of random noise. The "speed" of the market changes frequently, and observations are disturbed by large moves caused by information shocks or discontinuities in time owing to limited trading periods.

The basic time series is the value of the stock index DAX observed per minute in November 1994. The volatility is computed by averaging the last 60 absolute differences in the index. The results are plotted in Fig. 7. In this figure, a strong nonstationary evolution of the significance as a function of time is observed. Remarkable is the existence of a strong predictability over the entire time. The dimension of the embedding vector is $m = 5$ and the time lag is $\Delta = 60$. The DAX significances were computed for a slicing window with $N_W = 440$ and $N_S = 8$.

Electroencephalogram Data. A relevant application of nonlinear time series analysis is the investigation of EEG signals. For instance, considerable progress in the therapy of patients who suffer from epileptic seizures could be made, if the existence of a strange attractor in the signal space were established as remarked in Rapp *et al.* [47]: "Small changes in parameter values that could be effected pharmacologically could produce a reverse transition to ordered, physiologically acceptable behavior." In this section, we analyze the existence of nonlinear correlations in the EEG time series measured using intracranial electrodes in the entorhinal cortex (EC) of a male Wistar rat during provoked epileptic seizures

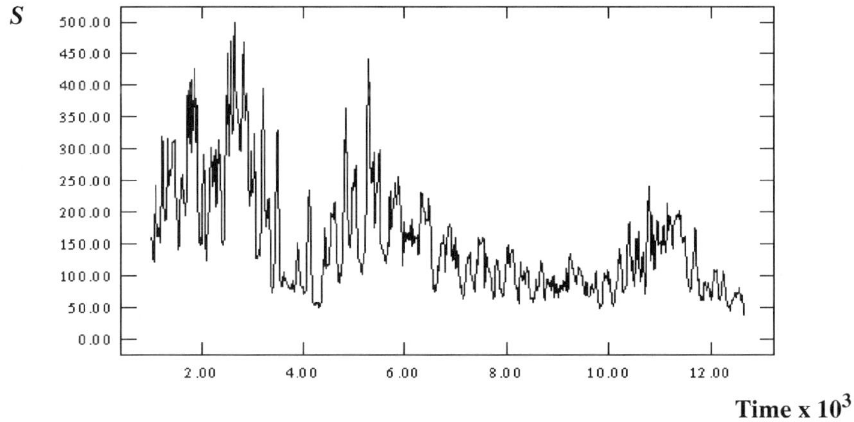

Figure 8 Significance as a function of time for the time series corresponding to the EEG signal measured using intracranial electrodes in the entorhinal cortex of a male Wistar rat during an epileptic seizure. The dimension of the embedding vector is $m = 3$ and the time lag is $\Delta = 10$. All quantities were computed for a slicing window with $N_W = 1000$ and $N_S = 20$.

triggered by kindling stimulus applied to the entorhinal cortex (see [48]). The epileptic seizure took approximately 24 seconds sampled in 12700 data points.

Figure 8 shows the significance as a function of time corresponding to an embedding vector of dimension $m = 3$ and a time lag of $\Delta = 10$. All quantities were computed for a slicing window with $N_W = 1000$ and $N_S = 20$. In the figure, it is possible to observe a clear rejection of the independence hypothesis, indicating the existence of statistical correlations in time. The evolution of the significance is nonstationary and shows a decay of correlations over time. This last fact agrees with the results of Pjin [48] who also detected an increase in the dimension of the attractor during the last 8 seconds, indicating a loss of the dynamics and therefore of the temporal correlations as in our experiment.

IV. NONPARAMETRIC CHARACTERIZATION OF DYNAMICS: THE INFORMATION FLOW CONCEPT

Over the past decade, there has been an increasing interest in a statistical- and information-theoretic-based approach to chaotic dynamical systems [49–56]. In spite of the deterministic character of the equations that generate chaotic dynamics, there appears to be a stochastic aspect in such systems because of the essential problem of extreme sensitivity to the initial conditions. In fact, if one has no exact knowledge about the initial state of a chaotic dynamical system, the prediction of states in the future can be done only in a probabilistic sense. The exponential divergence in time of two neighboring trajectories in a strange attractor is usually quantitatively described by a positive Lyapunov exponent [50]. This stochastic property of chaotic systems can be globally specified by the concept of Kolmogorov–Sinai (KS) entropy [57, 58]. This concept is an information-theoretic-based measure of the rate of information production in dynamical systems and therefore characterizes the degree of randomness in one-step predictions based on the whole past of a chaotic, dynamical system. In some typical chaotic systems, the sum of the positive Lyapunov exponents yields the KS entropy [59]. In general, this sum is an upper bound of the KS entropy [59]. The KS entropy is usually defined in the context of symbolic dynamics [56]. The symbolic dynamics can be defined by a partition of the phase space owing to the uncertainty in the measurement device [51, 54]. If the state of a dynamical system is measured with finite precision, the KS entropy yields a measure of the uncertainty of a future state provided the whole past is known. Some analytical results of the calculation and of the behavior of the KS entropy in one-dimensional chaotic maps can be found in Szépfalusy [55].

On the other hand, several other statistics-based measures were introduced in order to characterize dynamical chaos, for example, the cumulants of the information loss in one-step predictions [60], Rényi entropies [55], correlation integrals [61], and the generalized mutual information [53]. A decisively important and essential generalization of the KS entropy is a measure of the information loss or of statistical correlations for predictions p steps into the future. Especially in the works of Pompe [53, 54] and Matsumoto and Tsuda [62–64], information-theoretic-based measures are formulated for the study of the information flow p steps into the future, but only the dependence of the prediction point on the present point and not on the entire past is considered therein.

The aim of this section is to formulate a new entropy measure of the statistical correlations between the entire past of a dynamical system and a state p steps into the future. This measure, which is based on the concept of mutual information, characterizes the information loss through time for chaotic systems, providing us with a quantitative measure of the prediction horizon and the evolution of the uncertainty in p-step predictions. In the context of symbolic dynamics generated by a chaotic attractor, the introduced entropy measure yields a consistent instrument for the study of the evolution of the information loss during time.

The information loss provides us with a mean measure of sensitivity on the initial conditions, which is the main characteristic of chaotic, dynamical systems. In order to characterize the dynamics on a fine scale and to describe the true and intrinsic information flow of chaotic systems, infinitesimal partitions were used. In the framework of the thermodynamical formulation, a new characterization of information flow is formulated based on the behavior of the conditional mutual information between the points a single and p steps ahead in the future, given the entire past. The zeta-function formalism is applied. The result obtained offers a rigorous definition of chaotic systems in the language of information theory. A dynamical system is chaotic if the generalized Kolmogorov–Sinai entropy for a point p steps ahead in the future increases linearly with p, which is equivalent to say that the aforementioned conditional mutual information is always constant and equal to the Kolmogorov–Sinai entropy.

A. Information Flow for Finite Partitions

This section deals with sequences consisting of a finite number of symbols chosen from a finite alphabet $\aleph$. A sequence of n symbols will be referred to as a block of length n. If $\aleph$ consists of m different symbols, there are $k_n = m^n$ different blocks of length n. A measure of uncertainty of blocks of length n is given by the *block entropy* H_n defined by

$$H_n = -\sum_{i=1}^{k_n} p_i \log(p_i). \tag{87}$$

Here, p_i is the probability of finding a block of type i if a block of length n is randomly chosen. The sum extends over all types of blocks of length n. Another relevant quantity is the *entropy per step* h_n given by

$$h_n = H_{n+1} - H_n, \tag{88}$$

which is a measure of the uncertainty or of the predictability of an added symbol given the previous n symbols.

A measure of the predictability of an added symbol independent of the block length is defined by the *source entropy* h_s (also called the entropy rate) given by

$$h_s = \lim_{n\to\infty} H^{(n)}, \tag{89}$$

where $H^{(n)} = H_n/n$ is the uncertainty per symbol of a block of length n. It is easy to demonstrate [19] that

$$h_s = \lim_{n\to\infty} h_n. \tag{90}$$

Using the chain rule for the entropy [19], we obtain

$$h_s = \lim_{n\to\infty} H_{n|n-1,\ldots,1}, \tag{91}$$

with $H_{n|n-1,\ldots,1}$ being the conditional entropy of the nth symbol given the previous $n - 1$ symbols. This can be written as

$$H_{n|n-1,\ldots,1} = -\sum_{n=1}^{k_{n-1}} \sum_{j=1}^{m} p_{j,i} \log p_{j/i}, \tag{92}$$

where $P_{j,i}$ and $p_{j/i}$ are the joint and conditional probabilities, respectively, of observing the symbol j after the observation of the block of type i of a randomly chosen block of length $n-1$. The source entropy is a measure of the uncertainty of an added symbol given the whole past and therefore is an absolute measure of the predictability of the next symbol. For Markov processes of order p, Shannon [17] proved that the source entropy reaches its limit for $n = p$.

Now the same concepts can be used in the ergodic theory of chaotic systems [49, 50]. To do so, it is necessary to apply the concept of symbolic dynamics. Let us define a time series $\{x_t\}$, $t = 1, 2, \ldots,$ and a partition β of an attractor A which is a set of disjoint boxes B_i, that is,

$$\beta = \{B_i\}_{i=1}^{m}, \qquad \bigcup_{i=1}^{m} B_i = A, \qquad \text{and} \qquad B_i \cap B_j = \emptyset, \qquad i \neq j. \tag{93}$$

By interpreting each box B_j as a symbol i, we transform the original, real-valued time series $\{x_t\}$, $t = 1, 2, \ldots,$ into a sequence $\{i_t\}$, $t = 1, 2, \ldots,$ of m different symbols and thereby define the symbolic dynamics. Let $p_1^{i,\beta}$ denote the probability of observing symbol i of the partition β. Then the entropy of the symbolic sequence is defined by

$$H^{\beta} = H^{\beta}(1) = -\sum_{i=1}^{m} p_1^{i,\beta} \log p_1^{i,\beta}. \tag{94}$$

The block entropy for the symbolic dynamics is now given by

$$H^{\beta}(n) = -\sum_{i=1}^{k_n} p_n^{i,\beta} \log p_n^{i,\beta}, \tag{95}$$

where $p_n^{i,\beta}$ is the probability of finding a block of type i if a block of length n is randomly selected from the symbolic sequence. Let us now introduce the joint entropy

$$H^{\beta}(n, p) = -\sum_{i=1}^{k_n} \sum_{j=1}^{m} p_{n,p}^{i,j,\beta} \log p_{n,p}^{i,j,\beta}, \tag{96}$$

which is the entropy of the set of patterns which are the concatenation of subsequent symbols and the symbol which is p steps ahead. In Eq. (96), $p_{n,p}^{i,j,\beta}$ is the probability of observing a block of type i randomly chosen from blocks of length n and the symbol j which is p steps ahead with respect to the originally chosen block. Evidently, $H^{\beta}(n, 1) = H^{\beta}(n+1)$. The source entropy in this case is given by

$$\begin{aligned} h^{\beta} &= \lim_{n\to\infty} \frac{H^{\beta}(n)}{n} = \lim_{n\to\infty} \left(h^{\beta}(n+1) - H^{\beta}(n)\right) \\ &= \lim_{n\to\infty} \left(H^{\beta}(n, 1) - H^{\beta}(n)\right) = \lim_{n\to\infty} \left(H^{\beta}((n+1)|n, \ldots, 1)\right). \end{aligned} \tag{97}$$

In the last equation, $H^{\beta}((n+1)|n, \ldots, 1)$ is the conditional entropy of the one-step prediction given previous symbols. In this context, it is now possible to introduce the *Kolmogorov–Sinai* (KS) or *metric entropy* [57, 58] of the underlying dynamical system as the supremum over all possible partitions, that is,

$$h = \sup_{\beta} h^{\beta}. \tag{98}$$

Each partition for which the supremum is reached is called a *generating* partition.

In this section, we are interested in studying the flow of information, that is, in measuring the predictability not only one step ahead (KS entropy) but eventually arbitrary p steps into the future. Therefore, a natural extension of the concept of KS entropy is given by

$$I_p^{\beta} = \lim_{n\to\infty} I^{\beta}(n, p), \tag{99}$$

where $I^{\beta}(n, p) = H^{\beta} + H^{\beta}(n) - H^{\beta}(n, p) = H^{\beta} - H^{\beta}((n+p)|n, \ldots, 1)$ is the mutual information between a word of n subsequent symbols and the symbol which is p steps ahead. The mutual information is a measure of the statistical correlation between two random variables. In this case, we are therefore measuring the statistical correlation between the symbol p steps ahead and the whole past $n \to \infty$. It is easy to see that

$$0 \leq I^{\beta}(n, p) \leq H^{\beta}, \tag{100}$$

where the minimal value (0) corresponds to statistical independence and the maximal value to perfect predictability. For chaotic time series, we expect $I^{\beta}(n, p) > I^{\beta}(n, p+1)$ which expresses the loss of information in the prediction horizon. For the special case of $p = 1$ and if β_g is a generating partition, we get

$$\begin{aligned} I_1^{\beta_g} &= \left(H^{\beta_g} - \lim_{n\to\infty} \left(H^{\beta_g}(n+1) - H^{\beta_g}(n)\right)\right) \\ &= H^{\beta_g} - h, \end{aligned} \tag{101}$$

where H^{β_g} is the entropy corresponding to the symbolic dynamics associated with the generating partition β_g. We see that for the case of one-step predictions we recover the concept of KS entropy as the amount of lost information. As a result, information flow is an extension of the concept of KS entropy to predictions p steps into the future. It is important to notice the differences between this definition of information loss and the one given by other authors. Pompe [53, 54] and Matsumoto and Tsuda [62–64] formulate mutual-information-based measures for the study of the information flow p steps into the future, but they consider only the dependence of the prediction point on the present point and not on the entire past; that is, they consider basically $I^{\beta}(1, p)$ instead of our definition in Eq. (99).

B. Intrinsic Information Flow (Infinitesimal Partition)

In the case of infinitesimal partitions, it is not possible to study the information flow by the mutual information $I_p^{\beta} = \lim_{n\to\infty} I^{\beta}(n, p)$ owing to the fact that the partition β which is used in this case corresponds to a partition with infinitesimal diameter. Therefore, the first term in the definition of the information flow, that is, H^{β}, is always infinite yielding an infinite (and, consequently, meaningless) I_p^{β} for all p. These infinitesimal partitions are essential for the study of the effect of sensitivity on the initial condition in the present information-theoretic framework. The relevant term for the evolution of statistical dependences between future and past is therefore the second term of the mutual information. We denote this term h_p^{β} and call it the generalized conditional entropy for a point p steps into the future, that is,

$$h_p^{\beta} = \lim_{n\to\infty} H^{\beta}(n + p|n, \ldots, 1). \tag{102}$$

The *generalized Kolmogorov–Sinai entropy* for a point p steps ahead is defined by taking the supremum over all possible partitions, that is,

$$h_p = \sup_{\beta} h_p^{\beta}, \tag{103}$$

which yields a partition-independent result and therefore characterizes the intrinsic information flow. The supremum can be replaced by the limit [56]

$$h_p = \lim_{\varepsilon\to 0} h_p^{\beta}, \tag{104}$$

where $\varepsilon = \max(\text{diameter}(B_i))$ corresponding to the partition β. The evolution of the uncertainty in the case of infinitesimal partitions, that is, the information-theoretic measure for the sensitivity to initial conditions, is therefore described by h_p.

The increase of h_p as a function of p can be studied by analyzing the evolution of the *information flow*

$$I_p \lim_{\varepsilon \to 0} \lim_{n \to \infty} I^{\beta}(n+1; n+p|n, \ldots, 1) \tag{105}$$

as a function of p. Here, the conditional mutual information $I^{\beta}(n+1; n+p|n, \ldots, 1)$ is given by

$$\begin{aligned} I^{\beta}(n+1; n+p|n, \ldots, 1) = {} & H^{\beta}(n+p|n, \ldots, 1) \\ & - H^{\beta}(n+p|n+1, \ldots, 1). \end{aligned} \tag{106}$$

I_p is the supremum of the asymptotic value of the mutual information between points and p ahead given n previous symbols. In fact,

$$h_p = \sum_{j=1}^{p} I_j, \tag{107}$$

which can be easily derived using following relations:

$$\begin{aligned} H^{\beta}(n+p|n, \ldots, 1) = {} & \sum_{j=2}^{p} \big(H^{\beta}(n+j|n, \ldots, 1) - H^{\beta}(n+j-1|n, \ldots, 1)\big) \\ & + H^{\beta}(n+1|n, \ldots, 1), \end{aligned} \tag{108}$$

$$\begin{aligned} & \lim_{n \to \infty} I^{\beta}(n+1; n+p|n, \ldots, 1) \\ & \quad = \lim_{n \to \infty} \big(H^{\beta}(n+p|n, \ldots, 1) - H^{\beta}(n+p|n+1, \ldots, 1)\big) \\ & \quad = \lim_{n \to \infty} \big(H^{\beta}(n+p|n, \ldots, 1) - H^{\beta}(n+p-1|n, \ldots, 1)\big). \end{aligned} \tag{109}$$

Using the fact that conditioning reduces the entropy leads to $h_p \geq h_{p-1}$ and therefore $h_p \geq h$. On the other hand, using

$$\begin{aligned} & I^{\beta}(n+1, \ldots, n+p-1; n+p|n, \ldots, 1) \\ & \quad = H^{\beta}(n+p|n, \ldots, 1) - H^{\beta}(n+p|(n+p-1, \ldots, 1)) \end{aligned} \tag{110}$$

yields

$$h_p = ph - \lim_{\varepsilon \to 0} \lim_{n \to \infty} H^{\beta}(n+1, \ldots, n+p-1|n+p, n, \ldots, 1) \tag{111}$$

and because of the positivity of the entropy $h_p \leq ph$. To summarize, it was proven that

$$h \leq h_p \leq ph \tag{112}$$

holds.

It immediately follows that

$$I_p \leq h \tag{113}$$

holds. Using the zeta-function formalism, we will prove that the intrinsic information flow is given by

$$I_p = h \qquad \forall p, \tag{114}$$

which is equivalent to

$$h_p = ph \qquad \forall p. \tag{115}$$

This means that a chaotic system is characterized by an intrinsic information flow which is constant and therefore produces a linear increase as a function of p of the prediction uncertainty of a point p steps ahead given the whole past independent of the used partition.

It is important to notice that for a Markov process of order p the mutual information $\lim_{n\to\infty} I^{\beta}(n+1; n+p|n, \ldots, 1)$ decreases by increasing p, meaning that the points p steps ahead are more and more independent of the past points. In other words, the particularity of a chaotic system is that I_p is always constant and equal to the KS entropy and does not decay to zero as for Markov processes. The essential point in the characterization of the information flow is the fact that we study the supremum over all partitions (i.e., $\lim_{\varepsilon\to 0}$); only in this manner can we obtain an "intrinsic" evolution of the information flow. For partitions that do not maximize $\lim_{n\to\infty} I^{\beta}(n+p; n, \ldots, 1)$ for all p, a spurious Markov process may be observed (e.g., a tent map where the classical generating bipartition seems to have the structure of a first-order Markov process, i.e., $I_p = 0$ for $p > 1$; see the example in the preceding section). This means that a generating partition which maximizes only the KS entropy is not sufficient for a characterization of the information flow.

To summarize, the "true" information flow in chaotic systems is given by the fact that $I_p = h\ \forall p$. This is the principal result of this section. For the particular case of the tent map, this may easily be verified by refining the partitions. This results in a profile of I_p as a function of p that looks like a higher-order Markov process, that is, $I_p = 0$ for $p > p_\varepsilon$ with p_ε increasing when ε decreases (i.e., $I_p = h\ \forall p$ for $p_\varepsilon \to \infty$ and $\varepsilon \to 0$).

The proof of this will be given now for one-dimensional hyperbolic chaotic systems. Let us consider a decimated map $x_{(n+1)p} = f^{(p)}(x_{np})$. The KS entropy of this map is evidently

$$\begin{aligned} h^{(p)} &= \lim_{\varepsilon\to 0} \lim_{n\to\infty} H^{\beta}(np+p|np, np-p, \ldots, p) \\ &= \lim_{\varepsilon\to 0} \lim_{n\to\infty} H^{\beta}(n+p|n, n-p, \ldots, p) \\ &= ph. \end{aligned} \tag{116}$$

A sufficiently fine partition is also a generating partition of $f^{(p)}$. An infinite string of past symbols of the decimated sequence allows us to pin down x_{np} arbitrarily precisely, so that the other values of x_k (with $k \neq np, np - p, \ldots, p$) do not carry independent information and therefore $h_p = h^{(p)} = ph$.

An alternative proof based on the thermodynamical formalism is summarized in the following. The proof demonstrates that $I_p = h \; \forall p$ by decomposing the attractor in unstable periodic orbits. The associated zeta function (see [56]) is defined in such a way that its zeros result in I_p and it is shown that all zeros are identical and yield the same value h.

This means that I_p is equal to h for all p; that is, the process loses always, even for $p \to \infty$, information. A system can lose always information only if it has infinite information, and this is the case for deterministic chaos ($I_{n,p}^{\beta} = \infty$). The deterministic character of the dynamical equations is reflected by the fact that the process contains infinite information, and therefore it can lose always information with increasing p (i.e., $I_p = h$). In other words, the memory of the process is infinite. Moreover, the fact that $h_p = ph \; \forall p$ for chaos implies that the KS entropy h fully characterizes the information flow in the system. For a deterministic and nonchaotic process, the information is also infinite, that is, $I_{n,p}^{\beta} = \infty$, but the system never loses information ($I_p = 0 \; \forall p$ because $h = 0$).

It is important to notice that for chaos the most relevant fact is the study of the "intrinsic" information flow, that is, for the case of infinitesimal partitions, because it describes the sensitivity to the initial conditions on a fine scale and therefore characterizes the intrinsic evolution of statistical correlations.

V. CONCLUSIONS

In this chapter, an information-theoretic-based theory has been presented which, in a unifying fashion, poses the problem of structure extraction from observed data in a solid mathematical framework. Two approaches were analyzed: parametric and nonparametric.

The parametric formulation establishes a new statistical theory of neural learning which corresponds to the biologically inspired theory of redundancy minimization of Barlow. Barlow describes the process of cognition as a preprocessing of the sensorial information performed by the nervous system in order to extract the statistically relevant and independent features of the inputs without losing information. This means that the brain should statistically decorrelate the extracted information. As a learning strategy, Barlow formulated the principle of redundancy reduction. The term redundancy refers to the statistical dependence between the components involved, and therefore the principle of learning by redundancy reduction tries to find a transformation such that the transformed components are as statistically independent as possible. The basic idea is to define a constrained, general, nonlinear, parametric model (e.g., a neural network) that ensures the perfect transmission of information by performing a volume-conserving input–output transformation, and to achieve by learning a redundancy minimization between the output components, that is, to statistically decorrelate the output components by factorizing the joint output probability. This kind of transformation is called independent component analysis (ICA). Our theory formulates ICA in an information-theoretic framework and therefore generalizes previous works on ICA in two ways: our architecture performs a general nonlinear transformation without loss of information, and the statistical decorrelation is performed without assuming Gaussian distributions. Several architectures and paradigms are presented which handle the cases of nonlinear and linear ICA with Gaussian and non-Gaussian input distributions. Principal component analysis (PCA) is formulated as the very restricted, particular case of ICA when the transformation involved is linear and orthogonal and the input distribution is Gaussian.

An ensemble model of unsupervised learning defined by redundancy reduction at the output components and entropy conservation from inputs to outputs has also been derived. We have obtained an approximate expression for the probability distribution of the output components which is essentially determined by the probability distribution given by the best network and by the square root of the ratio between the determinants of the Fisher information with and without including the new point. In this framework, the problem of supervised learning has been posed as an unsupervised one. The ensemble theory derived for unsupervised learning then results in one for supervised learning by using ensemble theory based on the maximum-likelihood principle. An upper bound for the prediction probability of a new point not included in the training data is given. This upper bound is essentially determined by the ratio between the Fisher information given the training set and the one given a set which includes both the training data and the new point. This upper bound can be used as a mechanism to actively decide on the novelty of new data, and therefore it is a mechanism of query learning. Query learning aims to improve the generalization ability of a network that continuously learns by actively selecting optimal, nonredundant data, that is, data that contain new information for the model.

A very interesting perspective of this theory is its extension to the case of biologically based neural networks, that is, networks of spiking neurons. The ideas formulated in this work should be extended in order to understand what a real brain is actually doing. Information maximization through a network of spiking neurons instead of information preservation would be surely required together with the same goal of redundancy minimization. The essential problem here is the definition of the information channel and the identification of how information is coded in the spiked responses. Work of the author in this direction is now in progress.

In the nonparametric formulation, we address an *a priori* requirement for the parametric extraction of statistical structures, namely the detection of their existence and their characterization. For time series, for example, it is useful to know if the dynamics that originate the observed values are stationary or nonstationary, if the time series is deterministic or stochastic, and to be able to distinguish between white noise, colored noise, Markov processes, and chaotic and nonchaotic determinism. The detection and characterization of such dependences should therefore be previously performed in a nonparametric fashion in order to be able to model the process *a posteriori* in a parametric form. Another important application is the following one. When a parametric model of a nonstationary time series is constructed based on data, as in neural networks, it is very important to train the model with a set of data which contains the underlying structure to be discovered. The regions where only noisy behavior is observed should be ignored. Information about the predictability can be used, for example, to select regions where a temporal structure is visible in order to select data for training a neural network for prediction. The basic problem is of a statistical nature, and therefore information theory offers the ideal theoretical framework for a mathematical formulation.

In this work, we introduced a nonparametric, cumulant-based, statistical approach for detecting statistical dependences in nonstationary data. The statistical dependence is detected by measuring the predictability which tests the null hypothesis of statistical independence, expressed in Fourier space, by the surrogate method. The formulation of statistical independence in Fourier space leads automatically and consistently to a cumulant-based measure. In this formulation, linear and nonlinear as well Gaussian and non-Gaussian temporal dependences can be detected in time series. Contrary to box-counting-based methods for the estimation of temporal redundancy, this cumulant-based predictability offers a method for the estimation of temporal redundancy which is reliable even in the case of a small number of data. This fact permits us to avoid the assumption of stationarity and to use the slicing-window-based method for measuring the nonstationarity that can only be used reliably when the estimation of predictability can be done with a small number of data points. Therefore, the predictability is defined as a higher-order cumulant-based discriminating significance between the original data and a set of scrambled surrogate data which correspond to the null hypothesis of a noncausal relationship between past and present, that is, of a random process with statistical properties equivalent to the original time series. We demonstrated the efficiency of the methods for several artificial and real-world time series (financial and medical), which include chaos, stochasticity, and nonstationarity.

When structure is detected in a nonparametric fashion, additional information about the underlying dynamics can be extracted by the herein introduced concept of information flow. The information flow I_p characterizes the loss of information in dynamical systems and describes the decay of statistical dependences between the entire past and a point p steps into the future as a function of p. We studied this concept in the framework of symbolic dynamics for the cases of finite and infinitesimal partitions. In the case of chaotic systems, the most relevant case is the study of the "intrinsic" information flow, that is, for the case of infinitesimal partitions, because it describes the sensitivity to the initial conditions on a fine scale and therefore characterizes the intrinsic evolution of statistical correlations. The intrinsic information flow of hyperbolic chaotic systems is characterized by a constant value of I_p equal to the KS entropy h. This fact offers an interpretation of chaos based on the information flow: a deterministic chaotic process can permanently lose a constant amount of information ($I_p = h$) because of an infinite information content and because of stochasticity. This means that I_p is equal to h for all p; that is, the process loses always, even for $p \to \infty$, information. The deterministic character of the dynamical equations is reflected by the fact that the process contains infinite information, and therefore it can lose permanently information. In other words, the memory of the process is infinite. Moreover, the fact that $h_p = ph \; \forall p$ for chaos implies that the KS entropy h fully characterizes the information flow in the system. For a deterministic and nonchaotic process, the information is also infinite, but the system never loses information ($I_p = 0 \; \forall p$) because $h = 0$.

The information flow can be used to classify the dynamic behavior of complex systems, guiding therefore the selection of the most convenient form of the parametric model that should be used for modeling of the data.

A possible extension would be the formulation of the concept of information flow for the multivariate case in order to handle problems such as multivariate EEG or spatiotemporal turbulences as in transit flows or weather patterns. In this case, should not only present–past structures be detected but also statistical cross-correlations between the different time series could give us information about the local or global character of the involved dynamics. This could be crucial for the analysis case of EEG data where it is suspected that, in the case of epilepsy or by the influence of medicaments, the global character of the dynamics is affected. Also in weather-forecasting studies about the global character of the dynamics, information about the range of local corrections, which should be considered in local forecasting, could be gained. Work in this direction is also actually in progress.

We conclude that the theory exposed in this work not only presents a mathematically consistent solution of the problem of neural learning and of *a priori* nonparametric detection and characterization of structures, but also offers a concrete framework for the development and exploration of methods and theories for one of the most fascinating scientific fields: complex systems.

REFERENCES

[1] G. Zipf. *Human Behavior and the Principle of Least Effort*. Addison-Wesley, Cambridge, MA, 1949.

[2] F. Attneave. Informational aspects of visual perception. *Psychol. Rev.* 61:183–193, 1954.

[3] H. Barlow. Sensory mechanism, the reduction of redundancy, and intelligence. In *National Physical Laboratory Symposium 10. The Mechanization of Thought Processes*. Her Majesty's Stationery Office, London, 1959.

[4] H. Barlow. Unsupervised learning. *Neural Comput.* 1:295–311, 1989.

[5] H. Barlow, T. Kaushal, and G. Mitchison. Finding minimum entropy codes. *Neural Comput.* 1:412–423, 1989.

[6] G. Deco and D. Obradovic. Linear redundancy reduction learning. *Neural Networks* 8:751–755, 1995.

[7] R. Linsker. Self-organization in a perceptual network. *Computer* 21:105, 1988.

[8] R. Linsker. How to generate ordered maps by maximizing the mutual information between input and output signals. *Neural Comput.* 1:402–411, 1989.

[9] R. Linsker. Local synaptic learning rules suffice to maximize mutual information in a linear network. *Neural Comput.* 4:691–702, 1992.

[10] D. Obradovic and G. Deco. Generalized linear features extraction: an information theory approach. *Neurocomputing* 12:203–221, 1996.

[11] G. Deco and W. Brauer. Higher order statistics with neural networks. In *Advances in Neural Information Processing* (G. Tesauro, D. Touretzky, and T. Leen, Eds.), Vol. 7, pp. 247–254. MIT Press, Cambridge, MA, 1994.

[12] G. Deco and W. Brauer. Nonlinear higher-order statistical decorrelation by volume-conserving neural architectures. *Neural Networks* 8:525–535, 1995.

[13] G. Deco and B. Schürmann. Learning time series evolution by unsupervised extraction of correlations. *Phys. Rev. E* 51:1780–1790, 1995.

[14] G. Deco and B. Schürmann. Statistical ensemble theory of redundancy reduction and the duality between unsupervised and supervised learning. *Phys. Rev. E* 52:6580–6587, 1995.

[15] L. Parra, G. Deco, and S. Miesbach. Redundancy reduction with information-preserving nonlinear maps. *Network* 6:61–72, 1995.

[16] L. Parra, G. Deco, and S. Miesbach. Statistical independence with information preserving maps. *Neural Comput.* 8:262–271, 1996.

[17] C. E. Shannon. A mathematical theory of communication. *Bell System Tech. J.* 27:379–423, 623–656, 1948.

[18] S. Kullback and R. A. Leibler. On information and sufficiency. *Ann. Math. Statist.* 22:79–86, 1951.

[19] T. Cover and J. Thomas. *Elements of Information Theory*. Wiley, New York, 1991.

[20] P. Comon. Independent component analysis, a new concept? *Signal Process.* 36:287–314, 1994.

[21] A. Papoulis. *Probability, Random Variables and Stochastic Processes*, 3rd ed. McGraw-Hill, New York, 1991.

[22] C. W. Gardiner. *Handbook of Stochastic Methods*, 2nd ed. Springer-Verlag, New York, 1990.

[23] G. Deco and D. Obradovic. *An Information-Theoretic Approach to Neural Computing*. Springer-Verlag, New York, 1996.

[24] M. G. Kendall and A. Stuart. *The Advanced Theory of Statistics*. Charles Griffin, London, 1969.

[25] B. C. Y. Wong and I. F. Blake. Detection in multivariate non-Gaussian noise. *IEEE Trans. Comm.* 42, 1994.

[26] M. Casdagli. Nonlinear prediction of chaotic time series. *Physic. D* 35:335–356, 1989.

[27] G. Deco and B. Schürmann. Recurrent neural networks capture the dynamical invariance of chaotic time series. *IEICE Trans. Fund. Electron., Comm. Comput. Sci.* 77-A:1840–1845, 1994.

[28] F. Takens. Detecting strange attractors in turbulence. In *Dynamical Systems and Turbulence. Lecture Notes in Mathematics* (D. A. Rand and L. S. Young, Eds.), Vol. 898, pp. 366–381. Springer-Verlag, 1980.
[29] T. Sauer, J. Yorke, and M. Casdagli. Embedology. *J. Statist. Phys.* 65:579–616, 1991.
[30] W. Liebert and H. G. Schuster. Proper choice of the time delay for the analysis of chaotic time series. *Phys. Lett. A* 142:107–111, 1989.
[31] W. Liebert, K. Pawelzik, and H. G. Schuster. Optimal embedding of chaotic attractors from topological considerations. *Europhys. Lett.* 14:521–526, 1991.
[32] M. Mackey and L. Glass. Oscillation and chaos in physiological control systems. *Science* 197:287–291, 1977.
[33] A. M. Fraser and H. L. Swinney. Independent coordinates for strange attractors from mutual information. *Phys. Rev. A* 33:1134, 1986.
[34] N. Tishby, E. Levin, and S. Solla. Consistent inference of probabilities in layered networks: predictions and generalization. In *Proceedings of the International Joint Conference on Neural Networks*, Vol. 2, pp. 403–409. IEEE Press, Washington, DC, 1989.
[35] R. Meir and F. Fontanari. Data compression and prediction in neural networks. *Phys. A* 200:644–654, 1993.
[36] J. Rissanen. Modeling by shortest data description. *Automatica* 14:465–471, 1978.
[37] J. Rissanen. Stochastic complexity and modeling. *Ann. Statist.* 14:1080–1100, 1986.
[38] N. Abraham, A. Albano, A. Passamante, and P. Rapp. *Measures of Complexity and Chaos*. Plenum, New York, 1989.
[39] H. Tong. *Non-linear Time Series:A Dynamical System Approach.* Clarendon, Oxford, 1990.
[40] J. Theiler, S. Eubank, A. Longtin, B. Galdrikian, and J. Farmer. Testing for nonlinearity in time series: the method of surrogate data. *Phys. D* 58:77–94, 1992.
[41] M. Palus, V. Albrecht, and I. Dvorák. Information theoretic test for nonlinearity in time series. *Phys. Lett. A* 175:203–209, 1993.
[42] M. Palus. Testing for nonlinearity using redundancies: quantitative and qualitative aspects. *Phys. D* 80:186–205, 1995.
[43] D. Prichard and J. Theiler. Generalized redundancies for time series analysis. *Phys. D* 84:476–493, 1995.
[44] H. Herzel, A. Schmitt, and W. Ebeling. Finite sample effects in sequence analysis. *Chaos, Solitons Fractals* 4:97–113, 1994.
[45] L. Breiman. *Statistics*. Houghton Mifflin, Boston, 1973.
[46] M. Hénon. A two-dimensional mapping with a strange attractor. *Comm. Math. Phys.* 50:69, 1976.
[47] P. E. Rapp, I. D. Zimmerman, A. M. Albano, G. C. deGuzman, N. N. Greenbaum, and T. R. Bashore. *Experimental Studies of Chaotic Neural Behavior. Cellular Activity and Electroencephalographic Signals. Nonlinear Oscillations in Biology and Chemistry.* Springer-Verlag, Berlin/New York, 1986.
[48] J. P. Pijn. Quantitative evaluation of EEG signals in epilepsy: nonlinear associations, time delays and nonlinear dynamics. Ph.D. Thesis, University of Amsterdam, 1990.
[49] R. S. Shaw. Strange attractors, chaotic behavior, and information flow. *Z. Naturforsch.* 36a:80–112, 1981.
[50] J. P. Eckmann and D. Ruelle. Ergodic theory of chaos and strange attractors. *Rev. Modern Phys.* 57:617–656, 1985.
[51] B. Pompe, J. Kruscha, and R. Leven. State predictability and information flow in simple chaotic systems. *Z. Naturforsch.* 41a:801–818, 1986.
[52] B. Pompe and R. Leven. Transinformation of chaotic systems. *Phys. Scripta* 33:8–13, 1986.
[53] B. Pompe. Measuring statistical dependences in a time series. *J. Statist. Phys.* 73:587–610, 1993.
[54] B. Pompe. On some entropy methods in data analysis. *Chaos, Solitons Fractals* 4:83–96, 1994.

[55] P. Szépfalusy. Characterization of chaos and complexity by properties of dynamical entropies. *Phys. Scripta* 25:226–229, 1989.
[56] C. Beck and F. Schlögl. *Thermodynamics of Chaotic Systems. Cambridge Nonlinear Science Series.* Cambridge University Press, Cambridge, 1993.
[57] A. Kolmogorov. A new metric invariant of transient dynamical system and automorphism in Lebesgue spaces. *Dokl. Akad. Nauk SSSR* 119:861–864, 1958.
[58] Ya. G. Sinai. On the concept of entropy for a dynamic system. *Dokl. Akad. Nauk SSSR* 124:768–771, 1959.
[59] E. Ott. *Chaos in Dynamical Systems*. Cambridge University Press, Cambridge, 1993.
[60] F. Schlögl. The variance of information loss as a characteristic quantity of dynamical chaos. *J. Statist. Phys.* 46:135–146, 1987.
[61] P. Grassberger and I. Procaccia. Characterization of strange attractors. *Phys. Rev. Lett.* 50:346, 1983.
[62] K. Matsumoto and I. Tsuda. Information theoretical approach to noisy dynamics. *J. Phys. A* 18:3561–3566, 1985.
[63] K. Matsumoto and I. Tsuda. Extended information in one-dimensional maps. *Phys. D* 26:347–357, 1987.
[64] K. Matsumoto and I. Tsuda. Calculation of information flow rate from mutual information. *J. Phys. A.* 21:1405–1414, 1988.

Radial Basis Function Network Approximation and Learning in Task-Dependent Feedforward Control of Nonlinear Dynamical Systems

Dimitry Gorinevsky
Honeywell-Measurex
North Vancouver
British Columbia V7J 3S4, Canada

I. INTRODUCTION

Motor control in humans and animals is believed to use feedforward widely instead of relying only on feedback. One hypothesis on feedforward control organization is that control programs of complex motions are learned, memorized, and then just extracted from the memory when needed. This hypothesis provided an inspiration for this chapter, which presents a paradigm for task-level feedforward control. The practical value of this paradigm is demonstrated by applying it to solve a difficult control problem.

Classical automatic control theory heavily concentrates on the issues of low-level feedback control or feedforward compensation of disturbances. This is because most controlled systems used in industrial and other applications in the past two decades and many of those used now are characterized by a low level of computational power and relatively simple logic of operation. In addition to feedback control, classical control theory paid much attention to the issues of open-loop (programmed) control, in particular, optimal control. This was initially motivated by space flight and ballistic missile control applications, where the control program for a mission can be accurately computed in advance.

Many modern advanced control systems are integrated systems with high-performance computers, multiple sensors, and actuators. These advanced systems possess automated operation, adaptation and self-tuning features, built-in identification, and fault diagnosis capability. The trend in control practice is toward development and deployment of intelligent control systems performing increasingly complex tasks with minimal or no operator supervision and adapting to changing operation conditions. The setup, commissioning, and operation regime change for such systems should be also automated. Development of intelligent systems requires comprehensive multilayered control and information processing architectures, which are more complicated than classical feedback loops.

The task-level control algorithms considered in this chapter are assumed to be executed one level above the traditional planning and servo feedback level. To define this new control level, we assume that the overall motion planning and feedback tracking of this planned motion is a sequence of separate (but possibly related) tasks. We assume that each task and a control problem associated with this task are completely defined by a few task parameters. The algorithms considered in this chapter operate in discrete time: from task to task. In this respect, the proposed controllers can be considered as hybrid systems.

There are a few groups of papers in the literature more specifically related to the topic and technical approaches of this chapter. Recently, a *learning control* paradigm has been considered by a number of authors, starting with Arimoto [1], mostly with regard to the manipulator path tracking. The paradigm regards a feedforward control program as a high-dimension array, which stores a time history of the feedforward control input. Some authors [1–9] assume that a dynamic model of the system is to some extent known and consider an iterative method for improving the performance by updating a feedforward control program in the course of repeated motion trials. The iterations converge for a *single* given motion, but the learned control would not work for another trajectory.

Another direction of work related to the technical approach of this chapter is associated with approximation-based control. In a sense, any practical linear control approach is based on a model, which is but an approximation of a real plant. Herein, we consider nonlinear gray-box approaches that use generic computational architectures to approximate unknown nonlinear mappings. Such approximation-based approaches are usually applied within the neural network or fuzzy logic framework.

Many authors use neural networks to acquire knowledge of the controller plant dynamics that allow us to compute feedforward control at each instant, given the planned motion, velocity, and acceleration (see, e.g., [10–12]). Some related papers consider different techniques for computing the feedforward through approximation of the dynamics mapping of the system by using polynomial associative memories [13–15] or fuzzy control. Such an approach to learning is capable of generalization, which means that the knowledge of inverse dynamics acquired for one trajectory allow us to compute feedforward control for other motions. Yet the robustness of such an approach to high-order dynamics is not always acceptable and real-time implementation of the approach requires enormous resources, especially if the order of the system dynamics is high.

A distinct learning control paradigm combining features of the two previously described approaches was proposed in the author's earlier papers [16–18]. According to this paradigm, *a mapping* from the task parameters to the feedforward control program is learned (approximated). An example of the task parameters might include initial and the desired final state of the system in a transient motion control. The described approach is also related to the *task-level control* paradigm proposed in [19, 20].

This chapter surveys the author's work on approximation and learning in task-level control problems [16, 17, 21–23]. Herein, we mostly concentrate on the paradigms and algorithm ideas. The convergence results and applications of the approach are referenced but not presented in any detail. The learning process, which we consider in this chapter, can be regarded as an adaptive control of a linearly parametrized nonlinear system. In [24] we demonstrate an application of the algorithms considered in this paper to the adaptive control of an unstructured nonlinear system.

The proposed learning control paradigm has several fundamental advantages that make it potentially very useful for many applications. First, our approach is well suited for real-time implementation, because the entire control program is computed before the motion begins. Second, unlike sophisticated feedback controllers, the approach is robust to high-order unmodeled dynamics. Third, high sensitivity to the system parameter variations, which is inherent in open-loop control, is compensated by the on-line feedback update of the controller parameters.

Our approach is based on using a connectionist network approximation of the control program dependence on the task parameter vector. Such a network has a fixed number of weights that can be updated based on the system operation results. This chapter uses a radial basis function (RBF) network architecture for approximating dependences such as feedforward program dependence on the task parameters. It has recently been acknowledged that in many problems the RBF networks possess superior spatial filtration and approximation accuracy properties, as compared to the multilayered perceptron networks [25–31]. Even more important for this study is that the RBF network output is *linear* in the network weights. This property makes the powerful tools of the linear system theory applicable to system estimation (identification) with an RBF network. So designed algorithms converge much faster than usual neural network algorithms based on gradient descent (backpropagation). Furthermore, the computational complexity of the RBF network approximation does not grow with the dimension of the output variable, which is usually large for problems of this type.

The task-level algorithms considered in this chapter are expected to be useful for many applications. Currently, these algorithms have been applied to problems of robotics, process control, automotive, and flexible spacecraft control. In this chapter, we demonstrate the efficiency of the proposed approach in the control of flexible arm motions. We simulate very fast arm motions that take only about 1.5 periods of the lowest eigenfrequency oscillations. The system is oscillatory and very nonlinear, and, therefore, difficult to control using classical approaches even if the system dynamics are known exactly. For this system, we achieve a high control performance by using a learning control algorithm without any *a priori* information on the system dynamics.

The outline of the chapter is as follows. Section II considers a general statement of the task-level control problem. One of the goals of this section is to show how such problems relate to the classical feedback and programmed control problems. A concise formal mathematical problem statement of the task-level control is related to discrete-time on-line optimization of a static multivariate vector-valued mapping. We formulate four different problem statements, which assume different degrees of uncertainty about the system. The problems stated in Section II are then considered in four subsequent sections of the chapter.

Section III presents some background facts on the RBF approximation. The formulation of this section constitutes the basis for the development of Sections V and VI. Section III also discusses a basic architecture of task-level controller based on the RBF network approximation. Section IV discusses learning control algorithms applicable for a fixed parameter vector. The problem statement is similar to the standard learning control papers cited previously, but the suggested algorithms differ from most of these papers. The advantage of the algorithms considered is that they can be conveniently generalized to the case of changing task parameter vector.

Sections V and VI propose task-level feedforward control algorithms with learning (adaptive) capabilities. In Section V, we assume that the system sensitivity information is available prior to the system operation, similar to system gain information in feedback control. We then derive an algorithm for on-line update of the feedforward control approximation based on the task completion errors. In Section VI, the sensitivity information is assumed to be inaccurate and is updated on-line in the adaptive control manner. Section VI also presents an application of the algorithm to the terminal control of a two-link flexible manipulator. The manipulator performs a random sequence of point-to-point motion tasks.

II. PROBLEM STATEMENT

This section presents a formal statement of the learning control problem. The problem is formulated in a general form, so that the statement is applicable to a large class of control systems and tasks. To make the exposition more transparent, Section II.C introduces an example problem—one of feedforward control of a flexible planar arm—and illustrates how the general formulation considered can be applied to this example. Further, in Section VI, the algorithms developed in the following sections are applied to the same example.

A. CONTROL FORMULATION

This chapter considers task-level control problems and algorithms. Such algorithms perform computations and updates of control variables and internal models from task to task. They are essentially discrete-time algorithms evolving with the performed task number. The goal of this and the two following subsections is to explain how these task-level algorithms relate to more classical issues of feedback, feedforward, and programmed control. To this end, we start with a continuous-time controlled system in a state-space form and then arrive at the task-level control formulation for such a system. The task-level control problems and algorithms considered further, however, are generic discrete algorithms (in the same sense as classical optimization algorithms) and can be applied to a wider class of practical problems.

Let us consider a parametric family of nonlinear time-invariant dynamical systems of the form

$$\dot{x} = f(x, u; \mu), \tag{1}$$

$$y = h(x), \tag{2}$$

where $x \in \Re^{n_x}$ is a state vector, $y \in \Re^{n_y}$ is an observation vector, $u \in \Re^{n_u}$ is a control input vector, $\mu \in \mathcal{M} \subset \Re^{n_\mu}$ is a vector of system parameters, and $\mathcal{M}$ is a given domain. We further assume that the nonlinear mappings $f(\cdot, \cdot; \cdot)$ and $h(\cdot)$ are smooth and the nonlinear system (1), (2) is known to be observable and reachable for each value of the parameter $\mu \in \mathcal{M}$. We will discuss the assumptions about the smoothness, controllability, observability, and nature of available information about the system in more detail later on as they will be needed.

The system (1), (2) describes a task being executed. This system evolves in a local time, which is the time since the beginning of the task. In the sections to follow, we will consider issues of *task-level* control, that is, the control and estimation evolving from one task to another. It is assumed that a local time t can be used for the description of the system dynamics in each task.

In each task, we consider a controlled motion of the system (1), (2) on the given interval $[0, T]$ of the local time t and assume that the initial state vector belongs to a smooth manifold of the form

$$x(0) = \psi(\lambda), \tag{3}$$

where $\lambda \in \Re^{n_\lambda}$ is a vector that defines the initial conditions.

The control problem for the task is to find a control input $u(\cdot)$ defined on the time interval $[0, T]$ that allows us to achieve the control goal formulated as the minimization of the performance index of the form

$$J_1\big(y(\cdot), y_d(\cdot; \lambda, \nu); u(\cdot)\big) \to \min. \tag{4}$$

Here $\nu \in \Re^{n_\nu}$ is a vector that defines the task goal, for instance, in the form of the desired system state at the end time T, trajectory of motion, etc.; and $y_d(t; \lambda, \nu) \in \Re^{n_y}$ is the preplanned desired output of the controlled plant in the task. We will call ν a vector of the control goal parameters. The preplanned output y_d depends on both the initial condition vector λ and the control goal vector ν.

Let us introduce a task parameter vector p comprising the initial condition vector λ (3), the vector of the system parameters μ (1), and the vector of the control goal ν (4) for this task

$$p = \begin{bmatrix} \lambda \\ \mu \\ \nu \end{bmatrix} \in \Re^{N_p}, \qquad N_p = n_\lambda + n_\mu + n_\nu. \tag{5}$$

The optimal feedforward control input $u(\cdot)$ that solves the problem (1)–(4) depends on the vector p (5). We assume that the vector p belongs to a given compact domain $\mathcal{P} \subset \Re^{n_p}$.

We further assume that we can repeatedly apply the computed feedforward control $u(\cdot)$ to the system (1), (2) in different tasks, observe its output $y(\cdot)$ on the time interval $[0, T]$, and, possibly, use the obtained observations to update the control. For each task, we suppose that the local time t is reset to zero and the initial condition has form (3). We assume that the parameter vectors λ, μ, and ν can take different values for different runs (system motions). The control objective is to minimize the performance index (4) for each value of the parameter vector $p = [\lambda^T\ \mu^T\ \nu^T]^T$ (5) in the given domain: $p \in \mathcal{P} \subset \Re^{N_p}$.

At this stage, we do not specify how the parameter vector p (5) changes from one task to another. This issue is discussed further on in Section IV. The task-level learning control algorithms, which we are going to discuss, are more general than those considered in previous papers on learning and repetitive control. Standard learning control formulations (e.g., [2]) correspond to assuming the parameters λ, μ, and ν to be fixed. Repetitive control setting [5] assumes that the initial conditions for each task coincide with the final system state at the previous task, and the variation of the initial conditions is small. Furthermore, the prior learning control work mostly confines itself to the trajectory tracking problem. The task-level learning paradigm of [19, 20] is related to the preceding general problem statement, but covers a somewhat different class of problems. Herein, we formulate a control problem as a minimization of a general performance index, which includes most of the previous learning control formulations as special cases. In particular, as shown in the following discussion, such formulations can be used for both trajectory tracking and terminal control problems. A few recently published papers study control approaches related to some aspects of the considered formulation [32, 33].

As already mentioned in Section I, the learning control papers usually consider a trajectory tracking problem. This problem can be described with a performance index of the form (4)

$$J_2(y, y_d, u) = \int_0^T \left\| y(t) - y_d(t; \lambda, \nu) \right\|^2 dt + \rho \int_0^T \left\| u(t) \right\|^2 dt \to \min, \quad (6)$$

where $\| \cdot \|$ denotes the Euclidean norm of a vector. The first term in (6) is a penalty for the deviation of the system output from its desired value. Because the trajectory tracking problem is generally ill posed, the second term in the performance index (6) is very important. It regularizes the solution in accordance with the technique of [34]. For $0 < \rho \ll 1$, a solution to (6) provides good quality of the trajectory tracking (see [17] for additional discussion on the subject).

Similarly, the point-to-point control problem can be described with the performance index (4) of the form

$$J_{2f}(y, y_d, u) = \int_T^{T_f} \left\| y(t) - y_d(T; \nu) \right\|^2 dt + \rho \int_0^T \left\| u(t) \right\|^2 dt \to \min, \quad (7)$$

where we assume that the output y of the plant (1), (2) can be monitored on the time interval $[T, T_f]$, and $T_f > T$. The first term in (7) gives a measure of the overshoot, the second term, of the control effort. We assume that on the interval $T \le t \le T_f$ the desired output y_d is a constant defined by the vector ν, and $u(t) \equiv 0$, which corresponds to the system being in the desired final steady state. Unlike (6), the performance index (7) penalizes only overshoot of the system output *after* the end of the desired motion. Under appropriate conditions of controllability and observability, a solution $u(\cdot)$ of (7) approaches quadratic-optimal terminal control as $\rho \to 0$. For linear systems, this is studied in more detail in [35] and, generally, a smooth nonlinear system can be linearized in the vicinity of the optimal solution to make the linear system results applicable.

Unlike most other work on learning control, we will be interested in obtaining a *dependence* $u(\cdot\,; p)$ of the feedforward control (1) on the parameter vector (5), rather than in learning the feedforward control for a *single given value* of the parameter vector p.

The mapping $p \mapsto u(\cdot)$ defined by the optimal control problem (1)–(3), (4), and (5) is generally complicated and nonlinear, even for a linear system (1), (2). We suppose that the system (1), (2) and the performance index (4) are such that this mapping is continuous. For instance, it is continuous, if the right-hand-side mappings in (1), (2) are smooth, the system is reachable, $\rho > 0$, and we are considering a quadratic performance index (6) or (7). Yet, we would like to note that study of the properties of the mapping from the coordinates into control for a general nonlinear system is a complicated problem. In particular, it is known that for certain systems with smooth right-hand sides, such as nonholonomic systems, no continuous stabilizing state feedback exists [36]. Some discussion regarding approximation of the feedforward shape dependence on the task parameter vector for a nonholonomic system can be found in [21].

B. Example: Control of Two-Link Flexible Arm

Let us consider a two-link articulated arm shown in Fig. 1. We assume that inertial drives placed in the arm joints are connected to the links through lumped elastic elements and all motion is in a horizontal plane. Our goal is to demonstrate an application of the approach of this paper to a standard problem. Therefore, we employ commonly made assumptions about the elastic-joint manipulator dynamics [37]. In particular, we assume that the damping in the elastic elements is negligible, and that angular motion of the drive rotors is decoupled from the arm structure motion. The latter assumption holds for the drives with high transmission ratios.

Let $q \in \Re^2$ be the vector of the arm joint angles; $\gamma \in \Re^2$, the vector of the rotation angles for the output shaft of the drive; and $\mu \in \Re^1$, a lumped mass (payload) attached to the arm tip. Under the assumptions made, the equations of

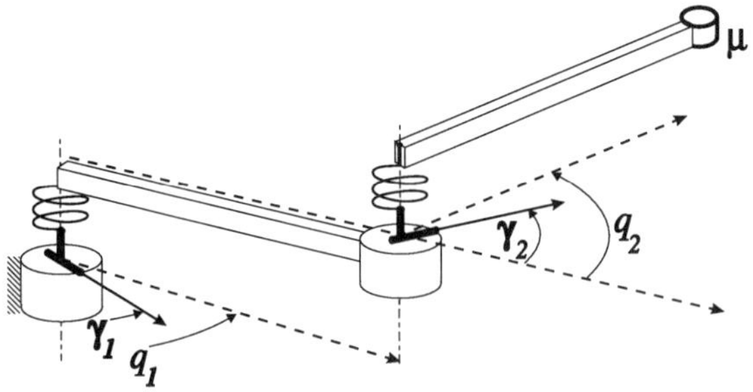

Figure 1 Two-link planar arm with flexible joints. Reproduced with permission from D. M. Gorinevsky, *IEEE Trans. Automat. Control* 42:912–927, 1997 (©1997 IEEE).

motion of the system have the form

$$M(q)\ddot{q} + C(q,\dot{q}) + K(q-\gamma) = 0, \tag{8}$$

$$J\ddot{\gamma} + B\dot{\gamma} + K(\gamma - q) = \tau, \tag{9}$$

where $M(q) \in \Re^{2,2}$ is the inertia matrix of the arm; $C(q,\dot{q}) \in \Re^2$ is the vector of Coriolis and centrifugal forces; the matrix $K = \text{diag}\{k_1, k_2\} \in \Re^{2,2}$ defines the elastic element stiffnesses k_1 and k_2; $J = \text{diag}\{j_1, j_2\} \in \Re^{2,2}$ is the diagonal matrix of the drive rotor inertias; $B = \text{diag}\{\beta_1, \beta_2\} \in \Re^{2,2}$ is the matrix of the internal drive damping; and $\tau \in \Re^2$ is the drive torque vector, which we consider as a control variable. The exact form of the nonlinear functions $M(q)$ and $C(q,\dot{q})$ for a planar arm with uniform links and a lumped payload can be found elsewhere (e.g., in [38]).

We further assume that the control torque vector τ applied to the drives is computed in the usual way, as a sum of a proportional and a derivative (PD) drive position feedback and feedforward compensation $u(t)$:

$$\tau(t) = K_*\big(q_d(t) - \gamma(t)\big) + B_*\big(\dot{q}_d(t) - \dot{\gamma}(t)\big) + u(t), \tag{10}$$

where $q_d(t)$ is the reference drive position, $K_* = \text{diag}\{k_{*1}, k_{*2}\} \in \Re^{2,2}$ is the matrix of the proportional drive position feedback gains; and $B_* = \text{diag}\{\beta_{*1}, \beta_{*2}\} \in \Re^{2,2}$ is the matrix of the velocity feedback gains.

We consider the feedback gains K_* and B_* as fixed parameters and the vector of the feedforward joint torques $u(t) \in \Re^2$ as an external (control) input to the system. We assume that the observations used in the learning include the drive rotation angles q and the elastic element deformations $q - \gamma$. We suppose that the drive velocities $\dot{\gamma}$ are available to the low-level servocontrol (10), but not to a higher-level controller implementing the learning algorithm. Such a situation is very common in robot control practice, where servocontrollers of the drives usually do not provide the upper level of the control system with the velocity information.

For the system (8)–(10), the state vector has the form $x = [q^T\ \dot{q}^T\ \gamma^T\ \dot{\gamma}^T]^T \in \Re^8$ and the observation vector, $y = [q^T\quad (q-\gamma)^T]^T \in \Re^4$. The system (8)–(10) is a special case of the system (1), (2).

Let us introduce a vector $\lambda \in \Re^2$ of the initial condition parameters, so that the initial state vector (3) has the form

$$x(0) = \begin{bmatrix}\lambda^T & 0 & 0 & \lambda^T & 0 & 0\end{bmatrix}^T, \tag{11}$$

where $\lambda \in \Re^2$ defines the initial joint angles of the arm and the drive angles. The initial condition (11) means that $q(0) = \gamma(0) = \lambda$ and $\dot{q}(0) = \dot{\gamma}(0) = [0\quad 0]^T$.

Let us consider the control goal parameter vector $\nu \in \Re^2$ which is equal to the desired joint angle vector $q_d(T)$ after the motion of the arm. The preplanned output $y_d(\cdot;\lambda,\nu)$ describes the desired path of the arm motion, which is used as a reference in the PD controller (10) and the planned joint deformations. The path planning method commonly used in robotics is to compute the reference path as a straight line in the joint angle space which can be written in the form

$$q_d(t) = \lambda\bigl(1 - s(t)\bigr) + \nu s(t), \tag{12}$$

where $s(t)$ is a smooth scalar-valued function (e.g., a third-order polynomial) such that $s(0) = s'(0) = s'(T) = 0$, and $s(t) \equiv 1$ for $t \geq T$. We assume that the planned values of the joint deformations $q_d(t) - \gamma_d(t)$ and their derivatives are always zero.

Let us consider a problem of point-to-point control of the arm stated as a so-called "cheap control problem," for example, the problem of minimizing the performance index of the form (7), where $0 < \rho \ll 1$. For $t > T$, we assume that the feedforward is zero and the PD controller uses the constant set point $y_d(T)$. As mentioned in Section II.A, a solution of the control problem (7) approaches a quadratic optimal solution of the terminal control problem as $\rho \to 0$.

C. Discretized Problem

Section II.A states the problem of approximating the optimal feedforward control u over a domain of the parameter vector p. This approximation problem involves a mapping from a domain $\mathcal{P} \subset \Re^{N_p}$ of the parameter vector p (5) into the Banach space $\mathcal{L}_2(\Re^{n_u}; 0, T)$, where the feedforward control input $u(\cdot)$ belongs. An implementable computational algorithm that solves the stated problem has to use a truncation of Banach space vectors such as $u(\cdot)$. Furthermore, a control algorithm implemented with a digital computer introduces the truncation in the form of input and output signal sampling. The problem treatment based on the truncation is acceptable, because, in practice, only a restricted approximation accuracy is required. After the truncation, the time histories of the feedforward input $u(\cdot)$ to the system (1), and the system output $y(\cdot)$ (2) will be represented by finite-dimensional vectors, and the approximation problem will include mappings between finite-dimensional vector spaces.

In order to truncate the formulated Banach-space problem, let us introduce a set of the shape functions $\phi_j(\cdot)$: $[0, T] \mapsto \Re$ $(j = 1, \ldots, N)$. The shape functions $\phi_j(\cdot) \in \mathcal{L}_2(\Re; 0, T)$ make a basis of the linear manifold $\Pi_N \subset \mathcal{L}_2(\Re; 0, T)$. By using the Galerkin–Rietz approach to the solution of the problem (1)–(3), we consider a projection of the control $u(\cdot) \in \mathcal{L}_2(\Re^{n_u}; 0, T)$ onto the linear manifold $\Re^{n_u} \otimes \Pi_N$, where $\otimes$ denotes a direct product of two vector spaces. This projection, also known as an assumed mode expansion, has the form

$$u(\cdot) = \sum_{j=1}^{N} U_j \phi_j(\cdot), \qquad U = \begin{bmatrix} U_1 \\ \vdots \\ U_N \end{bmatrix} \in \Re^{N_u}, \tag{13}$$

where we have collected the weight vectors U_j into the vector U of dimension $N_U = n_u N$. The vector U (13) is a coordinate vector on the linear space $\Re^{n_u} \otimes \Pi_N$. It is well known that for many choices of the shape functions—such as a trigonometric or polynomial Fourier series, B-spline approximations, or wavelet expansions—the Galerkin–Rietz method converges to the exact solution of the continuous-time problem, as $N \to \infty$.

We do not discuss a particular choice of the shape function set or the expansion order here, because this is a well-studied problem addressed elsewere. We assume that the expansion of the form (13) gives an acceptable solution of the problem. An additional insight can be obtained from the papers [33, 35, 39, 40], among many others, where the applications of expansions of the form (13) to related continuous-time control problems are discussed in more detail.

In many applications, it is advantageous to use B spline in the expansion (13). The B-spline shape functions with the same bases (support) differ only by translations. It is possible to use B splines of various orders in (13). In particular, zero-order B splines will give a piecewise constant feedforward (13), whereas cubic B splines will result in a twice continuously differentiable feedforward. In the application example of Section VI, first-order B-spline functions are used.

We further assume that the shape function set is given and the vector U defines the feedforward control on the interval $[0, T]$, that is, for the task in question. We will call U a *control input vector* or a *control program.*

Similarly, we describe the system output with a finite-dimensional output vector Y. We introduce a sampling time sequence $\{t_j\}_{j=1}^{L}$ and *output vector* Y and a desired output vector Y_d that have the form

$$Y = \begin{bmatrix} y(t_1) \\ \vdots \\ y(t_L) \end{bmatrix} \in \Re^{N_Y}, \qquad Y_d = \begin{bmatrix} y_d(t_1) \\ \vdots \\ y_d(t_L) \end{bmatrix} \in \Re^{N_Y}, \tag{14}$$

where $y \in \Re^{n_y}$ is the system output (2), $y_d(t) = y_d(t; \lambda, \nu)$ is the desired output as in (6) and (7), and $N_Y = n_y L$. Under appropriate observability conditions imposed on the sampling time sequence and on the controlled plant, the desired output of the system (1), (2) can be evaluated by monitoring only the sampled output (14). Owing to the measurement sampling in a digital computer, the measured output has the form (14) in most practical cases.

The input vector U defines the output vector Y in accordance with Eqs. (1), (2), (5), (13), and (14). We will write this dependence in the form

$$Y = S(U, p) \tag{15}$$

and assume that the system (1) is such that S is a smooth mapping. Let us also consider a modified form of the performance index (4)

$$J(Y, Y_d; U; p) \to \min, \tag{16}$$

where Y_d is obtained by sampling the preplanned output $y_d(t; \lambda, \nu)$ in the same ways as $y(t)$ in (14). The vector p in (15) is presented explicitly in order to show the dependence of the control problem on the task parameters λ, μ, and ν that influence the output according to (1)–(3), (13), (14), and (16). For a broad class of controlled systems, it is shown in [35] that, under appropriate conditions, a solution of the discretized problem as introduced in this subsection approaches the solution of the original continuous-time problem. Because of space limitations, we do not discuss this issue in more detail here and further will limit the consideration with the discretized problem (14), (15).

For the performance index (16), one can write the condition of the extremum in the form

$$\frac{\partial J}{\partial U}(U_*, p) + G^T(U_*, p)\frac{\partial J}{\partial Y}(U_*, p) = 0,$$
$$G(U, p) = \frac{\partial Y}{\partial U} = \frac{\partial S}{\partial Y}(U, p), \tag{17}$$

where $G \in \Re^{N_Y, N_U}$ is an input–output sensitivity matrix of the system—a Jacobian matrix of the mapping (15). If the mapping (15) is known analytically, one can use (17) to find analytically or numerically an optimal control input U_* for each value of p. For the performance index of the form (6) or (7), we can write the modified performance index (16) in the form

$$J = \|Y - Y_d\|^2 + \rho\|U\|_R^2 \to \min, \tag{18}$$

where $\|\cdot\|$ denotes the Euclidean norm in $\Re^{N_Y}$ and $\|U\|_R^2 = U^T RU$, where $R \in \Re^{N_U, N_U}$ is a positive-definite matrix defined through a Grammian of the shape function set $\{\phi_j(\cdot)\}_{j=1}^N$. Performance index (18) can approximate either (7) or (6), assuming that the sampling of the output $y(t)$ (14) is uniform on the interval $[T, T_f]$ or $[0, T]$, respectively. To make the presentation more transparent, we further assume that in (18) $R = I_{N_U}$ is a unity matrix and both norms in (18) are the regular Euclidean norms. In particular, this is valid if the shape functions (13) are orthonormal. In general, the orthonormality of the shape functions can be achieved with a simple linear transformation.

For problem (18), the extremum condition (17) has the form

$$\rho U + G^T\left[S(U_p) - Y_d\right] = 0 \tag{19}$$

and could be solved analytically or numerically, once the form of the mapping (15) is known. The solution to (19) has the form

$$U = U_*(p), \qquad p \in \mathcal{P}, \tag{20}$$

where we assume that the mapping $U_*(p)$ is smooth, which is valid for a broad class of systems (15).

D. Problems of Task-Dependent Feedforward Control

In what follows, we consider a few different control architectures and problems related to (15), (16). In particular, we will concentrate on the performance index (16) of the form (18).

The first problem we are going to consider is how to build a controller implementing the solution (20) to the problem (15), (16). The key issue here is to design a *practically implementable* controller that is able to compute the optimal feedforward vector U_* in real time with limited computational resources. This could be achieved by computing $U_*(p)$ *approximately*, rather than exactly. Such an approach is perfectly justified from a practical viewpoint because, anyway, real-life systems always differ from their computational models, however accurate the latter are.

Problem 1 (Computing task-dependent approximation for the feedforward). Design a practically implementable controller computing an accurate approximation $\widehat{U}(p)$ of the optimal solution $U_*(p)$ of the problem (15), (16) for any parameter vector p (5) in the given domain $\mathcal{P}$.

The design of an approximation-based controller solving Problem 1 is discussed in Section III.D. This design employs a radial basis function network for approximating the mapping $U_*(p)$.

As will be discussed further in more detail, Problem 1 can be solved by first obtaining an optimal control U_* for selected task parameters p and then interpolating these data. As an alternative to numerical model-based optimization, an optimal control vector U_* for a fixed given task parameter vector p can be directly learned in the course of repeated execution of the same task. Though repeated experiments are only possible in some problems, the following learning problem is of didactic importance to us. The RBF network learning problems considered further can be seen as generalizations of this problem.

Problem 2 (Iterative learning of feedforward for a given task). Let us assume that p in (15) is given and fixed. We assume that the mapping (15) is unknown, but we can repeatedly execute the same task defined by (1)–(3), (5), (13), (14). For each task repetition, an input vector U is applied and the output vector Y is observed. The problem is to design a learning control algorithm that iteratively updates the input vector U in order to optimize the performance index (18).

A learning procedure solving Problem 2 is studied in Section IV.

For many real problems, designing a feedforward controller by interpolating solutions of Problem 2 is not practical, because this would require multiple repetitions of the same task. A more practical controller can be obtained by first using an available model of the system for off-line design and then updating the feedforward control approximation based on output errors registered as it operates. Such an approach follows the usual practice of general feedback controller design where setpoints can be calculated off-line or on an upper level of control and then tracked using an error feedback. The difficulty in updating the control approximation $\widehat{U}(p)$ is that, unlike Problem 2, in the course of normal operation of the system, the sequence of the task vectors p is defined by a higher control level. Thus, it is unreasonable to make any assumptions about this sequence when designing the controller. The formal statement of this problem is as follows.

Problem 3 (On-line learning update in task-dependent feedforward approximation). Assume that an imprecise approximation $\widehat{U}(p)$ of the control input optimal in the sense of (16) is available for each value of p. Assume further that an approximation $\widehat{G}(p)$ of the system output Y sensitivity (16) to input U is available for each value of p. For a generic sequence of the parameter vectors p (5) and a corresponding sequence of the optimal control input vectors $\widehat{U}(p)$ (1), (13) applied to the system (1), (2), (15), observe the sequence of the output vectors Y (2), (14) and update an approximation $\widehat{U}(p)$ of the control input to optimize it in the sense of (16) for each value of p.

Problem 3 is studied in Section V.

As mentioned previously, Problem 3 is essentially about a discrete-time feedback controllers design. In some cases, such model-based controllers may not perform well enough in practice, in particular, if an available model of the system sensitivity (gain) is not sufficiently accurate. In such cases, one may want to use an *adaptive* or self-tuning controller, which estimates the system gain based on the closed-loop operation data (possibly with self-excitation added). The problem of designing such an adaptive controller is as follows.

Problem 4 (Adaptive learning of task-dependent approximation for feedforward). Given a sequence of the parameter vectors p (5), apply a sequence of the control input vectors U (1), (13) to the system (1), (2), (15) and by observing the sequence of the output vectors Y (2), (14) estimate for each p a local model of the input–output mapping $U \mapsto Y$ including the system sensitivity. Using this model, update an approximation $U_*(p)$ of the optimal control input (20).

Section VI presents an iterative adaptive learning procedure that updates an approximation of the mapping $p \mapsto U_*(p)$ for an *a priori* unknown mapping (15), as required in Problem 4.

III. RADIAL BASIS FUNCTION APPROXIMATION

This section considers a generic approximation problem, and introduces an RBF network architecture suitable for solving such problems. We then show how a controller using an RBF network approximation can be used to solve Problem 1 stated in Section II.D. The material of this section us used as a base for developing algorithms in the subsequent sections.

A. Exact Radial Basis Function Interpolation

Let us consider an auxiliary problem of approximating a smooth nonlinear mapping $G(\cdot)$: $\Re^{N_p} \mapsto \Re^{N_y}$ over a compact domain

$$Y = g(p), \qquad Y \in \Re^{N_y}, \qquad p \in \mathcal{P} \subset \Re^{N_p}, \tag{21}$$

where p is an input parameter vector, $\mathcal{P}$ is a compact domain, and Y is an output vector. We assume that a scattered (irregularly placed) set of N_t input–output pairs is available and call this set *the training data set*

$$\{Y^{(j)} = g(p^{(j)}), p^{(j)}\}, \qquad j = 1, \ldots, N_t. \tag{22}$$

The problem is to find an approximation $\widehat{Y} = \hat{g}(p)$ of the mapping (21) that can be used for any argument $p \in \mathcal{P}$.

A computationally convenient way of representing an unknown nonlinear function is to present it as an expansion. The expansion is linear in parameters that are assumed to be unknown. Let us consider an approximation of the mapping (21) that has the form

$$\widehat{Y} = \hat{g}(p) = \sum_{j=1}^{N_a} Z^{(j)} w_j(p), \tag{23}$$

where N_a is the order of the expansion, $w_j(p)$ are scalar expansion shape functions, and $Z^{(j)} \in \Re^{N_y}$ are the expansion weights. The truncated Fourier series expansion, polynomial expansion, and B-spline expansion all have the form (23). In the artificial neural network literature, approximations of the form (23) are known as functional link networks [41, 42] and we further call vectors $Z^{(j)}$ the network weight vectors.

Given the expansion shape functions $w_j(p)$ (23), a standard way to solve the scattered data approximation problem is to choose the parameter vectors $Z^{(j)}$ by fitting the expansion (23) to the data (22) with the least error. In the special case $N_a = N_t$, when the number of expansion weight vectors (23) coincides with the number of training set data pairs (22), one can generally fit the training data exactly.

In this section, we consider an expansion of the form (23) with functions $w_j(\cdot)$ that depend on the radii $r_j = \|Q^{(j)} - p\|$, where $Q^{(j)} \in \Re^{N_p}$ are given vectors. Such an expansion is known as radial basis function (RBF) approximation. RBF approximation has been used in computer graphics and experimental data processing applications (e.g., geophysical data) for two decades and has been demonstrated to provide for a high quality of approximation. One can find further details and references in [43–48]. These papers employ the method recently referred to as an *exact RBF interpolation*. This method uses the radial functions centered at each of the data points (22). In this method, the radial function centers are $Q^{(j)} = p^{(j)}$, and the expansion functions in (23) have the form

$$w_j(p) = h\big(p - p^{(j)}\big), \tag{24}$$

where $h(\cdot)$ is a radial function; that is, $h(p - p^{(j)})$ depends on the radius $\|p - p^{(j)}\|$. The most commonly used radial basis functions are

$$\begin{aligned} h(p) &= \exp\big(-\|p\|^2/d^2\big), \\ h(p) &= \big(1 + \|p\|^2/d^2\big)^{1/2}, \\ h(p) &= \big(1 + \|p\|^2/d^2\big)^{-1/2}, \end{aligned} \tag{25}$$

where $\|\cdot\|$ denotes the Euclidean vector norm. The first radial function in (25) is Gaussian, and the last two are called Hardy multiquadrics and reverse multiquadrics, respectively [46]. Usually, the radial function width parameter d in (25) is chosen to be about an average distance between the neighboring node centers [43, 44, 48, 49].

Let us introduce the data matrix $\mathbf{Y}$ and the parameter matrix $\boldsymbol{\theta}$ built of vectors (22) and (23)

$$\begin{aligned} \mathbf{Y} &= \big[Y^{(1)}, \ldots, Y^{(N_t)}\big] \in \Re^{N_y, N_t}, \\ \boldsymbol{\theta} &= \big[Z^{(1)}, \ldots, Z^{(N_a)}\big] \in \Re^{N_y, N_a}. \end{aligned} \tag{26}$$

In the exact RBF interpolation, we have $N_a = N_t$. Substituting (23) and (24) into (22) and using (26), we can write the condition of the exact fit for the training data $\widehat{Y}(p^{(j)}) = Y^{(j)}$ in the matrix form

$$\mathbf{Y} = \boldsymbol{\theta} H, \qquad H = \big\{h\big(p^{(i)} - p^{(j)}\big)\big\}_{i,j=1}^{N_t} \in \Re^{N_t, N_t}. \tag{27}$$

The symmetrical matrix H in (27) is called *interpolation matrix*. This matrix has been proved to be invertible for the commonly used radial functions, if the vectors $p^{(j)}$ are distinct [50]. With (23), (26) and (27) we obtain the interpolation of the mapping (21) of the form

$$\widehat{Y} = \hat{f}(p) = \mathbf{Y} H^{-1} \bar{h}(p), \qquad \bar{h}(p) = \operatorname{col}\Big(\big\{h\big(p - p^{(j)}\big)\big\}_{j=1}^{N_t}\Big) \in \Re^{N_t}. \tag{28}$$

It has recently been acknowledged that RBF interpolation minimizes a certain regularization performance index that describes the interpolated surface roughness [43, 47, 51]. Different forms of radial functions (25) correspond to minimization of different regularization indexes.

Note that the approximation (28) is *linear* in the data vectors $Y^{(j)}$ (26). Thus, the computational complexity of the method remains moderate even for a large dimension N_y of the vector Y.

The exact RBF interpolation (28) is *global* in the sense that it has the same form for any $p \in \mathcal{P}$. One needs to complete the most computationally expensive part of (28)—inversion of the matrix H (27)—only once for any number of points, where the approximation (28) of the function (21) is to be computed. Yet this advantage cannot be used if the training set (22) grows with time.

B. Radial Basis Function Network Approximation

Recently, some authors have treated the RBF approximation in the connectionist network setting [28, 30, 51–55]. They consider the radial function centers $Q^{(j)}$ (we will call $Q^{(j)}$ the network node centers) that do not coincide with the training set points, so that the expansion functions in (23) have the form

$$w_j(p) = h\big(p - Q^{(j)}\big), \qquad j = 1, \ldots, N_a. \tag{29}$$

Suppose that the node centers $Q^{(j)}$ $(j = 1, \ldots, N_a)$ are given and fixed vectors. Let us fit the training set data (22) using the network (23). Employing the same notation (26) as in (27), we can represent the fitting problem in regression form

$$\mathbf{Y} = \theta\mathbf{\Phi} + \mathcal{E}, \qquad \mathbf{\Phi} = \big\{h\big(p^{(k)} - Q^{(j)}\big)\big\}_{j,\,k=1}^{N_a,\,N_t} \in \Re^{N_a,\,N_t}, \tag{30}$$

where $\mathcal{E} = [e^{(1)}, \ldots, e^{(N_t)}] \in \Re^{N_y,\,N_t}$ is a residual error matrix. Because we cannot be sure that $\mathbf{\Phi}$ is a well-conditioned or even a full-rank matrix, we will look for a regularized least square solution to (30) that minimizes

$$\|\mathcal{E}\|_F^2 + \alpha\|\theta\|_F^2 \to \min, \qquad 0 < \alpha \ll 1, \tag{31}$$

where $\|\cdot\|_F$ denotes a matrix norm equal to the square root of the sum of the squared matrix entries (the Frobenius norm). In (32), α is a scalar regularization parameter, introduced to obtain the solution of a possibly ill-conditioned problem following the regularization technique of [34]. The parameter α is small and does not influence the solution if the problem is well conditioned. Solving (30) and (31) for $\boldsymbol{\theta}$ gives

$$\boldsymbol{\theta} = \mathbf{Y}\mathbf{\Phi}^T\big(\alpha I_{N_a} + \mathbf{\Phi}\mathbf{\Phi}^T\big)^{-1}, \tag{32}$$

where I_{N_a} is the $N_a \times N_a$ identity matrix.

If $N_t \gg N_a$ and the training set inputs $p^{(j)}$ are uniformly distributed in the input domain, matrix the $\mathbf{\Phi}$ should have rank N_a, because the basis functions (29) are linearly independent. The condition that $\mathbf{\Phi}$ has full rank is called the persistency of excitation condition. As the exact RBF interpolation is known to yield very accurate results, one can expect that an RBF network with fixed centers can provide good approximation accuracy. Additional discussion of the properties of the RBF network with nodes placed on a uniform grid can be found in [31]. The idea discussed in [31] and theoretically studied in more detail in [48] is that an RBF interpolation on a uniform grid performs a *spatial* filtering of the approximated function. Thus, the RBF approximation error is small, if the function has small high-frequency contents.

C. Recursive Identification of the Radial Basis Function Model

Equation (32) describes the computation of the network parameter matrix $\boldsymbol{\theta}$ with the method called *batch learning* in the ANN literature. This method assumes that the whole training data set (22) is available at once. In many practical situations, however, the training data pairs arrive one by one, and a recursive weight updating procedure is desirable. The recursive weight update enables the user not to keep track of all upcoming data, but rather to modify the parameter matrix $\boldsymbol{\theta}$ as the new data arrive. This feature is especially important for RBF network–based nonlinear adaptive control applications, such as these considered further in this chapter.

We can apply a well-known recursive least square estimation method to update an already available estimate of the matrix $\boldsymbol{\theta}$ (26). Let us introduce a regressor vector for the expansion (23), (29)

$$\Phi(p) = \left[h\left(p - Q^{(1)}\right) \cdots h\left(p - Q^{(N_a)}\right)\right]^T \tag{33}$$

and denote $\Phi^{(k)} = \Phi(p^{(k)})$. Note that the vectors $\Phi^{(k)}$ are the columns of the regression matrix $\boldsymbol{\Phi}$ (30). The RBF approximation (28), (29) can be presented in the form

$$\widehat{Y} = \hat{g}(p, \theta) = \theta\Phi(p), \qquad \theta = \left[Z^{(1)} \cdots Z^{(N_a)}\right]^T, \tag{34}$$

where $\theta \in \Re^{N_Y, N_a}$ is an RBF network parameter matrix.

A recursive estimation technique commonly used in signal processing and adaptive control is projection estimation. The projection estimation is a special case of the least mean square algorithm, which is known as the Widrow–Hoff updating rule in the signal processing literature and as a delta rule in the ANN literature. To derive the projection update, instead of minimizing a mean error index (31), let us minimize a one-step error increment index similar to (31). Let $\hat{\theta}^{(k)}$ be an estimate of θ available at step k and $e^{(k)}$ be the kth step approximation error

$$\begin{aligned} \left\|e^{(k)}\right\|^2 + \alpha\left\|\hat{\theta}^{(k+1)} - \hat{\theta}^{(k)}\right\|_F^2 &\to \min, \\ e^{(k)} = Y^{(k)} - \hat{\theta}^{(k)}\Phi^{(k)}&. \end{aligned} \tag{35}$$

The solution of (35) for $\hat{\theta}^{(k)}$ has a form similar to (32)

$$\hat{\theta}^{(k+1)} = \hat{\theta}^{(k)} + a^{(k)} e^{(k)} \Phi^{(k)^T} / \left(\alpha + \|\Phi^{(k)}\|^2\right), \tag{36}$$

where $\|\Phi^{(k)}\|^2 = \Phi^{(k)^T} \Phi^{(k)}$ and the dead-zone parameter $a^{(k)}$ is 1. In the presence of the approximation error, $a^{(k)}$ should be chosen zero inside a dead zone in the usual way to ensure robust convergence of the algorithm despite the approximation error and, possibly, measurement noise (see, e.g., [56] for more details).

Let us discuss the algorithm convergence issue. We assume that the RBF network approximation (34) of the mapping (21) can be made accurate enough by appropriate choice of the network weight matrix θ; that is, for some $\theta = \theta_*$,

$$\left\| g(p) - \hat{g}(p, \theta_*) \right\| \leq \delta_Y, \tag{37}$$

where δ_Y is a "sufficiently" small approximation error. The convergence of the recursive estimation algorithm (36) in the presence of the approximation error and, possibly, measurement noise can be guaranteed according to the standard results of adaptive control and parameter estimation theory [56, Section III.D, pp. 88–91]. To ensure the convergence, the dead-zone parameter sequence $a^{(k)}$ should be chosen as

$$a^{(k)} = \begin{cases} 0, & \text{if } \|e^{(k)}\| \leq \delta_Y, \\ 1, & \text{otherwise.} \end{cases} \tag{38}$$

Projection estimation of RBF network weights in adaptive control of a nonlinear system is considered, for instance, in [57]. The papers [25, 41] consider an application of the orthogonal least square modification of the recursive least square (RLS) algorithm to RBF network approximation. The papers [25, 29] consider modifications of the RLS identification of an RBF network for the cases of dynamical node creation, update and clustering of the node centers. In this chapter, we consider the number of nodes and the node centers as fixed parameters.

The well-known condition for the estimation algorithm (36) to converge to the correct parameter matrix θ_* is the *persistency of excitation* condition [56]. The persistency of excitation requirement is that, for $N_p \geq 1$, $\delta > 0$ exists such that, for any n,

$$\underline{\sigma}\left(\sum_{k=n+1}^{n+N_p} \Phi^{(k)} \Phi^{(k)^T} \right) > \delta, \tag{39}$$

where $\underline{\sigma}(A)$ denotes a minimal singular value of the matrix A. According to (33), the persistency of excitation depends on the sequence of training inputs $p^{(k)}$. In [58], it is proved that in the RBF network identification, persistency of excitation is provided if the inputs $p^{(k)}$ are in certain neighborhoods of the network node centers $Q^{(k)}$.

It is a well-established fact that for projection update the prediction errors $e^{(k)}$ always converge into the $2\delta_Y$ dead zone [56]. The convergence result is valid for any regressor vector sequence, that is, for an arbitrary task parameter vector sequence $\{p^{(k)}\}$.

D. Radial Basis Function Approximation of Task-Dependent Feedforward

Let us return to Problem 1 as introduced in Section II.D. It is possible to design a task-dependent approximation $\widehat{U}(p)$ of the optimal feedforward vector on the task parameter vector p (5) by using the RBF network approximation technique discussed in the beginning of this section.

Let us assume that the optimal feedforward vectors $U_{*,k} = U_*(Q^{(k)})$ are known for certain (discrete) values $Q^{(k)}$ of the task parameter vectors p. Given N_a pairs of vectors $Q^{(k)} \in \mathcal{P}$ and $U_{*,k}$ as $\{Q^{(k)}, U_{*,k} = U_*(Q^{(k)})\}_{k=1}^{N_a}$, we can approximate the mapping $U_*(p)$ over the given domain $\mathcal{P}$ of the task parameter vector p by using an exact RBF interpolation as discussed in Section III.A. This approximation can also be represented by an RBF network of the form

$$\widehat{U}(p) = K\Phi(p), \tag{40}$$

where $\Phi(p) \in \Re^{N_a}$ is the RBF regressor vector (33) and $K \in \Re^{N_U, N_a}$ is the RBF network weight matrix.

It follows from (23), (24) and (28), (40) that the matrix K can be found from exact RBF interpolation conditions as

$$K = [U_{*,1} \cdots U_{*,k}]H^{-1}, \tag{41}$$

where $H \in \Re^{N_a, N_a}$ is the RBF interpolation matrix of the form (27)

$$H = \left\{h(Q^{(i)} - Q^{(j)})\right\}_{i,j=1}^{N_a}. \tag{42}$$

The RBF network task-level controller defined by (40) is schematically shown in Fig. 2. A sequence of tasks, each defined by a task parameter vector p, is generated externally with respect to the controller. The vector p is supplied to the input of the system mapping $Y = S(U, p)$ (15) and to the input of the controller, which computes the feedforward U by using the RBF network approximation. The task parameter vector p changes in discrete time (from task to task). This change cannot be predicted by the controller and can be considered as an external disturbance acting on the task completion process. The designed RBF controller provides a feedforward compensation for this (measurable) disturbance.

Design of the proposed RBF network controller is straightforward. We assume that the dependence (20) of the optimal feedforward vector on the task parameters is a smooth mapping. Such a mapping can be approximated by an RBF network

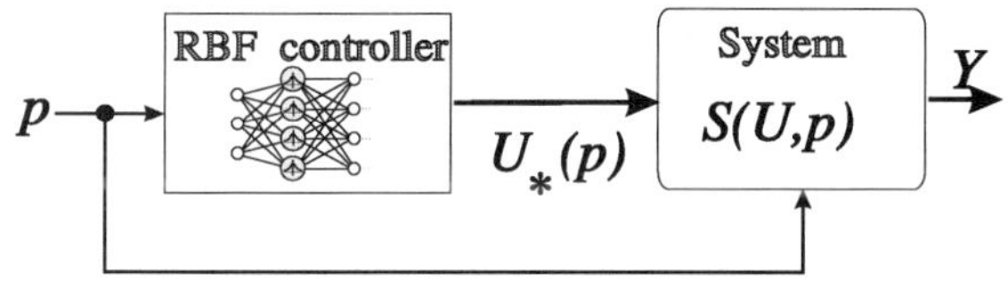

Figure 2 Schematics of feedforward controller using RBF approximation of dependence on task parameter vector.

to arbitrary accuracy, provided the grid of the RBF nodes $Q^{(k)}$ is dense enough. Thus, for $K = K_*$,

$$\left\| K_* \Phi(p) - U_*(p) \right\| \leq \delta_U, \tag{43}$$

where, as mentioned previously, δ_U can be made as small as needed by choosing a denser grid of the RBF node centers. The RBF network approximation weights K_* can be computed according to (41) or by another method.

The optimal feedforward vectors $U_{*,k}$ for the RBF interpolation (41) can be obtained in different ways. The vectors $U_{*,k}$ can be computed in the course of numerical optimization by using a detailed numerical model of the system. An advantage of an RBF network approximation in this case is that it can be used for a fast computation in real time, while the computationally expensive numerical optimization is done off-line. An application of this approach in the control of car parking is considered in [21], in the control of a free-flying space robot in [22], and in three-dimensional slewing maneuvers of a flexible spacecraft system in [59].

Alternatively, the optimal feedforward vectors can be learned in the course of iterative repetitive experiments with the system, where the task parameter vector p repeatedly takes the respective value $Q^{(k)}$. This approach, discussed in the next section, does not require detailed knowledge of the system dynamics. An experimental application of such an approach in the control of fast motions for a direct-drive robot is studied in [23].

IV. LEARNING FEEDFORWARD FOR A GIVEN TASK

This section is devoted to the problem of learning a single optimal shape vector U_*. We assume that the task parameter vector p is fixed. Therefore, we will not write explicitly dependencies on p unless this is needed to avoid ambiguity. The section has a didactic purpose and exposes some background ideas of learning control that are further elaborated in the more comprehensive approaches of Sections V and VI.

A. Learning Control as On-Line Optimization

Let us assume that the parameter vector p is fixed. In order to achieve the control goal, a feedforward shape vector U_* has to be found that minimizes the performance index (18). Let us first assume that an estimate $\widehat{G}$ for the Jacobian matrix $G = \partial S/\partial U$ of the mapping (15) is known for the optimal input U_*. Let $U^{(n)}$ and $Y^{(n)} = S(U^{(n)})$ be input and output vectors obtained at iteration n. By using the Levenberg–Marquardt algorithm [60], the next minimizing input guess can be computed as

$$U^{(n+1)} = U^{(n)} - \left(I_{N_u}(\rho + \mu_n) + \widehat{G}^T \widehat{G}\right)^{-1} \left(\rho U^{(n)} + \widehat{G}^T \left(Y^{(n)} - Y_d\right)\right), \qquad (44)$$

where $\mu_n \geq 0$ is a step length parameter, I_{N_U} is an $N_U \times N_U$ unity matrix, and $\widehat{G} = G(U^{(n)})$.

The motivation for the update (44) is as follows [60]. Let us consider a local affine model of the mapping (15) of the form

$$\widehat{Y} = Y^{(n)} + \widehat{G}\left(U - U^{(n)}\right). \qquad (45)$$

Let us also demand that the minimization step length does not exceed a given value $d_n > 0$:

$$U^{(n+1)} = U^{(n)} + s^{(n)}, \qquad \left\|s^{(n)}\right\| \leq d_n. \qquad (46)$$

By solving (18) and (45) with respect to $U^{(n+1)}$ and using the Lagrange multiplier method to satisfy the constraint (46), we arrive at (44). The Lagrange multiplier μ_n is nonnegative and can be computed once the Jacobian estimate $\widehat{G}$ and the allowed step length d_n are given. With an increase of μ_n in (44), all eigenvalues of the inverted matrix in (44) increase; hence, the step length $\|s^{(n)}\|$ decreases. Therefore, the dependence of μ_n on d_n is decreasing. In practice, instead of computing μ_n from d_n; usually μ_n itself is made a parameter of choice. More details on the method can be found in [60].

If μ is small, the method approximates the Newton–Gauss method; if μ is large, it approximates the downhill (gradient) method. Each iteration in (44) presumes a repeated execution of the same given task; each time using a different feedforward input $U^{(k)}$ and measuring the corresponding sampled output vector $Y^{(k)}$. Thus, (44) is a learning control iteration. In the commercially available software implementing the Levenberg–Marquardt method, the step-limiting parameter μ in (44) is chosen at each step. This, however, requires several evaluations of the minimized function. In learning control problems, each evaluation is a completion of the task for the controlled system and, thus, has a very large cost. Therefore, we are using the update (44) with a constant preselected parameter μ.

The schematics of the discussed learning update are shown in Fig. 3. At step k, the feedforward vector $\widehat{U} = U^{(k)}$ stored in the memory is applied to the system

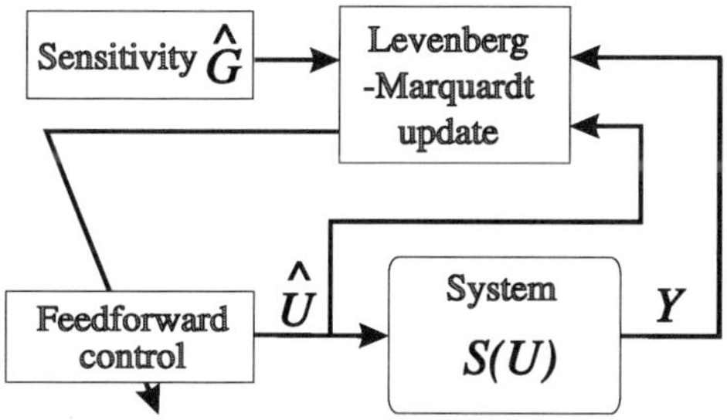

Figure 3 Schematics of learning update in a feedforward controller.

and the output $Y^{(k)} = S(U^{(k)})$ is obtained. This output us used in the Levenberg–Marquardt update (44) to compute an update for $\widehat{U}$. The updated value of $\widehat{U}$ is applied at the next iteration and so on. As shown in Fig. 3, the update uses an estimate $\widehat{G}$ for the sensitivity matrix $G(U_*)$.

The update rule (44) presumes that an estimate of the gradient (input–output sensitivity) matrix G is known and is sufficiently accurate. If unknown, this matrix can be estimated with a finite-difference method. The next subsection presents a simple result showing that the update (44) has certain robustness to error in the estimate $\widehat{G}$. Section IV.C discusses a finite-difference update for this estimate. Such an update would add an extra feedback loop for the gain $\widehat{G}$ in Fig. 3 and make the learning algorithm adaptive.

B. Robust Convergence of the Learning Control Algorithm

Analysis of the convergence of the Levenberg–Marquardt algorithm for a non-linear problem can be found in [60]. This analysis assumes that the Jacobian G is known exactly at each step. Herein, we consider the update (44) as a part of a discrete-time closed-loop control system. Following the established approach to analysis of such systems, let us study robust stability of the linearized loop. We assume that the system (15) is affine in U in the vicinity of the optimum. The affine model has the form

$$Y = GU + Z. \tag{47}$$

In the linear-quadratic setting (18), (47), the Levenberg–Marquardt algorithm (44) converges for any positive values of the parameters ρ and μ, and it is robust with respect to the error in estimating the matrix G. A sufficient condition for the convergence is given by the following theorem.

THEOREM 1. *Let us consider the update* (44) *of the system* (47) *input. The algorithm asymptotically converges for any initial condition* $U^{(1)}$ *if some* $k_0 \geq 1$ *exists such that for any* $k \geq k_0$ *the maximal singular value of the gradient estimation error satisfies the following inequality:*

$$\bar{\sigma}(\widehat{G} - G) < \frac{2\rho}{\sqrt{\rho + \mu}}. \tag{48}$$

Theorem 1 shows that the convergence robustness improves for larger values of the regularization parameter ρ and is absent if no regularization is performed. At the same time, increasing ρ increases the steady-state tracking error $\|Y - Y_d\|$ for the convergence achieved.

Proofs of Theorem 1 can be found in the papers [17, 23], which also present an analysis of static error for the learned feedforward depending on the error in the estimation of matrix G. The papers [17, 23] demonstrate experimental results in application of the learning control update of the form (44) to trajectory control of robotic arms.

C. FINITE-DIFFERENCE UPDATE OF THE GRADIENT

Let us proceed with a situation, where the Jacobian G is not known and we can only evaluate the function mapping (15) pointwise, by executing the respective task with the given input U and observing the output Y. In this case, a possible approach is to introduce an affine model (45) of the mapping (15) and update estimates of parameters of this model from the available input–output measurements.

The most common practically used method for estimating the Jacobian G is the *Broyden secant update*. Let $\widehat{G}^{(n)}$ be an estimate of the Jacobian at step n. Denote by $s^{(n)}$ the variation of input, and by $w^{(n)}$ the corresponding variation of the output at the previous minimization step. For a small step length $\|s^{(n)}\|$, the updated estimate should provide a fit to the observed data, that is,

$$Gs^{(n)} = w^{(n)}, \qquad s^{(n)} = U^{(n)} - U^{(n-1)}, \qquad w^{(n)} = Y^{(n)} - Y^{(n-1)}. \tag{49}$$

The Broyden update rule can be considered as an application to (49) of the projection estimation algorithm [56], which is very popular in adaptive control and signal processing applications. The Broyden update is used in conjunction with the input update (44) and has the form

$$\widehat{G}^{(n+1)} = \widehat{G}^{(n)} + \left(w^{(n)} - \widehat{G}^{(n)}s^{(n)}\right){s^{(n)}}^T / \left(c^2 + \|s^{(n)}\|^2\right), \tag{50}$$

where $c > 0$ is a scalar parameter used to avoid division by zero.

For a nonlinear mapping, local convergence of the Levenberg–Marquardt algorithm with the Broyden secant update can be proved using the *bounded deterioration* technique as considered in [60]. The idea of such a proof is that in the vicinity of the optimum and for a sufficiently small initial error of approximating the gradient G, the algorithm will converge before the gradient approximation error will have time to grow owing to the system nonlinearity.

Let us now consider another method for estimating the Jacobian G, which can be more appropriate if the measurements are corrupted with a noise. In the on-line learning algorithms discussed in the subsequent sections, approximation errors can be considered as such a noise. Let us write the affine model (47) for the mapping (15) in the form of a linear regression

$$\widehat{Y} = \widehat{G}U + \widehat{Z} = \hat{\theta}W; \qquad \hat{\theta} = [\widehat{Z}/c \; \widehat{G}], \qquad W = \begin{bmatrix} c \\ U \end{bmatrix}, \tag{51}$$

where c is a positive scaling constant, $\hat{\theta} \in \Re^{N_Y, N_U+1}$ is a regression parameter matrix, and W is a regressor vector. The Broyden gradient update (50) is a two-step estimation algorithm for the regression (51) that first sets $\widehat{Z} = \widehat{G}U^{(n-1)} - Y^{(n-1)}$ in (51) and then updates an estimate for G with the projection method. A more natural, one-step projection estimation algorithm for the model (51) has the form

$$\hat{\theta}^{(n+1)} = \hat{\theta}^{(n)} + a^{(n)}\big(Y^{(n)} - \hat{\theta}^{(n)}W^{(n)}\big)W^{(n)^T} / \|W^{(n)}\|^2, \tag{52}$$

where $\hat{\theta}^{(n)} = [Z^{(n)}/c \quad G^{(n)}]$, $W^{(k)} = [c \quad U^{(k)^T}]^T$, and $a^{(k)} \in \{0, 1\}$ is a scalar dead-zone parameter. Unlike the secant update (49), (50), which uses function values obtained on *two* consecutive steps, the update (52) uses only *one* function value. This makes it possible to generalize the update (52) for the optimization with changing task parameters, as shown in the next sections.

Let us write a step of the Levenberg–Marquardt algorithm for the affine model in the form (51). By solving (18) and (51) with respect to $U^{(n+1)}$ and using the Lagrange multiplier method to satisfy constraint (46), we obtain

$$\begin{aligned} U^{(n+1)} &= U^{(n)} - \big(I(\rho + \mu_n) + G^{(n)^T} G^{(n)}\big)^{-1} \\ &\quad \times \big(\rho U^{(n)} + G^{(n)^T}\big(\widehat{Y}^{(n)} - Y_d\big)\big), \end{aligned} \tag{53}$$

where $\widehat{Y}^{(n)} = G^{(n)}U^{(n)} + Z^{(n)}$. We have $\widehat{Y}^{(n)} = Y^{(n)}$ for the projection update (52), as long as $a^{(n)} = 1$. Therefore, (53) coincides with the Levenberg–Marquardt step (44). A more detailed theoretical study is presented in [61]. The experiments in applying learning control algorithms with an adaptive update of G as considered in this section to trajectory control of a direct-drive manipulator are described in [23].

V. ON-LINE LEARNING UPDATE IN TASK-DEPENDENT FEEDFORWARD

This section considers Problem 3 stated in Section II.D. We assume that a model of the system is available for the design of the task-dependent feedforward controller. This model is, however, imprecise and the approximation $\widehat{U}(p)$ of the optimal control input (1), (13), (16) is not satisfactorily accurate.

This section presents an algorithm that allows us to update (learn) the approximation $\widehat{U}(p)$ in the course of normal system operation, that is, assuming that the sequence $p^{(k)}$ of the task parameter vector is arbitrary. Upon completion of each task, characterized by the task vector $p^{(k)}$, the output vector $Y^{(k)}$ is used to compute an update of the controller approximation $\widehat{U}(p)$ available at this step. In what follows, we propose and study such an update.

A. APPROXIMATING SYSTEM SENSITIVITY

Similarly to the learning algorithm considered in Section IV, the algorithm of this section updates the guess of the optimal feedforward control based on the available estimate of the system input–output sensitivity matrix G in (17). Unlike Section IV, in this section we have to consider the dependence of this sensitivity matrix on the changing vector p. Let us introduce the matrix-valued function defining the system sensitivity at the optimal feedforward input

$$G_*(p) = \left.\frac{\partial S(U, p)}{\partial U}\right|_{U=U_*(p)}. \tag{54}$$

Though the mapping (54) is not known exactly, it can be approximated based on the available system model, in the same way as the approximation $\widehat{U}(p)$ for the optimal control is built in Section III.

Let us assume that for each of the RBF approximation nodes $p = Q^{(k)}$ used for building approximation of the optimal feedforward in Section III.D, the sensitivity matrix $G_{*,k} = G_*(Q^{(k)})$ is computed along with the optimal input vector $U_{*,k} = U_*(Q^{(k)})$. The sensitivity matrix would usually be computed as a by-product of a numerical optimization procedure (such as Levenberg–Marquardt) applied to the available system model to find $U_*(Q^{(k)})$. In the process of the numerical optimization, a matrix $G_*(Q^{(k)})$ can be obtained, for instance, by a finite-difference method.

Similarly to the approximation (40) for the mapping $U_*(p)$, let us use an RBF network approximation for the mapping $G_*(p)$. Unlike a *vector-valued* mapping $U_*(p)$, the mapping $G_*(p)$ is *matrix-valued*. To facilitate work with such mappings, let us introduce the vectorization operator vec$(\cdot)$. For a matrix $A \in \Re^{m,n}$, the vector vec$(A) \in \Re^{mn}$ is composed of all the entries of A, column by column.

An RBF network approximation $\widehat{G}(p)$ for $G_*(p)$ can be presented in the form, similar to (40),

$$\operatorname{vec}(\widehat{G}(p)) = \sum_{j=1}^{N_a} \operatorname{vec}(G_j) h(p - Q^{(j)}) = \Gamma \Phi(p), \qquad p \in \mathcal{P}, \tag{55}$$

where $G_j \in \Re^{N_Y, N_U}$ are the RBF network weights and $\Phi(p)$ is the RBF regression vector (33). Note that (55) can be also represented in the form

$$\widehat{G}(p) = \sum_{j=1}^{N_a} G_j h(p - Q^{(j)}). \tag{56}$$

The weights of the RBF network (55) can be computed so as to implement exact RBF interpolation of the matrices $G_*^{(k)} = G_*(Q^{(k)})$. In this case, the network node centers in (55) coincide with the data points $p = Q^{(k)}$ and the weight matrix Γ can be computed similarly to (41) as

$$\Gamma = \left[\operatorname{vec}(G_1^{(1)}) \quad \cdots \quad \operatorname{vec}(G_{(N_a)})\right] H^{-1}, \tag{57}$$

where H is the RBF interpolation matrix of the form (42).

Under general conditions, the function $G_*(p)$ has derivatives, which are uniformly bounded on $\mathcal{P}$, provided that the derivatives of $U_*(p)$ are bounded. Hence, the error of the RBF network approximation (55) can be made small for high-order N_a of the expansions (55).

Computing and storing the approximation $\widehat{G}(p)$ of the sensitivity matrix $G_*(p)$ is similar to the usual practice of storing the system gain information along with setpoints as a part of a feedback controller design. It is particularly related to gain scheduling methods, where gain and setpoint tables are stored in the controller.

B. Local Levenberg–Marquardt Update

In this subsection, we assume that the RBF approximation $\widehat{U}(p)$ (40) is not accurate enough and design an update for the feedforward, using the input–output data for the system (15). Let us assume that in task k an input vector $U^{(k)}$ (13) was applied to the system, and an output vector $Y^{(k)}$ (14) was obtained, while the task parameter vector was $p^{(k)}$. Note that the applied input will be $U^{(k)} = \widehat{U}(p^{(k)})$, where $\widehat{U}(p)$ is the approximation (40) of the optimal feedforward mapping as available at this step.

Following the usual practice of iterative optimization, as discussed in the derivation of the Levenberg–Marquardt algorithm in Section IV.A, let us consider an affine model of the mapping (15). A local model valid for the current task, that is, for $p = p^{(k)}$, can be obtained by using the input–output data $U^{(k)}$, $Y^{(k)}$ and the sensitivity estimate $\widehat{G}(p)$ (55). This local affine model has a form similar to (45)

$$\widehat{Y}\big(p^{(k)}\big) = Y^{(k)} + \widehat{G}\big(p^{(k)}\big)\big(U - U^{(k)}\big). \tag{58}$$

By substituting the affine model into the performance index (18) and finding the minimum, we arrive at the optimality condition

$$\widehat{G}\big(p^{(k)}\big)\big(\widehat{Y}\big(p^{(k)}\big) - Y_d\big) + \rho U = 0, \tag{59}$$

where $\widehat{Y}(p^{(k)})$ is defined by (58).

By solving (58), (59) and limiting the update step as in (46), we obtain the Levenberg–Marquardt update similar to (44). This update gives us $U = U^{(k|k+1)}$, the *a posteriori* optimal control input for task k (task parameter vector $p = p^{(k)}$),

$$\begin{aligned} U^{(k|k+1)} &= U^{(k)} + \Delta U^{(k)}, \\ \Delta U^{(k)} &= -\big(\widehat{G}^T\big(p^{(k)}\big)\widehat{G}\big(p^{(k)}\big) + (\rho + \mu)I\big)^{-1} \\ &\quad \times \big(\widehat{G}^T\big(p^{(k)}\big)\big(Y^{(k)} - Y_d\big) + \rho U^{(k)}\big), \end{aligned} \tag{60}$$

where $\Delta U^{(k)}$ is the update step for the feedforward U. Note that the update (60) is calculated assuming that the task parameter vector is fixed, $p = p^{(k)}$. In fact, our goal is to calculate an update for the RBF approximation controller (40) for all p. This can be done based on the update (60) as explained in the next subsection.

C. Update of Radial Basis Function Approximation in the Feedforward Controller

As mentioned previously, we assume that the feedforward vector U applied at step k is computed using the task-dependent RBF approximation controller (40) shown in Fig. 2. In accordance with (40), this control has the form

$$U^{(k)} = K^{(k)}\Phi\big(p^{(k)}\big), \tag{61}$$

where $K^{(k)}$ is the weight matrix of the feedforward RBF controller available at step k. Based on the *a posteriori* optimal feedforward solution (60), the weight matrix K should be modified so that the controller would yield this new optimal solution for $p = p^{(k)}$. The latter condition can be written as

$$U^{(k|k+1)} = K^{(k+1)}\Phi\big(p^{(k)}\big), \tag{62}$$

where $K^{(k+1)}$ is the updated RBF network weight matrix.

By using (60) and (62), we obtain a projection update for the controller (61) in the same way as (36) is obtained from (35). Modified to include a dead-zone compensation for the approximation error, this update has the form

$$K^{(k+1)} = K^{(k)} - a^{(k)} \Delta U^{(k)} \Phi^T\left(p^{(k)}\right) / \left\| \Phi\left(p^{(k)}\right) \right\|^2, \tag{63}$$

where $a^{(k)} = \{0, 1\}$ is a dead-zone parameter and $\Delta U^{(k)}$ is defined by (60). The update (60), (63) is analogous to the projection update (36). As one can easily check by substitution, for $a^{(k)} = 1$, (62) holds exactly. For $a^{(k)} = 0$, no update is performed.

Figure 4 illustrates the proposed design of the learning controller. The task parameter vectors p are generated externally to the diagram of Fig. 4 and are supplied to the controller (40) computing an RBF approximation for $U_*(p)$. The RBF approximation in (40) is updated in accordance with (60), (63) depending on the system output $U^{(k)}$. The update (60) uses the approximation for the Jacobian $\widehat{G}(p)$ in (60), which is computed by the RBF network (55). The weights of the network (55) are computed off-line, for example, according to (57).

The dead-zone parameter $a^{(k)}$ in (63) should be chosen similarly to (38) in order to ensure the algorithm convergence. The dead zone must compensate for the approximation error (43) and also for the initial error of the approximation (61). It is possible to demonstrate dead-zone convergence of the proposed algorithm and estimate the necessary dead zone by using a modification of standard convergence results in [56]. However, presenting such proof is beyond the scope of this

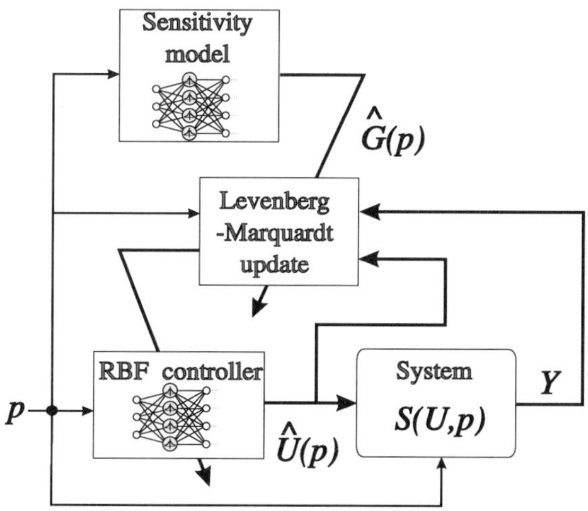

Figure 4 Schematics of learning update in a task-dependent RBF approximation feedforward controller.

chapter. The algorithms of this section have been applied by the author in control of slewing maneuvers of a flexible spacecraft system with nonlinear rotational dynamics.

VI. ADAPTIVE LEARNING OF TASK-DEPENDENT FEEDFORWARD

This section proposes a solution to Problem 4 stated in Section II.D. The problem is to learn (update) an approximation $\widehat{U}(p)$ (18) of the optimal feedforward input for an arbitrary sequence of the parameter vectors p (5) by using only input–output data. Unlike the previous section, an accurate approximation for the Jacobian $G_*(p)$ dependence on the task parameters is no longer assumed to be available in advance. The available approximation of the Jacobian is assumed to contain an error and is refined on-line as a part of the learning controller we are going to develop.

In order to solve the stated problem, we introduce an affine RBF network model of the system mapping and estimate this model on-line. The algorithm discussed in this section resembles the algorithm for on-line parametric nonlinear least square (NLS) optimization proposed and studied in [61, 63–65]. This algorithm can be viewed as a discrete-time adaptive algorithm for nonlinear system control.

A. Affine Radial Basis Function Network Model of the System Mapping

The on-line parametric optimization algorithm of [62] we are about to derive can be considered as an extension of the Levenberg–Marquardt algorithm. Similarly to a standard derivation of the Levenberg–Marquardt algorithm given earlier, our derivation here will be based on an affine model (45) of the mapping (15). Such an affine model can be written in the form

$$\widehat{Y} = \widehat{G}(p)U + \widehat{Z}(p), \tag{64}$$

where $\widehat{G}(p)$ and $\widehat{Z}(p)$ are (smooth) matrix and vector functions of p. For a fixed task parameter vector p, the model (64) is affine in U; at the same time, the model depends on p in a nonlinear way.

Let us introduce the functions:

$$Y_*(p) = S\big(U_*(p), p\big), \qquad Z_*(p) = Y_*(p) - G_*(p)U_*(p), \tag{65}$$

where $G_*(p)$ is given by (54). Similarly to (56), let us use RBF networks for approximating the functions $Z_*(p)$ and $G_*(p)$. Let us assume that these mappings

can be represented in the form

$$Z_*(p) = \sum_{j=1}^{N_a} Z_{*j}h\big(p - Q^{(j)}\big) + \delta Z(p),$$

$$G_*(p) = \sum_{j=1}^{N_a} G_{*j}h\big(p - Q^{(j)}\big) + \delta G(p),$$

where $Z_{*j} \in \Re^{N_Y}$ and $G_{*j} \in \Re^{N_Y,N_U}$ are the ex
errors δ_Z and δ_G can be made small for high-ord
(67), that is, by selecting sufficiently high-density F

We are now in a position to explain the basic al
are going to use for adaptive update of the controll
riving the algorithm, we neglect the approximation
errors are taken into account in the algorithm conve
the absence of the approximation error in (42) (for
sented in linear regression form

$$U_*(p) = K_*\Phi(p).$$

Similarly to (68), we can present (66) and (67) i
ear in the weights Z_{*j} and G_{*j}. Keeping these re
and (67) in mind, a model of the form (51) can be
regression:

$$\widehat{Y} = \Theta\overline{\Phi}(p, U), \qquad \Theta \in \Re^{N_Y, N_a(N_U+1)},$$

$$\overline{\Phi}(p, U) = \Phi(p) \otimes W, \qquad W =$$

where $\otimes$ denotes the Kronecker (direct) product
a scalar scaling parameter, $\Phi(p)$ is the regressor
extended regressor vector, and Θ is the RBF net
fixed parameter p, the model (69) has the form
$[Z_1/c \quad G_1 \quad \cdots \quad Z_{N_a}/c \quad G_{N_a}]$ into (69), one ob
form (64), where

$$\widehat{Z}(p) = \sum_{j=1}^{N_a} Z_j h\big(p - Q^{(j)}\big), \qquad \widehat{G}(p) = \sum$$

We assume that for $\Theta = \Theta_*$ and in the absence
that is, for $\delta y = \delta z = \delta u = 0$, the affine model (69) gi

$p^{(k+1)}, U^{(k+1|k)}$). By substitut-
J into (18) and optimizing the
$\|\cdot^{(k)}\| \le d_k$, we obtain

$^{(k+1)})(\widehat{Y}^{(k+1|k)} - Y_d)$

$\|$ (76)

$\|$ (77)

uced to comply with the step-
lassical Levenberg–Marquardt
d_k is monotone nonincreasing
omputing the Lagrange multi-
lf a parameter of choice. Recall
he weights of the RBF network
and (74), we can write

$^{(k+1)})$. (78)

ie RBF weight matrix update
obtain a step of the proposed

$^{1}\big(\widehat{G}^T(p^{(k+1)})$

$^{k)})\dfrac{\Phi^T(p^{(k+1)})}{\|\Phi(p^{(k+1)})\|^2},$ (79)

Φ $\widehat{U}^{(k+1|k)}$, by (73) and (33);
), (73), (75) as $\widehat{Y}^{(k+1|k)} =$

can be written in regression
update algorithm, we can use
neter matrix Θ in (69). This

$\overline{\Phi}^{(k)T}/\|\overline{\Phi}^{(k)}\|^2,$ (80)

p k and $\overline{\Phi}^{(k)} = \overline{\Phi}(p^{(k)}, U^{(k)})$
ate (80) can be considered as
0), $a^{(k)}$ is a scalar dead-zone
mpensate for the influence of
) is zero if the prediction error
defined by the approximation

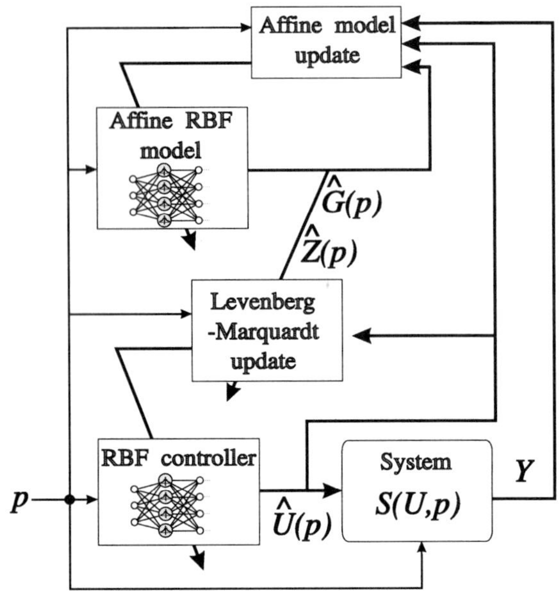

Figure 5 Schematics of adaptive learning update in a task-dependent RBF approximation feedforward controller.

Figure 5 illustrates the designed controller. Vector p is defined externally to the control diagram of Fig. 5 and acts as a disturbance for control (18). The Levenberg–Marquardt update (79) modifies the weight matrix K in the RBF network controller (18). The Levenberg–Marquardt update (79) uses estimates of the output $\widehat{Y}^{(k+1|k)}$ (74) and of the Jacobian (sensitivity) matrix $\widehat{G}(p)$ obtained from the affine model (69), (70). The external loop in Fig. 5 updates an estimate for this affine model according to (80).

The discussion on the choice of the dead-zone parameter $a^{(k)}$ in (80), as well as a proof for the algorithm convergence, are considered in [61, 62]. To ensure convergence of the estimation algorithm, a small self-excitation signal can be added to the computed control $U^{(k)}$ before it is applied to the system. This self-excitation is needed to make the regressor vector sequence in (80) persistently exciting as discussed in [57, 62].

C. Discussion

Equations (33), (70), (79), and (80) constitute the basic algorithm for on-line parametric NLS optimization first proposed in [62]. An analysis of the algorithm convergence is presented in [61, 62]. For a generic nonlinear mapping (15), it is only possible to prove *local* convergence of the algorithm. The locality here means that the initial approximation to the nonlinear feedforward control mapping (68), (33) should be sufficiently close to the optimum. The domain of the algorithm convergence theoretically ensured in [61, 62] depends on the degree of the system nonlinearity (second derivative bound).

The local convergence results of [61, 62] demonstrate that the algorithm of this section is consistent. It can be shown that, under certain persistence of excitation conditions (such as discussed in [58]), the parameter matrix Θ of the RBF affine model (69) converges into a dead-zone neighborhood of the matrix (72). At the same time, the approximation (61) of the feedforward converges into a neighborhood of the "best" approximation (43). The practical usefulness of the algorithm depends on its performance in a particular application. The next subsection considers one application of the proposed approach.

D. Application Example: Learning Control of Flexible Arm

This section applies the developed learning algorithm to the control of fast motions in a flexible-joint arm. No *a priori* knowledge of the system dynamics is assumed to be available. Let us consider a planar flexible-joint arm example as introduced in Section II.B, and the problem of point-to-point control for such an arm. We assume that the joint torques of the arm are computed according to (10), where the desired trajectory is planned as in (12).

The control problem is to compute feedforward $u(\cdot)$ so that the arm comes to the final position $q_d(T) = \nu$ at time $T = 1.5$ without oscillations. This is a difficult problem, because the oscillation period for the lowest eigenfrequency of the system is close to unity.

We divide the motion interval $[0, T]$ into seven subintervals $[\tau_j, \tau_{j+1}]$ ($j = 0, \ldots, 7$), $\tau_0 = 0$, $\tau_j = T$, and consider the feedforward input (13), $U \in \Re^{12}$, that is piecewise linear on these subintervals and zero at times 0 and T. In other words, the shape functions $\phi_j(\cdot)$ in (13) are the first-order (triangular) B splines. We monitor the arm motion on the interval $[T, T_f]$, $T_f = T + 0.5$ at $L = 14$ uniformly spaced output sampling instants $t_1 = T, \ldots, t_{14} = T_f$. The measurement vector y comprises drive angles and joint deformations, $y = [q^T \; (q - \gamma)^T]^T \in \Re^4$. By sampling the vector y at instants t_j, we obtain the output vector $Y \in \Re^{56}$.

The task mapping (15) in this problem depends on the task parameter vector p (5) that includes the initial and desired final configurations of the arm. Because the system is cyclic, the control depends only on the *variation* of the first joint angle. Thus, we can write the task parameter vector p in the form

$$p = \begin{bmatrix} q_{2d}(0) & q_{2d}(T) & \big(q_{1d}(T) - q_{1d}(0)\big) \end{bmatrix}^T. \qquad (81)$$

We assume that the task parameter vector (81) remains bounded inside the following domain:

$$\mathcal{P} = \left\{ p \in \mathcal{P}\colon \frac{\pi}{4} \le q_{2d}(0),\ q_{2d}(T) \le 3\pi/4; \right.$$
$$\left. -\frac{\pi}{2} \le q_{1d}(T) - q_{1d}(0) \le \pi/2 \right\}. \quad (82)$$

To implement the adaptive task-level learning algorithm of Section VI.B, we use a network of the Gaussian radial basis functions with the node centers $Q^{(j)}$ placed on a uniform mesh $5 \times 3 \times 3$ in the task parameter space. When *implementing* the control algorithm, we assume the dynamical model of the system to be completely unknown and set the initial estimate of the parameter matrix Θ in (80) to be zero.

The adaptive update algorithm of Section VI.B was implemented as a Matlab program on an 1.2 MFlops computer, and the arm motion simulation was coded in C. The task-level control algorithm does not exploit any initial knowledge of the controlled system dynamics available for simulation and uses just the input–output data for the control tasks. Given the input and output dimensions $U \in \Re^{12}$ and $Y \in \Re^{56}$ and the number of RBF network nodes $N_a = 45$, the sizes of the matrices in (70) are $\overline{\Phi}(p, U) \in \Re^{585}$ and $\Theta \in \Re^{56,585}$. These sizes cause no computational problems, as the updates (79) and (80) only include matrix multiplications, and the matrix inverted in (79) has the (15) size.

For our Matlab implementation of the algorithm, the control update (79) took 0.16 s, and the affine model update (80), 0.23 s. These computational delays could be acceptable even for the feedforward control of a real-life system, because the updates need to be done only once for each motion. The computation of control in accordance with (40) takes less than 25 ms, which suggests that the proposed algorithm is feasible for on-line control, especially if the updates (79) and (80) are scheduled outside time-critical feedback loops.

When *simulating* the planar arm motion, we assume that the arm links are uniform rods of unit mass and length. We take the moments of inertia of the drive rotors as $J = \mathrm{diag}\{2, 2\}$, the damping in drives as $B = \mathrm{diag}\{0, 0\}$, and the angular stiffnesses of the lumped elastic elements in the joints as $K = \mathrm{diag}\{200, 200\}$. We further assume that the angular position gain of the PD feedback controller (10) is $K_* = \mathrm{diag}\{100, 100\}$, and the angular velocity gain is $B_* = \mathrm{diag}\{40, 40\}$. Note that for the preceding parameters of the system, the period of oscillations with the lowest eigenfrequency is about 1, if the elbow angle is $3\pi/4$. The motion time $T = 1.5$ is close to this period, which makes the control problem very difficult. We have found that adding a small measurement noise to the simulated system output does not change the algorithm performance in any visible way. The reason is that for a random parameter vector sequence $p^{(k)}$, the error approximating the

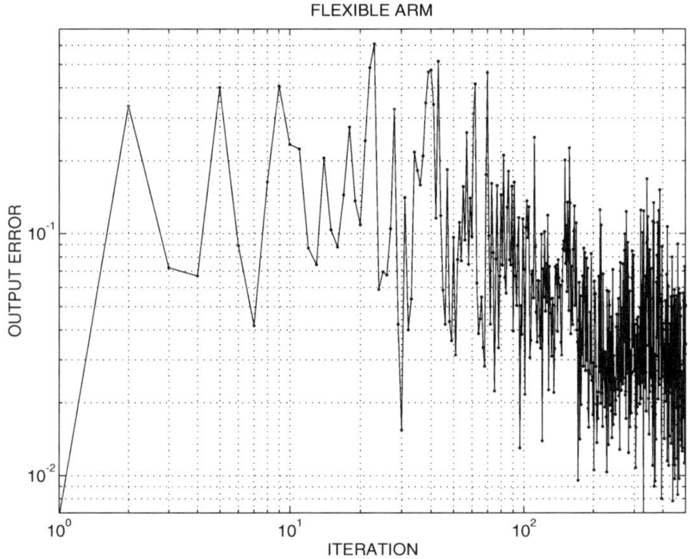

Figure 6 Progress of the terminal error $\|Y\|$ with the optimization iteration number. The arm moves through a randomly generated goal positions sequence. Reproduced with permission from D. M. Gorinevsky, *IEEE Trans. Automat. Control* 42:912–927, 1997 (©1997 IEEE).

system mappings with RBF networks in the algorithm already acts in the same way as an output noise.

In a numerical experiment, a sequence of the task parameter vectors p is generated so that the initial arm configuration coincides with the final arm configuration at the end of the previous task. Figure 6 shows the progress of the error $\|Y - Y_d\|$

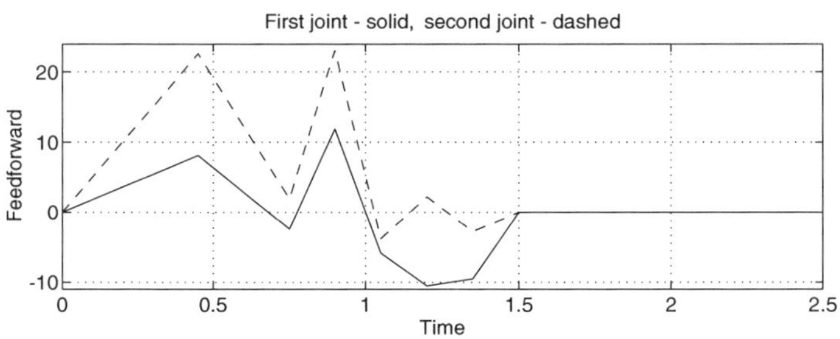

Figure 7 Feedforward for a test motion after algorithm convergence. Reproduced with permission from D. M. Gorinevsky, *IEEE Trans. Automat. Control* 42:912–927, 1997 (©1997 IEEE).

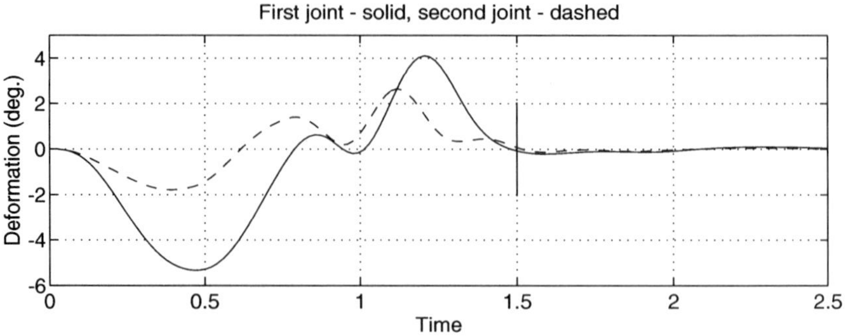

Figure 8 Joint deformations for the test motion with approximated feedforward. Reproduced with permission from D. M. Gorinevsky, *IEEE Trans. Automat. Control* 42:912–927, 1997 (©1997 IEEE).

with the optimization iteration number. One can see that the control error converges to a small acceptable value over the entire parameter vector domain. The error, which is achieved at the end of the optimization process, is about 20 times less than the initial error, which is obtained without feedforward. The oscillations of the motion error in Fig. 6 are related to the variation in the arm motion amplitude as new task parameter vectors p are randomly generated in the course of the learning.

Figure 7 illustrates the feedforward control computed as a result of the RBF network approximation after the algorithm convergence for the motion with initial joint angles $\lambda = [0° \quad 60°]^T$ and final angles $\nu = [70° \quad 105°]^T$. Figure 8 shows the joint deformations for the same motion. Owing to the computed feedforward, the deformation is small after time $T = 1.5$, which means the arm arrives at

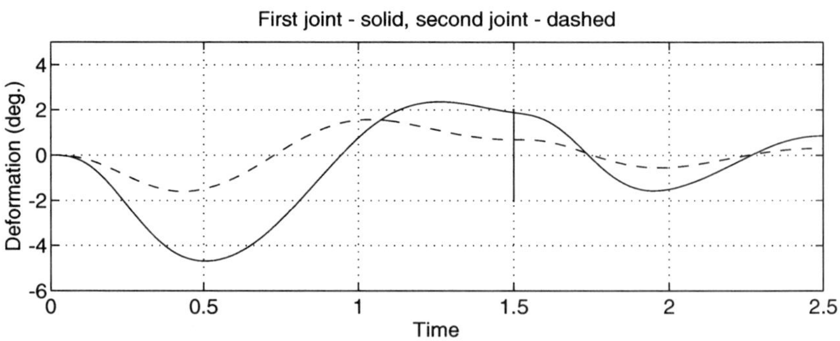

Figure 9 Joint deformations for the test motion with zero feedforward. Reproduced with permission from D. M. Gorinevsky, *IEEE Trans. Automat. Control* 42:912–927, 1997 (©1997 IEEE).

the final position without visible oscillations. An acceptable motion accuracy is achieved despite the high motion speed, low feedback gains, moderate network size, and large covered domain of the task parameters (82). For comparison, Fig. 9 shows the deformations for the same motion in the absence of feedforward.

VII. CONCLUSIONS

We have presented a new paradigm and RBF network architectures for task-level feedforward control of nonlinear systems. These algorithms belong to the realm of intelligent control and work at a higher hierarchical level compared to classical feedback or programmed control algorithms. We assume that the system operation can be considered as a sequence of clearly defined tasks and compute feedforward control for each task. The learning features of the proposed algorithms are aimed at optimizing the feedforward from one task to another based on the performance for a completed task. Dependence of the feedforward control on the task parameters is approximated using an RBF network.

The surveyed applications of the algorithms demonstrate their usefulness. This is illustrated by the example of point-to-point control of a flexible articulated arm. The computational resources required for practical implementation of the algorithms are moderate, especially as the algorithms need to run only once for each task. The application of the proposed paradigm can help to solve difficult practical control problems, for which other known methods are not adequate. The algorithms can also be extended to allow adaptive feedback control of nonlinear systems.

The main limitation of common approximation-based approaches to nonlinear control, such as neural networks, is the necessity of completing many learning trials in order to train the network. Generally, the number of examples required for network identification (training) grows exponentially with the input variable dimension. In the proposed paradigm, such an input variable is the task parameter vector p. Thus, the proposed task-level control technique offers the best advantage compared to state-space learning approaches if there are few task parameters and the state dimension is high. This advantage is achieved because the proposed algorithms do not attempt to approximate full nonlinear dynamics of the controlled system and limit themselves to optimizing performance just for a parametric family of control tasks.

REFERENCES

[1] S. Arimoto, S. Kawamura, and F. Miyazaki. Bettering operation of robots by learning. *J. Robotic Systems* 1:123–140, 1984.

[2] S. Arimoto. Learning control theory for robotic motion. *Internat. J. Adapt. Control Signal Process.* 4:453–564, 1990.

[3] Z. Geng *et al.* Learning control system design based on 2-D theory—an application to parallel link manipulator. In *Proceedings of the 1990 IEEE International Conference on Robotics and Automation*, Cincinnati, pp. 1510–1515, 1990.
[4] K. Gugliemo and N. Sadeh. Experimental evaluation of a new robot learning controller. In *Proceedings of the 1991 IEEE International Conference on Robotics and Automation*, Sacramento, pp. 734–739, 1991.
[5] S. Hara, Y. Yamamoto, T. Omata, and M. Nakano. Repetitive control systems: a new type of servo systems for periodic exogeneous signals. *IEEE Trans. Automat. Control* 33:659–668, 1988.
[6] R. Horowitz, W. Messner, and J. B. Moore. Exponential convergence of a learning controller for robot manipulator. *IEEE Trans. Automat. Control* 36:890–894, 1991.
[7] W. Messner *et al.* A new adaptive learning rule. *IEEE Trans. Automat. Control* 36:188–197, 1991.
[8] S. R. Oh, Z. Bien, and I. H. Suh. An iterative learning control method with application for the robot manipulator. *IEEE J. Robotics Automat.* 4:508–514, 1988.
[9] M. Togai and O. Yamano. Learning control and its optimality. In *Proceedings of the 1986 IEEE Conference on Robotics and Automation*, San Francisco, pp. 248–243, 1986.
[10] K. J. Hunt, D. Sbarbaro, R. Zbikowski, and P. J. Gawthorp. Neural networks for control systems—a survey. *Automatica* 28:1083–1112, 1992.
[11] M. Kawato. Adaptation and learning in control of voluntary movement by the central nervous system (tutorial). *Advanced Robotics* 3:229–249, 1989.
[12] V. D. Sanchez and G. Hirzinger. State-of-the-art robotic learning control based on artificial neural networks. An overview. In *The Robotic Review 2* (O. Khatib *et al.*, Eds.), Vol. 2. MIT Press, Cambridge, MA, 1991.
[13] C. G. Atkeson. Using locally weighted regression for robot learning. In *Proceedings of the 1991 IEEE International Conference on Robotics and Automation*, Sacramento, pp. 958–963, 1991.
[14] H. Tolle *et al.* Learning control with interpolating memories—general ideas, design lay-out, theoretical approaches and practical applications. *Internat. J. Control* 56:291–317, 1992.
[15] H. Tolle, J. Militzer, and E. Erzüe. Zur Leistungsfähigkeit lokal verallgemeinender assoziativer Speicher und ihren Einsatzmöglichkeiten in lernenden Regelungen. *Messen Steuern Regeln* 32:98–105, 1991.
[16] D. M. Gorinevsky. Learning and approximation in database for feedforward control of flexible-joint manipulator. In *ICAR '91: Fifth International Conference on Advanced Robotics*, Pisa, pp. 688–692, 1991.
[17] D. M. Gorinevsky. Experiments in direct learning of feedforward control for manipulator path tracking. *Robotersysteme* 8:139–147, 1992.
[18] D. M. Gorinevsky. Modeling of direct motor program learning in fast human arm motions. *Biol. Cybernet.* 69:219–228, 1993.
[19] C. G. Aboaf, C. G. Atkeson, and D. J. Reikensmeyer. Task-level robot learning. In *Proceedings of the IEEE International Conference on Robotics and Automation*, Philadelphia, pp. 1311–1312, 1988.
[20] M. S. Branicky. Task-level learning: experiments and extensions. In *Proceedings of the 1991 IEEE International Conference on Robotics and Automation*, Sacramento, pp. 266–271, 1991.
[21] D. M. Gorinevsky, A. Kapitanovsky, and A. A. Goldenberg. Neural network architecture for trajectory generation and control of automated car parking. *IEEE Trans. Control Systems Technol.* 4:50–56, 1996.
[22] D. M. Gorinevsky, A. Kapitanovsky, and A. A. Goldenberg. Radial basis function network architecture for nonholonomic motion planning and control of free-flying manipulators. *IEEE Trans. Robotics Automat.* 12:491–496, 1996.
[23] D. Gorinevsky, D. Torfs, and A. A. Goldenberg. Learning approximation of feedforward control dependence. *IEEE Trans. Robotics Automat.* 12, 1997.

[24] D. M. Gorinevsky. Sampled-data indirect adaptive control of bioreactor using affine radial basis function network approximation. *Trans. ASME J. Dynam. Systems Measurement Control* 188, 1996.
[25] S. Chen, S. A. Billings, and P. M. Grant. Recursive hybrid algorithm for non-linear systems identification using radial basis function networks. *Internat. J. Control* 55:1051–1070, 1992.
[26] S. Chen, C. F. N. Cowan, and P. M. Grant. Orthogonal least squares learning algorithm for radial basis function networks. *IEEE Trans. Neural Networks* 2:837–863, 1991.
[27] D. M. Gorinevsky and T. H. Connolly. Comparison of some neural network and scattered data approximations: the inverse manipulator kinematics example. *Neural Comput.* 6:519–540, 1994.
[28] E. Hartman and D. Keeler. Predicting the future: advantages of semilocal units. *Neural Comput.* 3:566–578, 1991.
[29] V. Kadirkamanathan, M. Niranjan, and F. Fallside. Sequential adaptation of radial basis function neural networks and its application to time-series prediction. In *Advances in Neural Information Processing Systems* (J. E. Moody, R. P. Lipmann, and D. S. Touretzky, Eds.), Vol. 3, pp. 721–727. Morgan Kaufmann, San Mateo, CA, 1991.
[30] J. Moody and C. J. Darken. Fast learning in networks of locally-tuned processing units. *Neural Comput.* 2:281–284, 1989.
[31] R. M. Sanner and J.-J. E. Slotine. Gaussian networks for direct adaptive control. *IEEE Trans. Neural Networks* 3:837–863, 1992.
[32] O. Bock, G. M. T. D'Eleuterio, J. Lipitkas, and J. J. Grodski. Parametric motion control of robotic arms—a biologically based approach using neural networks. *Telematics and Informatics* 10:179–185, 1993.
[33] N. Sadeh. A perceptron network for functional identification. *IEEE Trans. Neural Networks*, 4:982–988, 1993.
[34] A. N. Tikhonov and V. Ya. Arsenin. *Methods for Solution of Ill-Posed Problems*, 2nd ed. Nauka, Moscow, 1979 (in Russian).
[35] D. M. Gorinevsky. On the approximate inversion of linear system and quadratic-optimal control. *J. Comput. System Sci. Internat.* 30:6–23, 1992.
[36] R. W. Brockett. Control theory and singular riemanian geometry. In *New Directions in Applied Mathematics* (P. Hilton and G. Young, Eds.). Springer-Verlag, Berlin/New York, 1981.
[37] M. Spong. Modeling and control of elastic joint robots. *Trans. ASME J. Dynam. Systems Measurement Control* 109:310–319, 1987.
[38] J. Craig. *Introduction to Robotics*, 2nd ed. Addison-Wesley, New York, 1989.
[39] C. Fernandes, L. Gurvits, and Z. X. Li. Foundations of nonholonomic motion planning. Technical Report 577-RR-253, Robotics Research Laboratory, Courant Institute of Mathematical Sciences, New York, 1991.
[40] J. Vlassenbroeck and Van D. Rene. A Chebyshev technique for solving nonlinear optimal control problems. *IEEE Trans. Automat. Control* 33:333–340, 1988.
[41] S. Chen and S. A. Billings. Neural networks for non-linear dynamic system modelling and identifications. *Internat. J. Control* 56:319–346, 1992.
[42] Y.-H. Pao. *Adaptive Pattern Recognition and Neural Networks*. Addison-Wesley, Reading, MA, 1989.
[43] N. Dyn. Interpolation of scattered data by radial functions. In *Topics in Multivariable Approximation* (L. L. Schumaker, C. K. Chui, and F. I. Utreras, Eds.), pp. 47–61. Academic Press, Boston, 1987.
[44] R. Franke. Scattered data interpolation: test of some methods. *Math. Comput.* 38:181–200, 1982.
[45] R. Franke. Recent advances in the approximation of surfaces from scattered data. In *Topics in Multivariable Approximation* (L. L. Schumaker, C. K. Chui, and F. I. Utreras, Eds.), pp. 79–98. Academic Press, Boston, 1987.

[46] E. J. Kansa. Multiquadrics—a scattered data approximation scheme with applications to computational fluid dynamics, I. *Comput. Math. Appl.* 19:127–145, 1990.
[47] M. J. D. Powell. Radial basis functions for multivariable interpolation: a review. In *Algorithms for Approximation* (J. C. Mason and M. G. Cox, Eds.), pp. 143–168. Clarendon, Oxford, 1987.
[48] M. J. D. Powell. The theory of radial basis function approximation in 1990. In *Advances in Numerical Analysis* (W. Light, Ed.), Vol. 2, pp. 102–205. Clarendon, Oxford, 1992.
[49] M. Botros and C. G. Atkeson. Generalization properties of radial basis functions. In *Advances in Neural Information Processing Systems* (R. P. Lipmann, J. E. Moody, and D. S. Touretzky, Eds.), Vol. 3, pp. 707–713. Morgan Kaufmann, San Mateo, CA.
[50] C. A. Micchelli. Interpolation of scattered data: distance matrices and conditionally positive definite functions. *Constr. Approx.* 2:11–22, 1986.
[51] T. Poggio and F. Girosi. Networks for approximation and learning. *Proc. IEEE* 78:1481–1497, 1990.
[52] C. Bishop. Improving the generalization properties of radial basis function neural networks. *Neural Comput.* 3:579–588, 1991.
[53] J. A. Leonard, M. A. Kramer, and L. H. Ungar. Using radial basis functions to approximate a function and its error bounds. *IEEE Trans. Neural Networks* 3:624–627, 1992.
[54] J. Park and I. W. Sandberg. Universal approximation using radial-basis-function networks. *Neural Comput.* 3:246–257, 1991.
[55] J. Platt. A resource-allocating network for function interpolation. *Neural Comput.* 3:213–225, 1991.
[56] G. C. Goodwin and K. S. Sin. *Adaptive Filtering, Prediction and Control*. Prentice-Hall, Englewood Cliffs, NJ, 1984.
[57] S. Mukhopadhyay and K. S. Narendra. Disturbance rejection in nonlinear systems using neural networks. *IEEE Trans. Neural Networks* 4:63–72, 1993.
[58] D. M. Gorinevsky. On the persistency of excitation in radial basis function network identification of nonlinear systems. *IEEE Trans. Neural Networks* 6:1237–1244, 1995.
[59] D. Gorinevsky and G. Vukovich. Control of flexible spacecraft using nonlinear approximation of input shape dependence on reorientation maneuver parameters. In *13th World Congress of IFAC*, San Francisco, 1996.
[60] J. E. Dennis and R. B. Schnabel. *Numerical Methods for Unconstrained Optimization and Nonlinear Equations*. Prentice-Hall, Englewood Cliffs, NJ, 1983.
[61] D. M. Gorinevsky. An approach to parametric nonlinear least square optimization and application to task-level learning control. *IEEE Trans. Automat. Control* 42:912–927, 1997.
[62] D. M. Gorinevsky. An algorithm for on-line parametric nonlinear least square optimization. In *33rd IEEE CDC*, Lake Buena Vista, FL, 1994.
[63] D. M. Gorinevsky. Adaptive learning control using radial basis function network approximation over task parameter domain. In *Proceedings of the 1993 IEEE International Symposium on Intelligent Control*, Chicago, 1993.
[64] D. M. Gorinevsky. Learning task-dependent input shaping control using radial basis function network. In *IEEE World Congress on Computational Intelligence*, Orlando, 1994.
[65] D. Gorinevsky and L. Feldkamp. RBF network feedforward compensation of load disturbance in idle speed control. *IEEE Control System Mag.*, 1996.
[66] T. Ishihara, K. Abe, and H. Takeda. A discrete-time design of robust iterative learning algorithm. *IEEE Trans. Systems Man Cybernet.* 22:74–84, 1992.

Index

A

Accelerating backpropagation learning, 147
Activation functions, 96
ADALINE, 10, 34
Adaptive learning control, 277, 382
Adaptive update algorithm, 384
Advance rate, 190
  control routine, 190
Algorithms, parallelization, 216
Associative network, 175
Autoassociator, 23

B

Backpropagation, 16, 25, 176, 198, 200, 281
  algorithm, 81, 82, 85, 126, 145
  forgetting, 200
  learning, 147, 148, 163, 166
  networks, 175
  time, 192
Batchmode, 32
Boltzmann machines, 9, 225, 234
Boolean function, 106, 118, 126, 128
Broyden–Fletcher method, 147
Broyden secant update, 376
Broyden update, 385

C

Carve algorithm, 111, 112
Cascade net, 94
Cerebellar model articulation controller, 277
  techniques, 277, 278, 284, 287, 290, 293, 296
CMAC, *see* Cerebellar model articulation controller
Complex training sets, 86
Computational complexity, 148
  theory, 213, 214
Computational cost, 81, 179
Conjugate gradient direction, 156, 160
Constrained minimization problem, 181
Constraint satisfaction, 181, 209, 213, 218, 221, 227, 230
Constructive training methods, 85
Constructive techniques, 87, 88
Continuous Hopfield network, 228, 235
Continuous winner-take-all neural networks, 252, 255
Control neurons, 31
Convergence
  assessment, 74
  in neural network training, 74
  rate, 148
Cost function, 5, 14, 17, 22, 26, 188

D

Data based learning, 305
Decoupling network assumptions, 3, 25, 27
Descent direction, selection, 160
Discrete Hopfield network, 229, 230, 232, 234, 238, 239
Divide and conquer approach, 45
Dominant modes, 53
Dominant neuron techniques, 249, 250
Dynamic adaptation, 87
Dynamic barrier neural network, 238
Dynamic learning rate, 146
  optimization, 154

E

Error propagation learning, 282
Extended feedforward propagation procedure, 147
Extended Kolman filter, 147, 148, 164, 168
Exhaustive learning, 128

F

Fast and robust learning algorithms, 145
Fast backpropagation training, 145
Fault tolerance, 201
Feedforward controller, 380
Feedforward neural network architectures, 61, 83
Feedforward neural network systems, optimization, 53, 54
Finite difference update, 376
Fixed weight networks, 249
Forgetting behavior, 43
Forgetting problem, 175, 176
Fuzzy control, 354, 355

G

Gauss–Newton direction, 148, 160
General sequential constructive method, 101, 103
Genetic algorithms, 211, 225
Global knowledge of the state, 146
Global search, 222
Gradient backpropagation, 192
Gradient contribution matrix, 27
Gradient descent
  algorithms, 2, 151, 190
  optimization techniques, 32, 224, 240
Gradient of error function, 200
Growing networks, 44

H

Hamming clustering, procedure, 123, 125, 128
Hessian matrix, 19, 147, 179
Hidden layer, 14, 95
Hidden neurons, 17, 81, 95, 100, 126
Hidden nodes and links, optimum number, 60, 63
Hidden sublayer, 18
Hidden unit networks, 22, 191
Hidden units, 96
High speed training, 87
Homogeneous feedforward neural network, 63
Homogeneous network, 67
  optimization, 60
Hopfield network model, 240
Hopfield networks, 9, 228, 233, 234
Hybrid learning, 21
Hyperplane choice set, 117

I

Independent component analysis, 308, 311, 312
Information theoretic formulation, 309, 339, 342
Input nodes, optimum set selection, 59
Input preprocessing, 20
Irregular partitioning algorithm, 108, 110

J

Jacobian matrix, 147

K

K-winner-take-all, 235, 250, 251
K-winners-take-all neural networks, 268, 269, 270, 271

L

Lagrange multiplier method, 182
Large scale problems, 214
Layer weight matrix, 8
Learnable, 85
Learnability, 85
Learning
  algorithm, 98, 292
  as an optimization problem, 4
  control, 278, 374
    algorithm, 375
  environment, 5, 17, 26
  feedforward, 373
  from prior knowledge, 45
  mode, 32
  process, 192
  rate, 109, 146, 192, 200
    derivatives, 149, 151
    computation, 148
    search, 154
  rules, 299
  suboptimal solutions, 33
  trajectories, 42

with minimal degradation, 176, 183, 197, 199, 201
adaptation, 186
Levenberg–Marquardt method, 147, 379, 382, 384
Linear independent component analysis, 322
Linearly separable, 14
Local hidden unit, 204
Local minima, 10, 13, 14, 17, 26, 191
in neural networks, 34

M

Mackey–Glass series, 54, 62
Mapping local minima, 39
MAXNET, 259, 260, 261, 262, 264, 267
Memory and learning surface, 11
Minimal degradation with pruning, 204
Model reference adaptive control, 280
Modeling of the sunspot series, 65
Monk's problems, 134
Multiclass problems, 100
Multilayer backpropagation networks, 250
Multilayer feedforward neural network, 85, 147
Multilayer neural networks, 278
Multilayer perceptron, 86, 100, 355

N

Negative gradient, 148, 160
Network
architecture, 26, 83
with multiple outputs, 15
saturation, 41
size optimization, 58
Neural network
architecture, 87
learning of nonstationary processes, 175
supervised training, 82, 88, 249, 250, 323
systems, sequential constructive techniques, 81, 82
Neuron saturation, 41
Newton direction, 148, 160
Newton-like method, 148, 156, 164
Nonconstrained optimization, 177
Nonhomogeneous network, 68, 72
optimization, 60
Nonlinear autoassociators, 23
Nonlinear dynamical systems, feedforward control, 353
Nonlinear independent component analysis, 318
Nonlinear system identification, 161
Nonparametric characterization of dynamics, 337
Nonparametric statistical structure, 305, 306, 326
Normal batch training mode, 148
NP problems, 213, 233

O

Oil spot algorithm, 116
One-to-one mapping, 182
On-line learning, 378
Optical processor placement, 212
Optimal learning, 2, 44
in artificial neural networks, 1
rate, 145, 147, 148, 149, 154, 155
with autoassociators, 23
Optimal neural network architecture, 84
Optimal weight matrix, 86
Optimization
problem groups, 211
problems, 210
target function, 210
Orthogonality, principles, 217
Orthogonal transformation techniques, 53, 54, 55, 76
Output coding, 20
Output neuron, 9
Oversized neural network, 53
Overtraining effect, 201

P

Parametric statistical structure, 305, 306
Parity function, 128
Partial classifier, 98, 118, 122
Pattern
classification, 11
mode, 32
weight updating, 32
recognition, 12
Penalty-term methods, 44
Perceptron algorithm, 109, 118, 127
Perceptrons, 2, 16
Performance measure, 191
Physiological processes, 62
Premature saturation, 33, 41
Principle component analysis, 308

Problem solving techniques, 221
Projection matrix, 24
Pruning algorithms, 44, 54
Pruning methods, 85
Psuedo-Newton rule, 188
Pyramidal networks matching, 13

Q

Q R factorization, 55, 56
Quadratic optimization problem, 226

R

Radial basis functions, 16, 176, 297, 298, 353, 355, 365, 380
  approximation, 366, 367, 369, 372
  controller, 386
  interpolation, 367, 368
  model, 370
  units, 186
Random Boolean functions, 126, 131
Recurrent neural networks, 8, 9, 25, 39, 40, 42, 226, 227, 235
Recursive identification, 370
Recursive weight updating, 370
Regular partitioning algorithm, 108
Ring structure 31

S

Selected hidden nodes, 72
Sequential constructive methods, 81, 88, 94, 98, 99, 101, 102, 105, 106
Sequential decision lists, 89, 96, 98
Sequential window learning algorithm, 118
Simulated annealing, 211, 224
Single hidden layer, 96
Singular value
  decomposition, 53, 55, 60, 74, 78
  ratio spectrum, 78
Spatial crosstalk, 45
Spurious local minima, 34
Static networks, 7
Statistical independence measure, 328
Statistical structure extraction, 307
Steepest descent, 82
Stochastic methods, 211
Stochastic neuron dynamic, 234
Stopping criterion, 191
Structural local minima, 34, 35
Suboptimal learning, 41
Subset selection, 57
Sunspot series model, 66
Symmetrical configurations, 40
Symmetry function, 128
Synaptic weights, 147, 163

T

Target switch algorithm, 112, 114
Task-dependent feedforward control, 365, 378
Task-level control, 354, 356, 357, 358
Technical neuron problem, 227
Temporal crosstalk, 45
Threshold cost functions, 10
Threshold error function, 6
Threshold neurons, 96, 97, 106
Trained patterns, 178
Training
  algorithms, 88
  network, 44, 161
  process, 86
  regimes, 176
  set, 11, 118, 175
    size, 81
Two-layer perceptrons, 97

U

Unconstrained minimization problem, 179, 181
Unsupervised neural learning, 307, 308, 323, 325

W

Weight-specific local variation information, 146
Window activation function, 122
Window neurons, 117, 118
Winner-take-all, 235, 250, 251, 255, 256

X

X O R Boolean function, 36